AF333071

MOLECULAR BIOLOGY
INTELLIGENCE
UNIT

Nuclear Import and Export in Plants and Animals

Tzvi Tzfira, Ph.D.

Vitaly Citovsky, Ph.D.

Department of Biochemistry and Cell Biology
State University of New York at Stony Brook
Stony Brook, New York, U.S.A.

LANDES BIOSCIENCE / EUREKAH.COM
GEORGETOWN, TEXAS
U.S.A.

KLUWER ACADEMIC / PLENUM PUBLISHERS
NEW YORK, NEW YORK
U.S.A.

NUCLEAR IMPORT AND EXPORT IN PLANTS AND ANIMALS

Molecular Biology Intelligence Unit

Landes Bioscience / Eurekah.com
Kluwer Academic / Plenum Publishers

Printed in the U.S.A.

Kluwer Academic / Plenum Publishers, 233 Spring Street, New York, New York 10013, U.S.A.
http://www.wkap.nl/

Please address all inquiries to the Publishers:
Landes Bioscience / Eurekah.com, 810 South Church Street, Georgetown, Texas 78626, U.S.A.
Phone: 512/ 863 7762; FAX: 512/ 863 0081
http://www.eurekah.com
http://www.landesbioscience.com

Nuclear Import and Export in Plants and Animals, edited by Tzvi Tzfira and Vitaly Citovsky, Landes / Kluwer dual imprint / Landes series: Molecular Biology Intelligence Unit.

ISBN: 0-306-48241-X

Library of Congress Cataloging-in-Publication Data

Nuclear import and export in plants and animals / [edited by] Tzvi Tzfira, Vitaly Citovsky.
    p. ; cm. -- (Molecular biology intelligence unit)
  Includes bibliographical references and index.
  ISBN 0-306-48241-X
  1. Nuclear membranes. 2. Biological transport. 3. Proteins--Physiological transport.
  [DNLM: 1. Nucleocytoplasmic Transport Proteins--genetics. 2. Active Transport, Cell Nucleus--physiology. QU 55 N9627 2005] I. Tzfira, Tzvi. II. Citovsky, Vitaly. III. Series: Molecular biology intelligence unit (Unnumbered)
  QH601.2.N843 2005
  571.6'6--dc22

                              2005003124

This book is dedicated to our families.

CONTENTS

Preface xi

1. Structure of the Nuclear Pore 1
   Michael Elbaum
   Structure and Assembly 3
   Molecular Dissection and Proteomics 11
   FG Repeats 14
   Transport Models in Relation to Structure 15
   The Minimal Pore 16
   Assembly Revisited 17

2. Integral Proteins of the Nuclear Pore Membrane 28
   Merav Cohen, Katherine L. Wilson and Yosef Gruenbaum
   Yeast POMs 28
   Vertebrate POMs 30
   Cell Cycle Dynamics of the NPC 30
   Membrane Fusion and Nuclear Pore Formation 31

3. Subnuclear Trafficking and the Nuclear Matrix 35
   Iris Meier
   Nuclear Matrix Targeting Signals 36
   Regulated Nuclear Matrix Interaction 43
   Compromised Subnuclear Localization and Disease 44

4. Nuclear Import and Export Signals 50
   Toshihiro Sekimoto, Jun Katahira and Yoshihiro Yoneda
   Definition of Nuclear Import and Export Signals 50
   Basic Type NLSs 51
   Non-Basic Type NLSs 53
   NESs Recognized by Importin β Related Proteins 53
   Sequences Acting As Both NES and NLS 55

5. Nuclear Import of Plant Proteins 61
   Glenn R. Hicks
   Protein Import in Animals and Yeast 61
   Nuclear Translocation in Plants 62
   Regulated Protein Import in Plant Development 71
   Recent Advances in Plant Nuclear Translocation 74

6. **Nuclear Import of Agrobacterium T-DNA** 83
Tzvi Tzfira, Benoit Lacroix and Vitaly Citovsky
   The Genetic Transformation Process 85
   T-Complex Export to Plant Cells 86
   Molecular Structure of the Mature T-complex 87
   T-Complex Nuclear Import 88
   Host Cell Proteins That Interact with VirD2 and VirE2 89
   VirE3, a Bacterial Substitute for the Host Protein VIP1 92
   A Model for T-DNA Nuclear Import and Intranuclear Transport 92

7. **Regulation of Nuclear Import and Export of Proteins in Plants
and Its Role in Light Signal Transduction** 100
Stefan Kircher, Thomas Merkle, Eberhard Schäfer and Ferenc Nagy
   Nuclear Import of Proteins 100
   Nuclear Export of Proteins 102
   The Regulatory GTPase Ran 102
   Plant Factors and Plant-Specific Features of Nuclear Transport 103
   Regulation of Nuclear Transport As a Tool to Regulate Signaling 104
   Nucleocytoplasmic Partitioning in Light Signal Transduction 104

8. **Nuclear Export: Shuttling across the Nuclear Pore** 118
John A. Hanover and Dona C. Love
   The Nuclear Pore Complex (NPC) 119
   Methods for Analyzing Nuclear Export 119
   Rev-GR-GFP: Nuclear Export in Vitro 121
   The Nuclear Export Receptors (Karyopherins): Importins
     and Exportins 121
   A Nonclassical Export Receptor: Calreticulin 125
   Calcium-Dependent Modulation of Nuclear Transport? 125
   Mechanism of Nuclear Protein Export and Shuttling 127
   Export of RNA: Ribosomes, tRNA, snRNA and mRNA 128
   Chromatin Organization and Transcriptional Repression 131
   Export Machinery, Pre-mRNA Splicing, and Nonsense
     Mediated Decay 131

9. **Nuclear Protein Import: Distinct Intracellular Receptors
for DifferentTypes of Import Substrates** 137
David A. Jans and Jade K. Forwood
   The Transport Process 138
   α Importins 138
   Importin β1 and Homologs 150
   Competition between Target Sequences/Receptors 151
   Distinct Nuclear Import Receptor for Different Types of TFs;
     Differential Regulation? 153
   Unanswered Questions 155

10. The Molecular Mechanisms of mRNA Export 161

Tetsuya Taura, Mikiko C. Siomi and Haruhiko Siomi

Ran Dependent Nucleocytoplasmic Transport 162

RanGTPase Dependent RNA Exports 163

Export of mRNA 163

From Gene to Nuclear Pore to Cytoplasm 165

TAP-Mediated mRNA Export: Ran Independent
  Nucleocytoplasmic Transport 165

An Adaptor Protein and Other Conserved mRNA Export Factors 166

Interactions between mRNA Export Machineries
  and Nucleoporins 167

Links between mRNA Quality Control and Nuclear Export 167

11. Nuclear Import and Export of Mammalian Viruses 175

Michael Bukrinsky

Transport to the Nuclear Envelope 176

Interactions at the Nuclear Pore 177

Export through the Nuclear Pore 179

12. Nuclear Import of DNA 187

David A. Dean and Kerimi E. Gokay

The Nuclear Envelope Is a Barrier to Gene Delivery 187

Nuclear Import of DNAs in Non-Dividing Cells 189

Plasmid Nuclear Import 189

Nuclear Import of Plasmids in Cell-Free Systems 195

Alternative Pathways for Plasmid Nuclear Uptake 196

Viral Nuclear Import 197

Nuclear Import of Single-Stranded DNA 199

Nuclear Import of Oligonucleotides 199

13. Research Methodologies for the Investigation of Cell Nucleus 206

Jose Omar Bustamante

Methods 207

Index 225

EDITORS

Tzvi Tzfira
Vitaly Citovsky
Department of Biochemistry and Cell Biology
State University of New York at Stony Brook
Stony Brook, New York, U.S.A.
Chapter 6

CONTRIBUTORS

Michael Bukrinsky
Department of Microbiology
 and Tropical Medicine
George Washington University
Washington, DC, U.S.A.
Chapter 11

Jose Omar Bustamante
The Nuclear Physiology Lab
 and Nanobiotechnology Group
Department of Physics, Universidade
 Federal de Sergipe
The Brazilian Millenium Institute
 of Nanosciences
The Brazilian Nanosciences
 & Nanotechnology Network
Sao Cristovac, Brazil
Chapter 13

Merav Cohen
Department of Genetics
The Institute of Life Sciences
The Hebrew University of Jerusalem
Jerusalem, Israel
Chapter 2

David A. Dean
Division of Pulmonary and Critical
 Care Medicine
Northwestern Universitiy Medical School
Chicago, Illinois,U.S.A.
Chapter 12

Michael Elbaum
Department of Materials and Interfaces
Weizmann Institute of Science
Rehovot, Israel
Chapter 1

Jade K. Forwood
Nuclear Signaling Laboratory
Department of Biochemistry
 and Molecular Biology
Monash University
Clayton, Australia
Chapter 9

Kerimi E. Gokay
Division of Pulmonary and Critical
 Care Medicine
Northwestern Universitiy Medical
 School
Chicago, Illinois,U.S.A.
Chapter 12

Yosef Gruenbaum
Department of Genetics
The Institute of Life Sciences
The Hebrew University of Jerusalem
Jerusalem, Israel
Chapter 2

John A. Hanover
Laboratory of Cell Biochemistry
 and Biology
NIDDK, National Institutes of Health
Bethesda, Maryland, U.S.A.
Chapter 8

Glenn R. Hicks
Department of Botany and Plant
 Sciences
Center for Plant Cell Biology
University of California
Riverside, California, U.S.A.
Chapter 5

David A. Jans
Department for Biochemistry
  and Molecular Biology
Monash University
Clayton, Australia
Chapter 9

Jun Katahira
Department of Frontier Biosciences
Graduate School of Frontier Biosciences
Osaka University
Suita, Osaka, Japan
Chapter 4

Stefan Kircher
Institut für Biologie II/Botanik
Universität Freiburg
Freiburg, Germany
Chapter 7

Benoit Lacroix
Department of Biochemistry and Cell
  Biology
State University of New York at Stony
  Brook
Stony Brook, New York, U.S.A.
Chapter 6

Dona C. Love
Laboratory of Cell Biochemistry
  and Biology
NIDDK, National Institutes of Health
Bethesda, Maryland, U.S.A.
Chapter 8

Iris Meier
Department of Plant Biology
Plant Biotechnology Center
Ohio State University
Columbus, Ohio, U.S.A.
Chapter 3

Thomas Merkle
Center for Biotechnology
Universität Bielefeld
Bielefeld, Germany
Chapter 7

Ferenc Nagy
Biological Research Center
Plant Biology Institute, Szeged
and Agricultural Biotechnology Center
Godollo, Hungary
Chapter 7

Eberhard Schäfer
Institut für Biologie II/Botanik
Universität Freiburg
Freiburg, Germany
Chapter 7

Toshihiro Sekimoto
Department of Cell Biology
  and Neuroscience
Graduate School of Medicine
Osaka University
Suita, Japan
Chapter 4

Haruhiko Siomi
Institute for Genome Research
Univertiy of Tokushima
Tokushima, Japan
Chapter 10

Mikiko C. Siomi
Institute for Genome Research
Univertiy of Tokushima
Tokushima, Japan
Chapter 10

Tetsuya Taura
Institute for Genome Research
Univertiy of Tokushima
Tokushima, Japan
Chapter 10

Katherine L. Wilson
Department of Cell Biology
Johns Hopkins University of Medicine
Baltimore, Maryland, U.S.A.
Chapter 2

Yoshihiro Yoneda
Department of Frontier Biosciences
Graduate School of Frontier Biosciences
Osaka University
Suita, Osaka, Japan
Chapter 4

PREFACE

The nucleus is perhaps the most complex organelle of the cell. The wide range of functions of the cell nucleus and its molecular components include packaging and maintaining the integrity of the cellular genetic material, generating messages to the protein synthesis machinery of the cell, assembling ribosome precursors and delivering them to the cell cytoplasm, and many more. As a complex machine, the nucleus maintains a constant two-way flow of information with the surrounding cytoplasm, such as import and export of ions, small and large proteins and protein complexes, and ribonucleoprotein particles. These transport processes occur through the nuclear pore complexes which represent the selective gateways through the nuclear envelope, a major barrier that isolates the nucleus from the cytoplasm.

More than one hundred and seventy years have passed since Robert Brown discovered the cell nucleus using his simple light microscope, and, since then, remarkable progress has been made, both technically and conceptually, in studying and understanding the structure and function of the cell nucleus. In these days of modern cellular and molecular biology, we are capable of employing a vast array of sophisticated technologies and approaches to image the nucleus and its substructures, to isolate and functionally characterize its molecular components, and to modify the nuclear genetic material. With the resulting knowledge, we have come to appreciate the complexity of the nuclear structure and function. In particular, the ability of various types of molecules to be actively transported through the well-guarded nuclear pore complexes is extremely intriguing. The chapters of this book provide insights into the intricate mechanisms of nuclear import and export. To better understand these processes, one must first elucidate the organization of the physical gateways into the nucleus. Thus, we begin this book with a detailed description of the nuclear pore structure and composition. The signal sequences that specify nuclear import and export of proteins are discussed next followed by eight chapters, each dedicated to a specific aspect of the nuclear import and export in plant and animal cells. Among these, special chapters are dedicated to nuclear import of *Agrobacterium* T-DNA during plant genetic transformation, nuclear import and export of animal viruses, and nuclear intake of foreign DNA. A chapter on research methods to study nuclear transport concludes the book. The result is a compact book which we hope the readers will find useful as a guide and a reference source for diverse aspects of nuclear import and export in plant and animal systems.

We would like to express our sincere gratitude to all the authors for their outstanding contributions, to the staff of Eurekah.com for their help and patience during the long period of the book production and, in particular, to Ms. Cynthia Conomos for assistance in all technical aspects of the chapter productions.

Tzvi Tzfira and Vitaly Citovsky
January 2005, New York

CHAPTER 1

Structure of the Nuclear Pore

Michael Elbaum

The nucleus is a defining hallmark in cells of all the higher organisms: yeast, animals, and plants. As the repository of the genome, it both encloses the chromatin and regulates its accessibility. It is also the site of nucleic acid synthesis, including replication of DNA, transcription and editing of messenger RNA, synthesis of ribosomal RNAs, and assembly of ribosomal subunits. By contrast, the cytoplasm is the site of protein synthesis, where functional ribosomes translate mRNA into polypeptides. The nuclear envelope defines the border between these two distinct biochemical worlds. The nuclear pores (or nuclear pore complexes, NPCs) serve as guardians of this border, acting as the gateway for molecular exchange between the two major cellular compartments. They are deeply integrated to the physiological function of every cellular pathway involving communication between enzymatic, signaling, or regulatory activities on one hand, and gene expression on the other. The nuclear pore complex is also a fascinating molecular machine, facilitating the passage of specific macromolecules in one direction while ferrying others in the opposite sense.

The nuclear envelope (NE) defines the boundary between nucleus and cytoplasm. It is formed by two juxtaposed lipid bilayer membranes, the outer one of which is contiguous with the endoplasmic reticulum. The outer and inner lipid bilayers are also connected continuously through the nuclear pores themselves, though their protein compositions differ. A matrix of filaments underlies the inner nuclear membrane, providing mechanical support and anchoring sites for the enclosed chromatin. In animal cells these filaments are composed largely of lamins, similar in structure to intermediate filaments. Aside from a few known exceptions associated with viral infection, all molecular exchange across the nuclear envelope takes place via the nuclear pores, whose number ranges from many tens to several thousand per nucleus. Thus RNAs and ribosomal subunits are exported to the cytoplasm, while proteins needed in the nucleus must be imported, and often reexported when their task there is done. Each pore is a large multi-protein complex, consisting of 30 or more distinct protein components in multiple copies. Its total molecular weight has been measured at 125 MDa for vertebrate cells, and about 60 MDa for yeast. Individual nuclear pores are thought to mediate traffic in both directions.

The functional task of the nuclear pore is to regulate entry to, and exit from, the nucleus. Specific pathways are discussed at greater depth in other chapters of this book. A degree of consensus has emerged in describing nuclear transport as a receptor-mediated translocation process.[1] Molecular cargo is marked for import (or export) by the presence of peptide signals,[2-4] which are then recognized by specific receptors that serve to usher the cargo across the pore.[5-7] Models of translocation can been categorized into those that anticipate some form of micromechanical movement (for example iris-like closures) of the pore itself on one hand,[8-11] or entirely biochemical sieves on the other.[12-15] While deep modulation of calcium levels has a

Nuclear Import and Export in Plants and Animals, edited by Tzvi Tzfira and Vitaly Citovsky.
©2005 Eurekah.com and Kluwer Academic / Plenum Publishers.

pronounced effect on nuclear pore structure in vitro,[11,16] calcium depletion does not appear to be coupled to nuclear transport regulation in intact cells.[17,18] The lack of intrinsic ATPase activity in the nuclear pore supports the second, nonmechanical class of models.

A rather minimalistic model for nucleocytoplasmic transport describes the nuclear pore and its associated soluble biochemistry as an affinity-regulated chemical pump.[14,19-22] Two apparently distinct modes of transport are identified: small molecules including water, ions, metabolites, and even small proteins (up to ~40 kDa molecular weight) can pass by simple diffusion so that their concentrations in solution equilibrate on the two sides of the NE; larger proteins and protein complexes are transported by an "active" mechanism that is able to pump the molecular cargo against a gradient in concentration, and so to accumulate it on one or the other side of the NE. In the latter case, proteins bearing nuclear localization signal (NLS) peptides associate with receptors of the importin/karyopherin family in the cytoplasm, and dissociate from them inside the nucleus. The canonical import receptor is importin β,[23] also known as karyopherin β,[24,25] or as p97.[26] This receptor interacts with NLS via an importin α (karyopherin α) adapter protein, so that a single cargo molecule enters the nucleus as a heterotrimer with the receptors. Their dissociation is governed by a competitive interaction with the small GTPase Ran,[27,28] which in its GTP form binds the importin β and releases the α molecule and the NLS-cargo.[5,12,29-32] A differential concentration of RanGTP across the nuclear envelope is maintained by the localization of the associated GTP exchange factor RanGEF (independently known as the chromatin condensation factor RCC1) loosely bound to chromatin within the nucleus, and the GTPase activating protein RanGAP associated with the peripheral cytoplasmic structures of the NPC.[33-38] Thus Ran is primarily in the GTP form within the nucleus, and in the GDP form in the cytoplasm.[39] Computer simulations support the assertion that receptor selectivity at the pore is sufficient for its function as a molecular pump, in combination with the Ran cycle; specific transport directionality is not required.[40,41] In some cases the directionality of transport could be inverted by artificially inverting the RanGTP gradient.[42] The same paradigm operates for export, except that the association of the cargo and RanGTP to the export receptor is synergistic rather than competitive.[43-50] Transfer RNAs make use of a specific receptor for export,[51,52] while export of other RNAs is thought to be governed by signals on associated proteins. In the case of large substrates a restructuring of the cargo itself may also be involved. A beautiful example was observed by electron microscopy for Balbiani ring mRNA export in *Chironomus* salivary glands. A series of snapshots shows the spiral ring unwinding and feeding progressively through the pore.[53]

The major role of the fixed structure of the NPC in such a model is to provide a selective translocation barrier, limiting passage to a rather short list of proteins. Those which are able to associate with signal-bearing molecular cargo, most notably importin β, are recognized as nucleocytoplasmic transport receptors, effectively opening the barrier to pass the complex where the cargo alone would be excluded. (It should not be overlooked that the transport receptors may have other roles in the cell as well.[54-56]) Within this picture the "active" transport is achieved primarily by the Ran switch, whose role is primarily to recycle the components of the chemical pump. No "moving parts" are required in the pore itself. A number of other proteins on the NPC's recognition list, particularly those involved in signal transduction such as β-catenin,[57,58] are able to pass pore autonomously. Their directionality and temporal accumulation are governed primarily by retention on nuclear or cytoplasmic structures, rather than by restriction of the reverse passage through the pore (reviewed in ref. 59). Perhaps the essential structural question is how the NPC can be so selective, passing relatively large cargo and complexes while blocking the passage of smaller ones. High selectivity normally implies a high and specific equilibrium affinity, but in the case of transport strong binding would of course be antithetical to translocation.

Structure and Assembly

Nuclear pores have been studied since the early days of biological electron microscopy.[60-69] Several approaches and techniques have been pursued to determine their structure. Traditional sectioning of embedded nuclei shows the juxtaposed lipid membranes pinched at the edges of a hole approximately 50 nm in diameter. Close observation reveals some poorly-resolved structure on both the cytoplasmic and nuclear faces. Scanning electron microscopy provides a detailed relief view of these surfaces[70] while the introduction of field-emission sources enabled imaging at high resolution.[71,72] Rotary shadowing in transmission electron microscopy can provide a similar level of detail.[73,74] These surface views show a characteristic eight-fold rotational symmetry, with eight fibrils protruding into the cytoplasm, and eight fibers collected into a ring on the nucleoplasmic side, forming the nuclear basket. Atomic force microscopy shows similar surface structures at somewhat lower resolution, though with the advantage at least in some cases that imaging can be performed in hydrated, near-native conditions.[11,16,75-78] Figure 1 shows a number of views of the nuclear pore.

Between these peripheral fibrilar structures lies the central framework of the NPC. This domain was examined extensively by transmission electron microscopy.[74,79-87] The favored sample has been the giant nucleus (germinal vesicle) present in oocytes of *Xenopus laevis*, though several studies have demonstrated a universality of the basic elements across many representative animal species.[64,70,88] Oocyte nuclei can be extracted by hand under a simple dissecting microscope, and the nuclear envelope spread flat on a microscope grid. Computerized image processing methods may be used to orient and average the images of individual pores, thereby improving signal to noise. Image averaging emphasizes the common underlying features, while intrinsic variability is lost along with the noise. Thus eight-fold symmetry is emphasized, with density appearing in a pattern of radial spokes. The hints to the protruding filament structures are lost, due to their intrinsic disorder. A dense object appears at the center of the pore. Because of its strategic location this object has often been called the "central transporter". Comparison of the average with individual images shows that this object is highly variable, on the other hand, leading to suggestions that it may not be a distinct structural feature of the pore itself but rather evidence of cargo caught in transit. Central protrusions appear with similar variability in scanning electron and atomic force microscopy imaging. They have been observed with particular regularity by atomic force microscopy under conditions of calcium depletion. Clearly this remains the most enigmatic part of the pore structure.

Tomographic methods have generated three-dimensional structural models. The thin, flat samples prepared by spreading *Xenopus* germinal vesicle nuclear envelopes are ideally suited for these studies. Full tomography involves the acquisition of a series of images where the sample is progressively tilted to steeper and steeper angles. A variant called random conical tilt involves the acquisition of a flat, normal incidence view and a single tilted image. The in-plane rotations of distinct (but ostensibly identical) objects are used to provide the multiple angular views required for three-dimensional reconstruction. Due again to the variability of peripheral structures, these studies have focused on the core region of the NPC.

The consensus three-dimensional structure of the vertebrate NPC is described as a three-layer sandwich, consisting of cytoplasmic and nucleocytoplasmic rings surrounding a set of eight spokes projecting inward from the lipid membrane pore. The spokes themselves have radial structure, with two lobes appearing within the diameter of the lipid membrane pore and one extending beyond it. The outermost diameter of the protein structure reaches ~ 120 nm, where the third lobes join circumferentially to form a lumenal ring in the space between the two membranes of the nuclear envelope. The latter join circumferentially to form a lumenal ring between the membrane layers of the NE. The surrounding cytoplasmic and nuclear rings are continuous, while eight internal voids appear between the spokes. This led to the suggestion that passive transport may take place through these spaces, rather than through the central

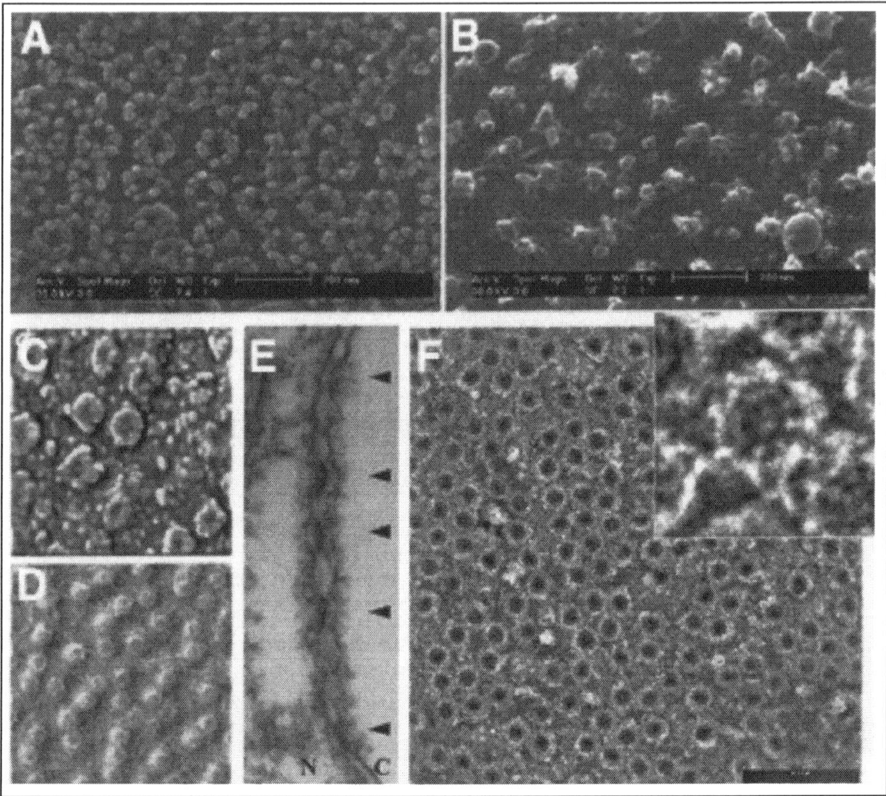

Figure 1. The nuclear pores as they appear in four different imaging methodologies. *top panels:* High-resolution, field-emission scanning electron microscopy of the *Xenopus laevis* germinal vesicle envelope shows surface topology of the nuclear pores, seen from the cytoplasmic (A) and nucleoplasmic (B) sides. C,D) Atomic force microscopy of the same sample; sample topography can be measured quantitatively. (Reprinted from ref. 77.) E) Typical view of the double lipid bilayer nuclear envelope, seen in cross-section through a nucleus reconstituted in *Xenopus* egg extract. Nuclear pores are marked by arrowheads, cytoplasmic and nucleoplasmic sides by C and N respectively. Note the equatorial cut in the lowermost pore, and the near-glancing section in the uppoermost pore, where peripheral structures are seen clearly. F) The *Xenopus* germinal vesicle is spread on a thin grid and observed in negative stain by transmission electron microscopy; protein density is white. scale bar = 500 nm. The inset shows a single pore at high magnification. (Reprinted from ref. 21.)

channel. Direct observation of small colloidal gold particles in transit puts the diffusional channel along the central axis, however, in the same location where signal-mediated translocation occurs.[89] Kinetic analysis of diffusive transport through individual nuclear pores also indicated that passage occurs through a single channel.[90]

An alternate viewpoint would regard the NPC as a large barrel of eight staves surrounding a central channel, with inward projections attached to the center of the staves. Bands corresponding to the two rings at the top and bottom close the staves. This central part of the NPC normally possesses clear eight-fold rotational symmetry. There appears to be some chiral character with a clockwise vorticity on the cytoplasmic side, becoming anti-clockwise on the nuclear side. Most models also include a central element connected to the staves by radial spokes, corresponding again to the "central transporter". A detailed view of the most recently-published tomogram of the nuclear pore appears in Figure 2.

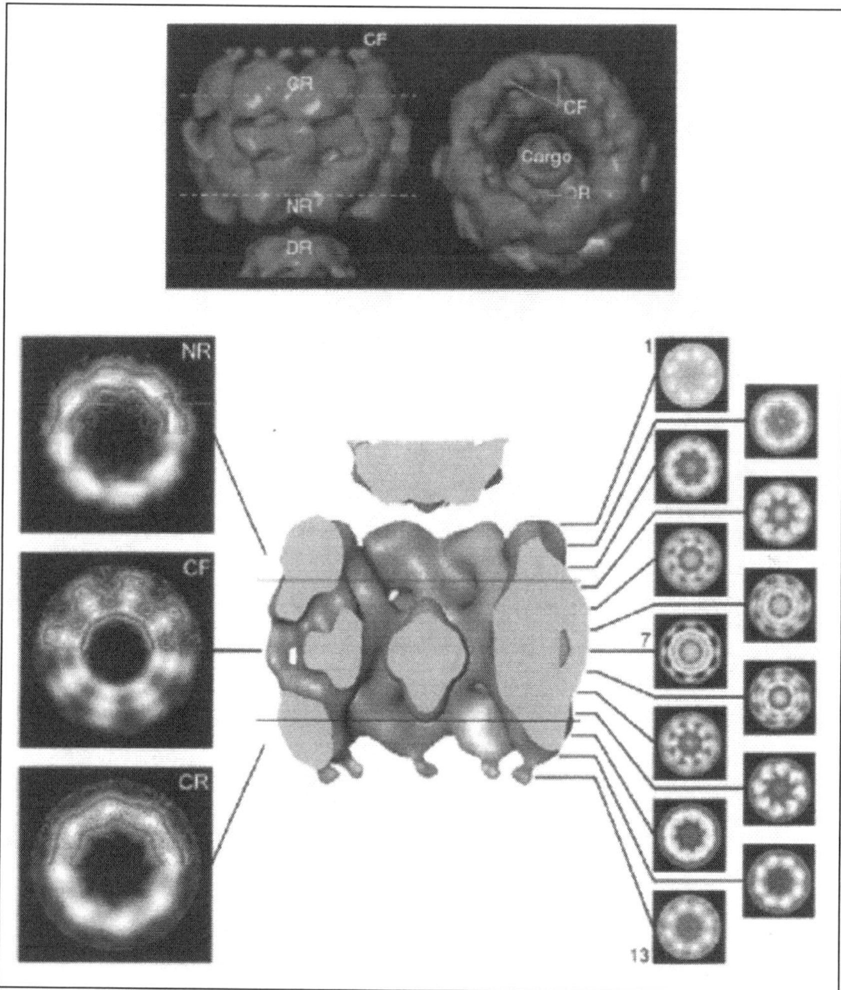

Figure 2. Tomographic reconstruction of the nuclear pore by energy-filtered cryo-electron microscopy. *Above:* isosurface representations. CF—cytoplasmic filaments; CR—cytoplasmic ring; NR—nucleoplasmic ring; DR—distal ring (or basket). (Reproduced with permission from Nat Rev Mol Cell Biol 2003; 4:757-66, ©2003 Macmillan Magazines Ltd.) *Below:* protein density shown in thirteen sections through the pore, with density contours at the cytoplasmic and nucleoplasmic rings (CR & NR) and through the central framework (CF). (Reprinted from: Staffler D et al. Cryoelectron tomography provides novel insights Into nuclear pore architecture: Implications for nucleocytoplansmic transport. J Mol Biol 2003; 328:110-130. ©2003, with permission from Elsevier.)

NPCs of yeast have a similar structure to those of *Xenopus*.[91-93] Overall the construction appears somewhat simpler, with only two lobes in the radial spokes, the outer one of which overlaps the membrane pore closely. Smaller radial arms cross the membrane, and there is no evidence for a lumenal ring. The yeast NPC also appears to lack the cytoplasmic and nuclear rings surrounding the major spoke complex, suggesting a weaker connection to peripheral structures than in the case of the vertebrate nuclear pore. The diameter of the membrane pore itself is similar for yeast and Xenopus, while the height of the NPC is approximately half: 30

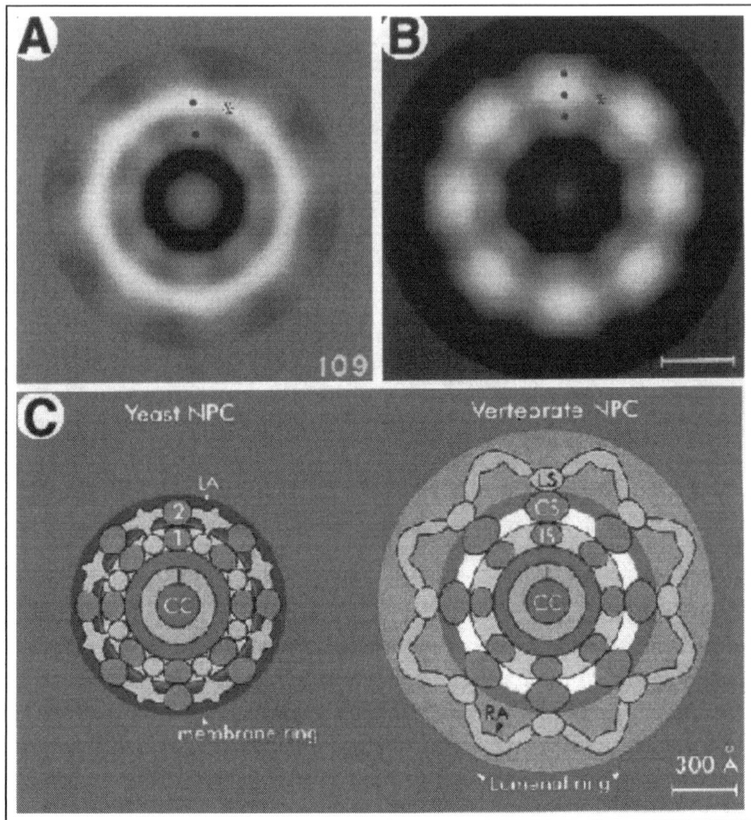

Figure 3. Comparison of yeast and vertebrate nuclear pores. Rotationally averaged density projections show two concentric rings for the yeast pore (A) and three for the *Xenopus* pore (B). The dots represent lobes along the radial spokes, as described in the text. C) a cartoon representation of the above. (Reprinted from: Yang Q, Rout MP, Akey CW. Three-dimensional architecture of the isolated yeast nuclear pore complex: Functional and evolutionary implications. Mol Cell 1998; 1:223-234. ©1998, with permission from Elsevier)

nm vs. 65 nm, respectively. A comparison of the vertebrate and yeast pore appears in Figure 3. Early works on plant nuclear pores revealed a generally similar structure,[94-96] though to date there have been no structural studies at a comparable level of detail.

The NPC can be disassembled into component sub-structures by detergent treatment or gentle proteolysis. This approach has been especially fruitful with the *Xenopus* oocyte NE. The breakdown may be sufficiently delicate that the products retain octagonal symmetry so that they can be identified with spoke complexes, or nuclear or cytoplasmic rings. The mass of each such component could then be measured quantitatively using scanning transmission electron microscopy (STEM).[97] The molecular weight of the spoke complex was found to be 52 MDa, while the heavier cytoplasmic ring and lighter nucleoplasmic ring were assigned masses of 32 and 21 MDa respectively. This leads to a total of 105 MDa, to be compared to a measured 112 MDa for NPCs where the central plug was apparently absent, or 124 MDa where it was present. An earlier biochemical estimate of the NPC mass found 110-148 MDa for unfixed preparations.[98] Also by STEM, the mass of the yeast nuclear pore was placed at 54.5 MDa, while light scattering and sedimentation gave estimates between 55 and 66 Mda.[91,92]

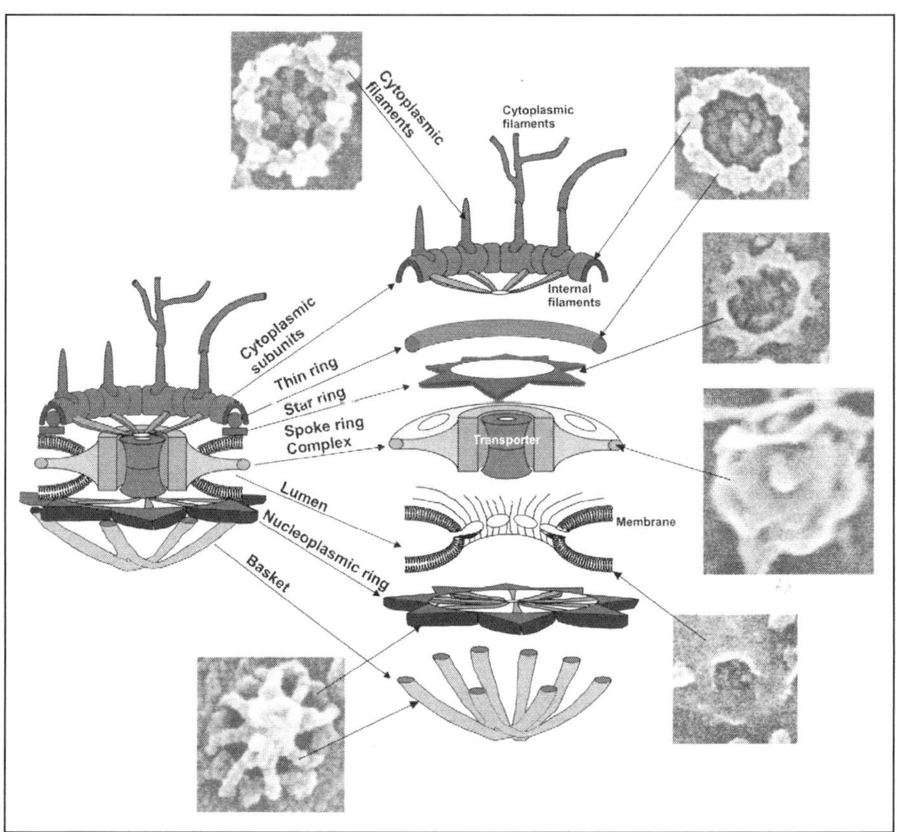

Figure 4. A sketch of the three-dimensional structure of the nuclear pore, inspired by high-resolution scanning electron microscopy of assembly intermediates. (Reprinted with permission from ref. 246.)

Gentle disassembly treatments have also been used to dissect the *Xenopus* NPC structurally using high-resolution, field-emission in-lens scanning electron microscopy (FEISEM).[99,100] This method shows the topography of the exposed surface, so the structural intermediates can be seen clearly as the layers are peeled away. The peripheral filaments are removed first, followed by the cytoplasmic and nucleoplasmic rings. In comparison with earlier studies, the cytoplasmic ring comes off in two steps, first as a "thin ring" that serves as a base for the protruding filaments, and secondly an underlying "star ring" that connects to internal elements. The central spoke ring complex lies underneath, in the plane of the nuclear envelope. From the nuclear side a similar picture emerges, with the basket filaments removed first, and then the nucleoplasmic ring. Unlike its cytoplasmic counterpart, though, this ring comes off as a single unit with the internal filaments. Once gone, the view of the central spoke ring is similar to that seen from the cytoplasmic side. Figure 4 shows a sketch of a hypothetical assembly model based on these observations.

Exceptions exist to the eight-fold symmetry of the NPC. Seven and nine-fold pores have been observed in in vitro reconstitutions of nuclei from *Xenopus* egg extract[100] (to be described in more detail below), while nine and tenfold pores were seen in germinal vesicle nuclear envelopes.[101] While rare, their existence yields important clues about pore assembly. A tomographic reconstruction of nine-fold NPCs shows that the basic radial units of spokes (or staves) is conserved. They contain roughly the same mass and occupy the same volume as the corresponding unit in the normal pore. In other words the larger pore structure is made of similar building blocks at the level of the stave. This would suggest that an early step in pore genesis might fix the symmetry. If a stable seed-pore forms with abnormal symmetry, this would be propagated to the final structure. It is interesting to speculate whether each assembly intermediate of the stave perpendicular to the membrane requires closure of the in-plane ring, or if each of the eight subunits is built autonomously. The symmetry exceptions suggest that local lateral interactions, which can suffer some strain, set the eight-fold symmetry, rather than some absolute requirement of eight units for closure. The same lateral interactions could establish a temporal asymmetry in assembly and disassembly. In the former case addition of NPC subunits might not depend on full closure of the n-fold ring, while in the latter the destabilization of one element of a given ring could lead to rapid loss of the entire substructure. This would cause accumulation of disassembly intermediates in stages where eight-fold substructures predominate.

A second very noteworthy exception is that the nuclear pore can form in membranes other than the nuclear envelope. These are known as "annulate lamellae"; they form as stacks of double-bilayer lipid membranes very similar to the NE.[102-111] They form under a wide range of conditions in vivo, though their biological function remains unknown. It was proposed that they may act as a storage medium for NPC components, or perhaps they represent dead-end assembly of pores in excess membranes. In any case the pores that form in them appear morphologically identical to those in the NE. In some examples they appear to be oriented all in the same direction, while in others they seem to face inward and outward at random. Figure 5 shows two views of nuclear pores in annulate lamellae.

Animal and plant cells employ an open mitosis, where the nuclear envelope breaks down and reassembles with every cell division. The nuclear pores are similarly broken down and reassembled in each cycle. The number of pores doubles during the G2-S transition.[112] In yeast, by contrast, the nuclear envelope remains intact throughout the closed mitosis. In early *Drosophila* embryogenesis the syncytial nuclear division occurs surrounded by a spindle membrane similar to the nuclear envelope but lacking nuclear pores.[113-115] These observations indicate two potentially different modes of pore assembly: one concomitant with nuclear envelope assembly and the other involving insertion into preexisting membranes.

Nuclear reconstitution in vitro affords a particularly powerful system for the study of NPC assembly. Extracts from amphibian,[116] sea urchin,[117,118] and fish eggs,[119] *Drosophila* embryos,[120,121] and even tissue culture cells[122,123] support nuclear assembly (reviewed in refs. 124, 125). Cell-free nuclear reconstitution was also achieved in plant extracts.[126,127] By far the most popular system has been based on extracts from eggs of *Xenopus leavis*. Such extracts imitate the normal process of rapid cell division following fertilization, wherein mRNA transcription and most protein expression are silenced during the first twelve cycles. A single egg therefore contains a stockpile of material sufficient for 4096 daughter cells and daughter nuclei. Egg extracts can be prepared and arrested at a variety of meiotic, mitotic, and interphase checkpoints.[128-133] Upon addition of a source of chromatin to an interphase extract, typically demembranated *Xenopus* sperm heads, nuclei assemble spontaneously. Important stages include a preliminary swelling of the chromatin, accumulation of membranes (as vesicles) to the chromatin surface, fusion of the membranes to form a smooth nuclear envelope, and finally swelling of the nuclei to their typical near-spherical shape. Nuclear pores appear on these nuclei, and they are functional for nucleocytoplasmic transport.[134] Figure 6 shows the course of a typical reconstitution assay.

Reconstituted nuclei and their cell-free cytosol can be manipulated biochemically. Addition of a variety of chemical and biochemical inhibitors to the extract inhibits nuclear envelope and pore assembly at distinct stages. The alkylating agent N-ethylmaleimide (NEM) and the nonhydrolyzable GTP analog GTPγS were shown to inhibit the membrane fusion events

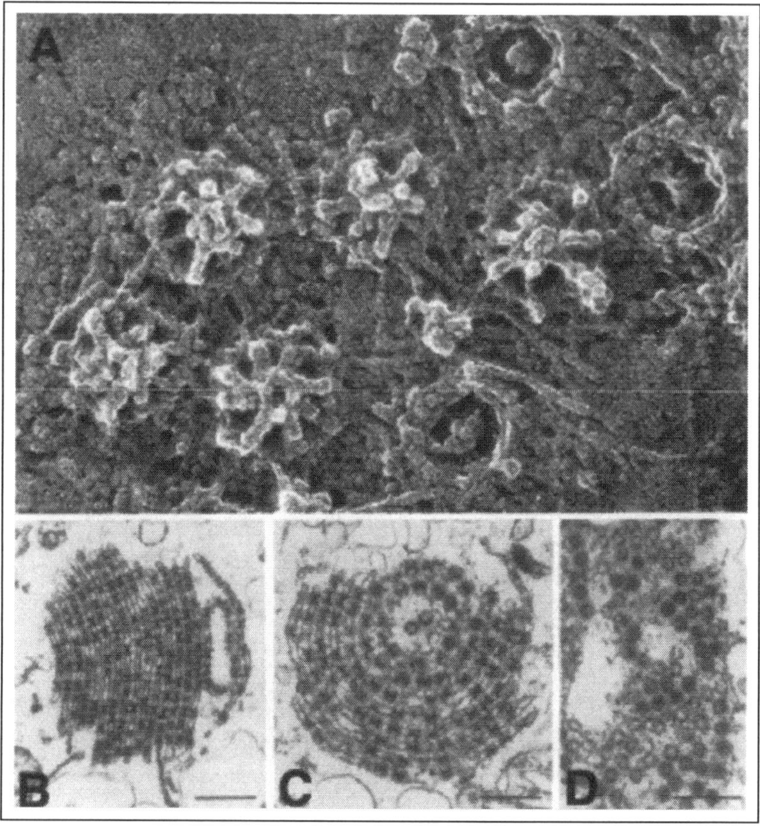

Figure 5. Nuclear pores in annulate lamellae (AL). A) rotary metal shadowing of frozen etched AL from *Dictyostelium.* (Courtesy of J. Henser; Reprinted from: Suntharalingam M, Wente SR. Peering through the Pore: Nuclear pore complex structure, assembly, and function. Developmental Cell 2003; 4:775-789. ©2003, with permission from Elevier.) AL in chromatin-free Xenopus egg extract, seen in transverse (B,C) and tangential (D) sections. (Reproduced from J Cell Biol 1991; 112:1073-1082, by copyright permission from The Rockefeller University Press.)

required for nuclear envelope formation, while the calcium chelator BAPTA permits nuclear envelope closure but completely blocks assembly of nuclear pores.[135] When pore-free nuclei were prepared in the presence of BAPTA and then transferred to a BAPTA-free cytosol, nuclear pores assembled into the preformed nuclear envelope. Pore assembly continued as well when the replacing cytosol contained GTPγS. A study using high resolution scanning electron microscopy showed that BAPTA itself has multiple effects. When added to a reconstitution assay from the start it blocked all pore formation. When added after sufficient time for assembly of preliminary pore structures, it led to accumulation of those intermediates, culminating with star rings at 40-45 minutes.[136] The same study showed a concentration-dependent inhibition of pore assembly by the known *transport* inhibitor wheat germ agglutinin (WGA). At low concentrations there was no effect, while at high concentrations pore assembly was blocked entirely. At intermediate concentrations there was an accumulation of stabilized but apparently empty pores, i.e., at a stage prior to formation of star rings. The observations permitted conjecture of a reasonable pattern of assembly stages, with membrane dimples followed by stabilized

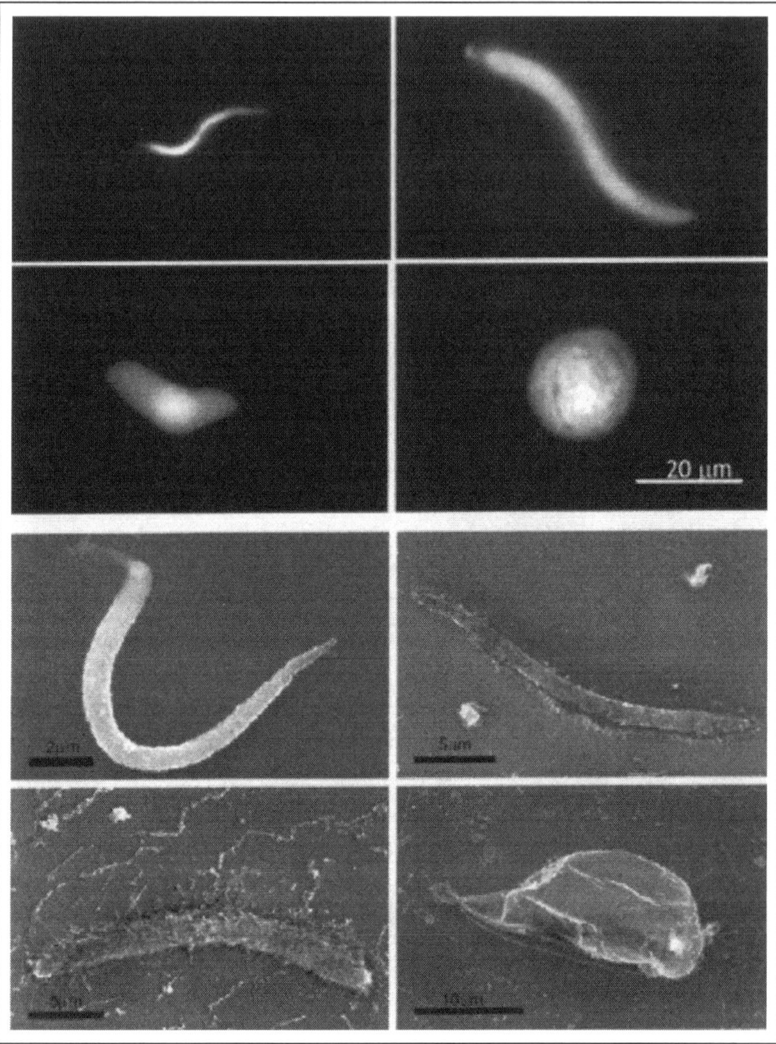

Figure 6. Cell-free reconstitution of nuclei in *Xenopus* egg extract. *Upper panels:* chromatin observed by fluorescent Hoechst stain. The demembranted sperm starts from an initial "corkscrew" shape, swelling quickly on exposure to the extract, and then gradually inflating as a nuclear envelope forms and the chromatin decondenses. The process typically takes 60-90 minutes. All images are shown at the same scale for comparison. *Lower panels:* scanning electron microscopy shows the progression: bare sperm, swelled chromatin, membrane vesicle condensation followed by fusion to a continuous nuclear envelope bearing nuclear pores. Scale bars as shown.

pores, followed by star rings (as viewed from the cytoplasmic side), thin rings, and finally cytoplasmic filaments. A similar pattern was subsequently observed in vivo in early *Drosophila* embryos.[115]

Other factors involved in transport have also been implicated in the nuclear envelope and pore assembly processes. It was shown that GTP hydrolysis by the major transport regulator Ran is required for envelope formation.[137] Addition of excess wild-type Ran in a reconstitution promotes nuclear assembly, while a mutant incapable of GTP hydrolysis, RanQ69L, inhibits it, promoting instead formation of annulate lamellae.[138,139] By contrast the RanT24N mutant, which blocks nucleotide exchange by RCC1 and therefore lets Ran accumulate in the GDP form,[140,141] inhibited annulate lamellae formation in assays without chromatin. Applied to nuclear reconstitution, RanT24N and RanQ69L both inhibit early stages of membrane vesicle fusion.[137] Importin β also has a number of striking effects. When added in excess to a reconstitution assay, membrane vesicles accumulate on the chromatin surface but fail to fuse.[138,139] This block can be reversed by excess Ran-GTP, suggesting that the balance of importin β to Ran is important in regulating membrane fusion. In contrast to full-length importin β, a mutant lacking both importin α and Ran binding sites (β 45-463) does not inhibit membrane fusion, but entirely blocks a later stage in pore assembly. Similar to BAPTA, closed nuclear envelopes form but these envelopes are devoid of nuclear pores. The β 45-463 block of pore assembly occurs downstream from the BAPTA block, and it is not reversible by Ran-GTP. This same importin mutant is also a powerful transport inhibitor.[142]

Molecular Dissection and Proteomics

A major interest in understanding the NPC structure is in identifying its molecular components, and then placing them within the assembly. Presumably the molecular structure of each pore component protein, or nucleoporin (Nup), should reveal clues to its role in the global assembly or transport functions. Its location and orientation could be equally revealing. Proteomic studies have achieved what is likely to be a complete catalog of Nups in yeast[13] and mammalian cells.[143] Biochemical preparations from annulate lamellae in *Xenopus* egg extracts yielded a very similar list.[110] Contrary to earlier suppositions that the nuclear pore should contain as many as 100 different proteins, roughly 30 were found in all three cases. With a consensus that the list is more or less complete, it becomes possible to categorize the nucleoporins and to try to build a map of their assembly into the NPC.

During open mitosis, the nuclear pore decomposes into a rather small number of stable sub-complexes of nucleoporins.[144-149] On reassembly, these same multi-protein sub-complexes organize as the basic architectural building blocks of the pore. This is perhaps the most important simplifying aspect in appreciating its molecular structure. Orienting the sub-complexes accurately within the overall pore structure, detecting the order of their accrual, and determining the spectrum of their functional roles, remain to a lesser or greater extent, open challenges. In addition, there exist a number of noncomplexed Nups. These are most notable in prominent locations, i.e., the transmembrane proteins and those making up the peripheral cytoplasmic filaments and nuclear baskets.

Transmembrane Nups anchor the protein assembly into the membrane pore. They may also act as fusogens, joining the two lipid bilayers and producing the incipient "empty" hole seen by FEISEM in assembly reactions.[146,150] These are gp210 and POM121 in vertebrates, and Ncd1, POM34, and POM152 in yeast. The yeast transmembrane nucleoporins, interestingly, show no sequence homology to the vertebrate ones. Gp210 contains a short cytoplasmic tail and a large domain protruding into the NE lumen, making it the obvious candidate for the lumenal ring seen by electron microscopy. POM121, on the other hand, has a large cytoplasmic domain and a short lumenal one, suggesting it as a primary anchor. Photobleaching of a green fluorescent protein fusion to POM121 in live cell cultures showed that it remains associated with the same NPC throughout the cell cycle, again consistent with an architectural role.[151] Moreover the pores were largely immobile within the NE. In yeast, on the other hand, it was shown that pores could move from one nucleus to another in haploid mating assays,[152] suggesting a very different mode of anchoring within the nuclear envelope.

Peripheral NPC structures are also associated with specific Nups. The cytoplasmic filaments of the vertebrate pore are composed primarily (or perhaps entirely) of Nup358.[153,154]

Also known as Ran binding protein 2 (RanBP2), Nup358 binds the Ran GTPase activating protein RanGAP and accelerates its promotion of Ran-bound GTP hydrolysis.[155] This would be a logical termination step for recycling of importin β-type receptors that should be released from Ran on return to the cytoplasm. Targetting of RanGAP to Nup358 depends on a ubiquitin-like SUMO modification.[36-38] The cytoplasmic filaments were also suggested as potential docking sites for import complexes, which would accumulate at the mouth of the pore before traversing it. Nup358 contains a Zn-finger domain, suggesting an interaction with oligonucleotides, perhaps in RNA export.

Xenopus egg extracts that had been depleted of Nup358 yielded reconstituted nuclei lacking cytoplasmic filaments. It came as a great surprise, given the biochemical richness of this protein, to find that the depletion had little effect on nuclear import.[156] Apparently the functions of the filaments are duplicated, or redundant, in spite of their prominent appearance. Yeast and plants have no sequence homolog to Nup358, indicating that its role in transport *per se* must not be entirely essential. High resolution scanning electron microscopy does show what appear to be cytoplasmic filaments decorating a cytoplasmic ring on the yeast nuclear pore.[93]

Nup153 and Tpr are found prominently on the nucleoplasmic face of the vertebrate NPC.[157,158] Nup153 is closely related to the nuclear basket structure, though its precise localization by immuno-labelling in the electron microscope has been controversial due to variability among the antibodies employed. This is exemplified by the finding that antibodies to the N-terminal peptides recognize the proximal nuclear rim, while antibodies to the Zn-finger epitopes place those at the distal ring, and the C-terminus appears to localize without preference.[159] This suggests that the basket filaments may be composed entirely of Nup153, and would logically place Tpr further inward to the nuclear interior. Tpr has a coiled-coil structure, and was associated by immunolabelling with intranuclear fibers.[160,161] Subsequent studies found Tpr more closely linked to the NPCs, particularly at the nuclear basket, while it also appears in a punctate rather than fibrous pattern within the nucleus.[162] A recent work goes so far as to locate Nup153 uniquely to the nucleoplasmic ring of the NPC, and to identify the fibers of the nuclear basket with the coiled-coils of Tpr.[163] Mechanical changes that might indicate some kind of gating in the nuclear basket were seen by atomic force microscopy. The baskets appeared to open and close reversibly on addition or removal of Ca^{++} ions.[11]

Many biochemical pathways converge on Nup153. Antibodies injected to *Xenopus* oocytes blocked snRNA, mRNA, and 5S rRNA, but not tRNA or importin β receptor recycling.[164] Like Nup358, Nup153 has a Zn-finger domain as well as protein-interaction domains. In vitro it interacts with poly(G) and poly(U) RNAs, as well as with importin α/β and transportin. Fragments of the protein containing these docking sites acted as dominant-negative inhibitors of the respective import pathways, when added in excess to in vitro import assays.[158] Immunodepletion of Nup153 from *Xenopus* egg extracts implicate the protein in immobilizing NPCs within the NE, and specifically in importin-mediated transport.[165]

Like Nup153, Tpr has binding sites to importin β, and binds importin α/β complexes in vitro.[158] In *Xenopus* egg extracts this binding is released by GMP-PNP, a nonhydrolyzable analog of GTP. Unlike Nup153, however, Tpr cannot bind importin α/β when the latter are complexed to NLS-bearing cargo. Microinjection of anti-Tpr antibodies to mitotic tissue culture cells blocked the protein's reassociation with the NPCs on return to interphase.[162] Nuclear protein export was inhibited, but import remained unaffected. The yeast homologs of Tpr, myosin-like proteins Mlp1 and Mlp2, form long coiled-coils that project into the nucleus.[166-168] Produced by alternative splicing, they have been implicated as anchors for transcriptionally silent telomeres,[169,170] and their deletion leads to suppression of double-strand break repair.[169,171] Mlp1 is also required for retention of immature, intron-containing mRNAs.[172] These proteins provide a clue to the coupling of nuclear transport with RNA processing and other intranuclear regulatory functions.

   Immobilized fragmients of Nup98 and Nup153 extract a number of other nucleoporins from *Xenopus* egg extracts, suggesting a stable subcomplex. These binding partners include Nup107, Nup133, Nup160, and Nup96.[148] The interaction between Nup107 and Nup133, as well as the interaction with Nup96, were seen independently in yeast two-hybrid screens and by immunoprecipitation.[149] These turn out to be members of a single large pore sub-complex, the so-called Nup107-160 complex, which includes as well Nup96, Sec13, Seh1, Nup37, and Nup43.[173-175] The corresponding yeast Nup84 complex contains Nup85 (homolog of vertebrate Nup85), Nup84 (homolog of vertebrate Nup107), Nup120 (homolog of vertebrate Nup160), Nup145C (homolog of vertebrate Nup96), Sec13, and Seh1.[176] The latter are alternately known as endoplasmic reticulum proteins associated with membrane fusion. The association with Nup153 (or Nup1 in yeast) indicates a contact with the nucleoplasmic face of the NPC. The Nup107-160 complex has a striking architectural role: without it, nuclei form with closed nuclear envelopes entirely lacking nuclear pores. This was hinted to by small inhibitory RNA knockdown, and demonstrated conclusively by quantitative immunodepletion from Xenous egg extracts.[173,174,177] Its association with the membrane pore must be a very early and pivotal step in pore assembly.

   A second subcomplex on the nuclear face or basket includes Nup93, Nup188, Nup205.[178] Nup93 was found by immunogold labeling in electron microscopy to be located at the nuclear face of the NPC and at the basket. Immunodepletion of this complex from *Xenopus* egg extract, via antibodies to Nup93, impaired the growth of nuclei in a reconstitution assay. The number of assembled pores was also greatly reduced. Import of an NLS-bearing substrate was not strongly affected, though. Yeast homologs are Nic96, Nup188, and Nup192. Pore assembly was similarly impaired by thermosensitive mutations in Nic96, at the restrictive temperature.[179] In yeast these Nups were found to be symmetrically distributed between the cytoplasmic and nuclear faces, however.[13]

   The Nup62 complex, consisting of Nup62, Nup58, Nup45, and Nup54,[180] was identified by its affinity for the lectin wheat germ agglutinin (WGA), which indicates glycosylation by O-linked GlcNAc moieties.[181,182] The vertebrate Nup62 complex has a yeast homolog comprising Nsp1, Nup49, and Nup57.[183] When nuclei are reconstituted in *Xenopus* egg extracts depleted of WGA-binding proteins, the classical NLS-based import pathway is blocked.[181] Substrates that normally accumulate in the nuclei are instead excluded. Similarly, microinjection of WGA into living cells blocked NLS-dependent nuclear import, while diffusive entry of 10 kDa dextran was unaffected.[184,185] Electron microscopy shows that internal structure in the pore may be lacking.[186] That the removal of internal structure leads to a block of passage, rather than a block of equilibration, emphasizes the specificity of interactions involved in transport. Ironically, the biochemical importance of the O-GlcNAc modification remains a mystery, even though it provided the first criterion for molecular dissection of the nuclear pore. Other O-GlcNAc bearing Nups are Nup358, Nup214, Nup153, and Nup43. Perhaps glycosylation serves as a moderator of phosphorylation on the same sites through the cell cycle.[187] Plant nucleoporins also show O-linked GlcNAc modification, though the sugars are oligomeric rather than monomeric.[188]

   The fourth major nucleoporin complex is that composed of Nup214/CAN[24,145,189] and Nup88,[190] alternately known as Nup84.[191] Yeast homologs are Nup159 and Nup82, respectively. Mutations in CAN are associated with acute myeloid and undifferentiated leukemia.[192,193] Loss of CAN in vivo, in knockout mouse embryos, affects both nucleocytoplasmic transport and cell cycle progression.[194] This complex lies at the cytoplasmic ring of the NPC. Nup214 interacts with the Crm1 export receptor[45,190] and is implicated as well in mRNA export.[195-197] Nup214 is exploited as a docking receptor for *Adenovirus* in preparation for nuclear import of its DNA.[198]

   Nup98 is perhaps the most enigmatic component of the nuclear pore. It is expressed in two routes of alternative splicing from a gene that includes Nup96 as well.[199] In one pathway

Nup98 is expressed directly as a 98kDa precursor. In the other pathway it is coexpressed with, and then cleaved from Nup96. In both cases an 8 kDa C-terminal fragment is cleaved from the 98 kDa precursor. The autoproteolytic cleavage domain was studied by X-ray crystallography, where the protein was found to interact noncovalently with the cleaved fragment.[200] A similar mechanism generates the homologous yeast nucleoporins N-Nup145p and C-Nup145p.

Injection of polyclonal antibodies against Nup98 to *Xenopus* oocytes inhibits RNA export, but has little effect on protein import.[201] A significant fraction of Nup98 is also found within the nucleus, in specific point-like locations.[202] These were called GLFG bodies because the accumulation of Nup98 there depended on the presence of its GLFG domain. They do not colocalize, however, with other known intranuclear structures. Photobleaching showed that Nup98 is mobile, with a rapid exchange between the pores and the intranuclear pool. Most interestingly, this exchange is blocked by inhibitors of RNA polymerases I and II, leading to the suggestion that Nup98 may actually accompany its cargo to the pore. Biochemically, Nup98 associates with both its genetic partner Nup96, and with Nup88.[203] These are members of distinct sub-complexes, Nup88/214 and Nup107-160. Consistent with this, Nup98 is found on both sides of the pore complex.

The comprehensive proteomics study in yeast included an attempt to localize each of the Nups within the NPC using antibody-labeled colloidal gold particles. Most Nups were found to be distributed symmetrically on both cytoplasmic and nuclear sides of the NPC, with a relative few displaying a strong bias to one side or the other. Only Nup159, Nup42, and Nup82 (homologs of vertebrate Nup214, Nup45/Nup58, and Nup88 respectively) were exclusively cytoplasmic, while Nup1 (homolog of vertebrate Nup153) and Nup60 were exclusively nuclear. Though such a study has yet to be performed in vertebrate pores, the result is still consistent with localization of the corresponding vertebrate peripheral complexes. Within the central framework of the vertebrate nuclear pore, the two-fold structural symmetry is also consistent with a largely symmetric spatial distribution of the major sub-complexes.

The total mass of the NPC could be estimated by quantifying relative fractions of the Nups and assuming population in integer multiples of eight.[13,143] Nup153, for example, appears to be present in a single copy per stave (i.e., 8 per NPC), associated with or possibly comprising the nuclear basket fibers, while Nup58 is six times more abundant. Tallying the total stoichiometry in yeast suggested a total mass of 44 MDa, rather close to the previous estimate of ~55 MDa. More surprisingly, the mass tally in the vertebrate pore comes to only 60 MDa, far less than the 125 MDa mass estimate made by quantitative scanning tunneling electron microscopy (STEM).[97] The discrepancy is currently unexplained. The former method probably represents a lower limit, accounting for the possibility of missed components or nonuniformities in sensitivity to different Nups, while STEM measures total diffracting mass and therefore would include transiently-associated components, cargo in transit, and perhaps a contribution from nonNPC structures such as lamins. The question is in fact crucial, as the contours and surfaces presented in the tomographic reconstructions enclose a certain total mass, and a lower estimate would imply a significantly more open structure.

FG Repeats

For the transport function, the most common and important peptide motifs among the nucleoporins are those containing a high proportion of phenylalanine (F) and glycine (G). Such "FG repeats" appear in about one third of the nucleoporins. In spite of the name they are always interspersed among other residues, and typically joined by hydrophilic linkers. The peripheral Nup153 and Nup358 both contain FG repeats, as do members of the Nup62 complex, and Nup98. In fact there is also considerable variety among the motifs, represented by FG, GLFG, FXFG (X being any amino acid), and likely others not yet recognized. The common feature is their specific interaction with transport receptors. These should satisfy the apparently contradictory requirements of high specificity and short lifetime. Exactly such an interaction was found experimentally for yeast Nup1 with Kap95p-Kap60p heterodimers, yeast homologs of importin β/α.[204] A hypothesis arises naturally that these interactions may be biased in the preferred direction direction of translocation, with the receptors hopping down a gradient in affinity.[12] Supporting this picture, the affinity of importin β for three nucleoporin FG regions increases from Nup358 (cytoplasmic) to Nup62 (central) to Nup153 (nucleoplasmic).[205] In yeast, a similar increase of affinity is observed between Kap95p-Kap60p and nucleoporins Nup42, Nup100, and Nup1.[206]

From the structural point of view, the most interesting aspect of the FG repeat regions is precisely their lack of secondary structure. Based on prediction, circular dichroism and Fourier transform infrared spectroscopy, as well as in situ protease digestion, they are described as natively unfolded domains,[207,208] or more humorously as "oily spaghetti".[14] This suggests a very different picture of the relevant protein-protein interactions than the conventional lock and key, which directly impact the central biophysical mystery of the NPC's function: how can it be so specific to the passage of transport receptors and their complexes, while faithfully excluding other, much smaller protein species.

A number of transport receptors have been cocrystallized with FG domains to highlight the molecular interactions. The structure of importin β shows a spiral wrap of 19 HEAT repeats, each formed by a pair of parallel α-helices.[209-211] Structures of the N-terminal residues 1-442 in combination with FxFG[212] and GLFG[213] show primary interactions with repeats 5 and 6, with the phenylalanine residue wedged between the α-helices in a hydrophobic pocket. An additional binding domain was found biochemically in the C-terminal region involvng HEAT repeats 14-16.[214] Other transport receptors show similar interactions with the essential phenylalanine, though the shape of the binding pocket is different.[213,215]

Transport Models in Relation to Structure

Early models considered an iris-like mechanical action for selective transport of large substrates,[8,9,82] and distinct peripheral channels for diffusion.[83] More recent work showed that both diffusion and active transport occur via a single channel,[89,90] and moreover that the translocation of cargo-receptor complexes does not require hydrolysis of any nucleotide triphosphate.[216,217] The latter is hard to reconcile with a motor-like mechanochemical mechanism. There is evidence, on the other hand, that large substrates may require GTP hydrolysis for translocation, perhaps to release adhesion to nucleoporins along the path.[218]

Recent models attempt to correlate between the special properties expected for FG repeats as natively unfolded structures, and the demands of molecular specificity for function of the nuclear transport apparatus. In general the NPC should appear as a size-selective sieve. Canonical numbers for the cutoff are ~8 nm diameter, as detected by passage of colloidal gold particles observed in electron microscopy,[219,220] or 40-60 kDa molecular weight.[221-226] The relatively large transport receptors (e.g., importin β, 97 kDa) can pass rather freely, and moreover can mediate the passage of smaller molecules whose translocation would otherwise be blocked.

Translocation through the nuclear pore has been likened, rather loosely, to the physical model known as the thermal ratchet,[227] suggesting diffusion between binding sites of progressively increasing affinity.[12] A terminating step should be required in order to liberate the cargo-transport receptor complex from the final, tightest binding site. The recognition of the FG repeats and of the importance of their interactions with the transport receptors led to more specific models, all of which attempt to explain the specificity of transport without invoking mechanochemical reactions that should be dependent on NTP hydrolysis. The term "virtual gating" was offered to describe passive mechanisms of selective translocation.[13]

In order to address selectivity, it was proposed that the unstructured FG repeats project into the NPC central channel, where thermal forces keep them in a constant flailing motion common to flexible polymers. This "entropic exclusion" mechanism leads naturally to a size-based cutoff for nonspecific translocation, since diffusion constants depend inversely on hydrodynamic radii. A protein arriving at the pore would see these filaments as a barrier if they sweep through a large space during the time it would take the arriving protein to diffuse across. If the protein's diffusion is faster than the movement of the filaments, on the other hand, it would slip unhindered through the sieve. Where the translocating species can interact with the FG motif, on the other hand, short-lived binding could dock the transport receptors and their complexes along the way, allowing them to hop from site to site in spite of repulsion due to the filament motion. A gradient in affinity from one site to the next might provide further selectivity and directionality.

An alternate view considers the FG repeats as a self-interacting mesh, or gel, of filaments.[228] Rather than flailing in Brownian motion, such a net would present a static sieve. The size cutoff relevant to noninteracting molecules would relate directly to the spacings in the mesh. If a transport receptor bears an affinity to the filaments, on the other hand, its interactions would replace those between the filaments themselves, and it could be thought to dissolve into the gel. Just as amphiphilic molecules can pass lipid bilayer membranes while polar moieties are excluded, transport receptors might partition to the gel of FG repeats, and once entrapped inside they could exit with equal ease in either direction. It was suggested that hydrophobicity plays an important role in determining the selectivity.[15] Quantitatively, however, such a model requires a delicate tuning of parameters to avoid retention of the receptors by the gel.[229] Also, while importin β is exceptionally hydrophobic, and phenylalanine might give the nucleoporin FG repeats a hydrophobic character, natively unfolded proteins tend to be hydrophilic due to polar interactions on the backbone that are hidden upon folding to α helices and β sheets.

A third model proposes the mechanism of a metastable gel.[230] The FG repeats are still considered to form a statically linked network, but one whose disruption can be catalyzed by the arrival of an interacting transport receptor. Rather than joining the gel, the receptor induces its temporary collapse. The challenge to explain selectivity lies in the requirement that the network should reclose fast enough to prevent random leakage. Undoubtedly the mechanistic proposals will continue to be refined. It is even possible that different substrates, e.g., small proteins or large mRNA, execute their specific translocation through the same pores by fundamentally distinct mechanisms, with different molecular and/or energetic requirements.

The Minimal Pore

A systemmatic deconstruction of the nuclear pore was undertaken in yeast, based on genetic deletions of 11 FG-containing Nups in various combinations.[231] The earlier comprehensive proteomic study of the *Saccharomyces cerevisciae* nuclear pore included an immunolocalization of all the individual nucleoporins by electron microscopy.[13] The general rule was a symmetric distribution of both FG-containing and non-FG Nups on the cytoplasmic and nucleoplasmic sides of the pore. Relatively few showed an exclusive bias to one face or the other. In the case of the FG-Nups, five are asymmetrically located (cytoplasmic: Nup159 and Nup42, nucleoplasmic: Nup1, Nup60, and the mobile nucleoporin Nup2[232-234]) and eight are symmetrically distributed or moderately biased: Nsp1, Nup49, Nup53, Nup57, Nup59, Nup116, Nup100, and Nup145N. Surprisingly, when the asymmetric FG-Nups (excepting Nup53 and Nup59, whose FG repeats are sparse) were removed, the cells remained viable and functional for transport. This included deletions of the entire set, or the cytoplasmic and nucleoplasmic sub-groups independently. Among the symmetric Nups, pairwise depletions showed that GLFG, but not FxFG nucleoporins, were essential. No correlation was found between lethality or transport defect and the amount of protein mass removed by the deletions. It was further possible to define a minimal nuclear pore and to test the function of various transport receptors. Kinetics of import in the classical NLS pathway mediated by yeast Kap60p/Kap95p (homologs to importin (karyopherin) α/β) were slower by a factor of 3 in all the viable mutants. In symmetric Nup mutants, on the other hand, import of the Nab2-NLS of histone H2B1, which uses primarily the transporter Kap114, was strongly inhibited. Thus different transport pathways may depend to some degree on different subsets of the symmetric Nups. The lack of sensitivity to depletion of the asymmetric Nups is especially surprising in light of the translocation models based on affinity gradients for the receptors.

Assembly Revisited

With the catalog of nucleoporins more or less completed and organized into subcomplexes, the questions of nuclear pore assembly can be addressed on a molecular basis. A study using green fluorescent protein labeling in mammalian cell cultures showed that nucleoporin recruitment follows membrane closure in the sequence: POM121, Nup62, Nup214, gp210, Tpr.[146] Nup153 was found to associate to chromatin even before membranes. Since Nup62 and Nup214 are representative members of stable subcomplexes, a plausible order of assembly begins to emerge, with core elements assembling prior to more peripheral ones. Import functionality apparently follows recruitment of the Nup62 complex.[235] In a different approach based on immunodepletion from Xenopus egg extracts, a critical role was found for the large, non-FG repeat Nup107-160 complex. In vitro reconstitution of nuclei in depleted extracts yielded closed nuclear envelopes that completely lacked nuclear pores.[138,139] Presumably the recruitment of this complex to the incipient pore structure takes place at a very early stage.

Summary and Outlook

Great progress has been made in understanding protein components of the nuclear pore complex. Their placement structurally within the pore has also seen progress, though ambiguities remain. Further studies will undoubtedly pinpoint this issue, and make a connection with the deconstruction approach that watches for loss of function. Hopefully these anticipated structural understandings will also help to clarify remaining mysteries surrounding the biophysical mechanism of molecularly selective translocation. Comparison of the animal, yeast, and eventually plant pore proteomes might explain the evolutionary origins and divergence of the NPC.

Misassembly or disfunction of the nuclear pore proteins can lead to severe human disease. CAN/Nup214 has long been associated with leukemia.[192,193] Similarly, rearrangements and fusions of the Nup98 gene are implicated in acute leukemia forms.[236] A novel nucleoporin called ALADIN, discovered in the human proteomic screen,[143] is associated with the genetically heritable triple A syndrome.[237,238] In common with proteins of the nuclear lamina, several nucleoporins are targeted in autoimmune diseases (reviewed in 239). The centrality of the nuclear pore in regulation of gene expression by cytoplasmic signaling mechanisms suggests the involvement of transport deficiencies in a wide range of signaling-related diseases (reviewed in ref. 240).

Finally, the plant nuclear pore has received relatively little attention, compared with its animal or yeast counterpart. The few electron microscopic images that appear in the literature suggest a generally consistent eight-fold symmetric structure. (Indeed the earliest images remain in many ways the most informative.[94]) Very little is known about its protein composition. Bioinformatic screens identify putative *Arabidopsis thaliana* homologs of four human nucleoporins: Tpr, Nup98, Nup155, and gp210.[241,242] While homologs of Ran and RanGAP are found in plants, the RanGEF RCC1 is still missing. As in yeast, there is apparently no Nup358/RanBP2 homolog. Among transport factors, it was shown that nuclear import depends, at least in tested cases, on importin α alone.[243] Most surprisingly, protein import to isolated plant nuclei in vitro does not require soluble factors.[244,245] All the required biochemistry apparently resides on the pore itself, or can be supplied from within the nucleus. Clearly, the plant nuclear pore provides a fertile ground for new discovery.

Acknowledgements

I would like to thank members of my lab and of the Weizmann Institute Electron Microscopy Unit for assistance with original figures, particularly Aurelie Lachish-Zalait, Ella Zimmerman, Ilana Sabanay and Sharon Wolf. This work was supported in part by the United States-Israel Binational Science Foundation, and the United States-Israel Binational Agricultural Research and Development Fund, and the Human Frontier Science Program.

References

1. Gorlich D, Kutay U. Transport between the cell nucleus and the cytoplasm. Annu Rev Cell Dev Biol 1999; 15:607-660.
2. Goldfarb DS, Gariepy J, Schoolnik G et al. Synthetic peptides as nuclear localization signals. Nature 1986; 322:641-644.
3. Kalderon D, Roberts BL, Richardson WD et al. A short amino acid sequence able to specify nuclear location. Cell 1984; 39:499-509.
4. Lanford RE, Butel JS. Construction and characterization of an SV40 mutant defective in nuclear transport of T antigen. Cell 1984; 37:801-13.
5. Gorlich D, Mattaj IW. Nucleocytoplasmic transport. Science 1996; 271:1513-1518.
6. Mattaj IW, Englmeier L. Nucleocytoplasmic transport: The soluble phase. Annu Rev Biochem 1998; 67:265-306.
7. Yoneda Y. Nucleocytoplasmic protein traffic and its significance to cell function. Genes Cells 2000; 5:777-87.
8. Pante N, Aebi U. Toward the molecular dissection of protein import into nuclei. Curr Opin Cell Biol 1996; 8:397-406.
9. Pante N, Aebi U. Molecular dissection of the nuclear pore complex. Crit Rev Biochem Mol Biol 1996; 31:153-99.
10. Kiseleva E, Goldberg MW, Cronshaw J et al. The nuclear pore complex: Structure, function, and dynamics. Crit Rev Eukaryot Gene Expr 2000; 10:101-12.
11. Stoffler D, Goldie KN, Feja B et al. Calcium-mediated structural changes of native nuclear pore complexes monitored by time-lapse atomic force microscopy. J Mol Biol 1999; 287:741-52.
12. Rexach M, Blobel G. Protein import into nuclei: Association and dissociation reactions involving transport substrate, transport factors, and nucleoporins. Cell 1995; 83:683-92.
13. Rout MP, Aitchison JD, Suprapto A et al. The yeast nuclear pore complex: Composition, architecture, and transport mechanism. J Cell Biol 2000; 148:635-51.
14. Macara IG. Transport into and out of the nucleus. Microbiol Mol Biol Rev 2001; 65:570-94, table of contents.
15. Ribbeck K, Gorlich D. The permeability barrier of nuclear pore complexes appears to operate via hydrophobic exclusion. EMBO J 2002; 21:2664-71.
16. Perez-Terzic C, Pyle J, Jaconi M et al. Conformational states of the nuclear pore complex induced by depletion of nuclear Ca2+ stores. Science 1996; 273:1875-7.
17. Strubing C, Clapham DE. Active nuclear import and export is independent of lumenal Ca2+ stores in intact mammalian cells. J Gen Physiol 1999; 113:239-48.
18. Wei X, Henke VG, Strubing C et al. Real-time imaging of nuclear permeation by EGFP in single intact cells. Biophys J 2003; 84:1317-27.
19. Moore MS. Ran and nuclear transport. J Biol Chem 1998; 273:22857-60.
20. Rout MP, Aitchison JD. The nuclear pore complex as a transport machine. J Biol Chem 2001; 276:16593-6.
21. Elbaum M. The nuclear pore complex: Biochemical machine or Maxwell demon? Paris: CR Acad Sci 2000; 2:681-807.
22. Weis K. Nucleocytoplasmic transport: Cargo trafficking across the border. Curr Opin Cell Biol 2002; 14:328-35.
23. Gorlich D, Prehn S, Laskey RA et al. Isolation of a protein that is essential for the first step of nuclear protein import. Cell 1994; 79:767-78.

24. Radu A, Blobel G, Moore MS. Identification of a protein complex that is required for nuclear protein import and mediates docking of import substrate to distinct nucleoporins. Proc Natl Acad Sci USA 1995; 92:1769-73.

25. Iovine MK, Watkins JL, Wente SR. The GLFG repetitive region of the nucleoporin Nup116p interacts with Kap95p, an essential yeast nuclear import factor. J Cell Biol 1995; 131:1699-713.

26. Chi NC, Adam EJ, Adam SA. Sequence and characterization of cytoplasmic nuclear protein import factor p97. J Cell Biol 1995; 130:265-74.

27. Moore MS, Blobel G. The GTP-binding protein Ran/TC4 is required for protein import into the nucleus. Nature 1993; 365:661-3.

28. Melchior F, Paschal B, Evans J et al. Inhibition of nuclear protein import by nonhydrolyzable analogues of GTP and identification of the small GTPase Ran/TC4 as an essential transport factor. J Cell Biol 1993; 123:1649-59.

29. Paschal BM, Delphin C, Gerace L. Nucleotide-specific interaction of Ran/TC4 with nuclear transport factors NTF2 and p97. Proc Natl Acad Sci USA 1996; 93:7679-83.

30. Moroianu J, Blobel G, Radu A. Nuclear protein import: Ran-GTP dissociates the karyopherin alphabeta heterodimer by displacing alpha from an overlapping binding site on beta. Proc Natl Acad Sci USA 1996; 93:7059-62.

31. Koepp DM, Silver PA. A GTPase controlling nuclear trafficking: Running the right way or walking RANdomly? Cell 1996; 87:1-4.

32. Avis JM, Clarke PR. Ran, a GTPase involved in nuclear processes: Its regulators and effectors. J Cell Sci 1996; 109(Pt 10):2423-7.

33. Coutavas E, Ren M, Oppenheim JD et al. Characterization of proteins that interact with the cell-cycle regulatory protein Ran/TC4. Nature 1993; 366:585-7.

34. Bischoff FR, Klebe C, Kretschmer J et al. RanGAP1 induces GTPase activity of nuclear Ras-related Ran. Proc Natl Acad Sci USA 1994; 91:2587-91.

35. Izaurralde E, Kutay U, von Kobbe C et al. The asymmetric distribution of the constituents of the Ran system is essential for transport into and out of the nucleus. EMBO J 1997; 16:6535-47.

36. Mahajan R, Delphin C, Guan T et al. A small ubiquitin-related polypeptide involved in targeting RanGAP1 to nuclear pore complex protein RanBP2. Cell 1997; 88:97-107.

37. Matunis MJ, Coutavas E, Blobel G. A novel ubiquitin-like modification modulates the partitioning of the Ran-GTPase-activating protein RanGAP1 between the cytosol and the nuclear pore complex. J Cell Biol 1996; 135:1457-70.

38. Matunis MJ, Wu J, Blobel G. SUMO-1 modification and its role in targeting the Ran GTPase-activating protein, RanGAP1, to the nuclear pore complex. J Cell Biol 1998; 140:499-509.

39. Kalab P, Weis K, Heald R. Visualization of a Ran-GTP gradient in interphase and mitotic Xenopus egg extracts. Science 2002; 295:2452-6.

40. Smith AE, Slepchenko BM, Schaff JC et al. Systems analysis of Ran transport. Science 2002; 295:488-91.

41. Gorlich D, Seewald MJ, Ribbeck K. Characterization of Ran-driven cargo transport and the RanGTPase system by kinetic measurements and computer simulation. EMBO J 2003; 22:1088-100.

42. Nachury MV, Weis K. The direction of transport through the nuclear pore can be inverted. Proc Natl Acad Sci USA 1999; 96:9622-7.

43. Ossareh-Nazari B, Bachelerie F, Dargemont C. Evidence for a role of CRM1 in signal-mediated nuclear protein export. Science 1997; 278:141-4.

44. Stade K, Ford CS, Guthrie C et al. Exportin 1 (Crm1p) is an essential nuclear export factor. Cell 1997; 90:1041-50.

45. Fornerod M, Ohno M, Yoshida M et al. CRM1 is an export receptor for leucine-rich nuclear export signals. Cell 1997; 90:1051-60.

46. Neville M, Stutz F, Lee L et al. The importin-beta family member Crm1p bridges the interaction between Rev and the nuclear pore complex during nuclear export. Curr Biol 1997; 7:767-75.

47. Fukuda M, Asano S, Nakamura T et al. CRM1 is responsible for intracellular transport mediated by the nuclear export signal. Nature 1997; 390:308-11.

48. Pasquinelli AE, Powers MA, Lund E et al. Inhibition of mRNA export in vertebrate cells by nuclear export signal conjugates. Proc Natl Acad Sci USA 1997; 94:14394-9.

49. Kudo NS, Khochbin K, Nishi K et al. Molecular cloning and cell cycle-dependent expression of mammalian CRM1, a protein involved in nuclear export of proteins. J Biol Chem 1997; 272:29742-51.

50. Kutay U, Bischoff FR, Kostka S et al. Export of importin alpha from the nucleus is mediated by a specific nuclear transport factor. Cell 1997; 90:1061-71.

51. Arts GJ, Fornerod M, Mattaj IW. Identification of a nuclear export receptor for tRNA. Curr Biol 1998; 8:305-14.

52. Kutay U, Lipowsky G, Izaurralde E et al. Identification of a tRNA-specific nuclear export receptor. Mol Cell 1998; 1:359-69.

53. Kiseleva E, Goldberg MW, Allen TD et al. Active nuclear pore complexes in Chironomus: Visualization of transporter configurations related to mRNP export. J Cell Sci 1998; 111(Pt 2):223-36.

54. Wiese C, Wilde A, Moore MS et al. Role of importin-beta in coupling Ran to downstream targets in microtubule assembly. Science 2001; 291:653-6.

55. Nachury MV, Maresca TJ, Salmon WC et al. Importin beta is a mitotic target of the small GTPase Ran in spindle assembly. Cell 2001; 104:95-106.

56. Gruss OJ, Carazo-Salas RE, Schatz CA et al. Ran induces spindle assembly by reversing the inhibitory effect of importin alpha on TPX2 activity. Cell 2001; 104:83-93.

57. Fagotto F, Gluck U, Gumbiner BM. Nuclear localization signal-independent and importin/karyopherin-independent nuclear import of beta-catenin. Curr Biol 1998; 8:181-90.

58. Yokoya F, Imamoto N, Tachibana T et al. Beta-catenin can be transported into the nucleus in a Ran-unassisted manner. Mol Biol Cell 1999; 10:1119-31.

59. Xu L, Massague J. Nucleocytoplasmic shuttling of signal transducers. Nat Rev Mol Cell Biol 2004; 5:209-19.

60. Callan HG, Tomlin SG. Experimental studies on amphibian oocyte nuclei. I. Investigation of the structure of the nuclear membrane by means of the electron microscope. Proc R Soc Lond B Biol Sci 1950; 137:367-78.

61. Gall JG. Observations on the nuclear membrane with the electron microscope. Exp Cell Res 1954; 7:197-200.

62. Watson ML. Further observations on the nuclear envelope of the animal cell. J Biophys Biochem Cytol 1959; 6:147-56.

63. Gall JG. Octagonal nuclear pores. J Cell Biol 1967; 32:391-9.

64. Daniels EW, McNiff JM, Longwell AC. Ultrastructure of oyster gametes. ANL-7635. ANL Rep 1969; 195-9.

65. Franke WW, Scheer U. The ultrastructure of the nuclear envelope of amphibian oocytes: A reinvestigation. I. The mature oocyte. J Ultrastruct Res 1970; 30:288-316.

66. Franke WW, Scheer U. The ultrastructure of the nuclear envelope of amphibian oocytes: A reinvestigation. II. The immature oocyte and dynamic aspects. J Ultrastruct Res 1970; 30:317-27.

67. Maul GG. On the octagonality of the nuclear pore complex. J Cell Biol 1971; 51:558-63.

68. Feldherr CM. Structure and function of the nuclear envelope. Adv Cell Mol Biol 1972; 2.

69. Markovics J, Glass L, Maul GG. Pore patterns on nuclear membranes. Exp Cell Res 1974; 85:443-51.

70. Kirschner RH, Rusli M, Martin TE. Characterization of the nuclear envelope, pore complexes, and dense lamina of mouse liver nuclei by high resolution scanning electron microscopy. J Cell Biol 1977; 72:118-32.

71. Ris H. The three dimensional structure of the nuclear pore complex as seen by high voltage electron microscopy and high resolution low voltage scanning electron microscopy. EMSA Bulletin 1991; 21:54-56.

72. Goldberg MW, Allen TD. High resolution scanning electron microscopy of the nuclear envelope: Demonstration of a new, regular, fibrous lattice attached to the baskets of the nucleoplasmic face of the nuclear pores. J Cell Biol 1992; 119:1429-40.

73. Whytock S, Stewart M. Preparation of shadowed nuclear envelopes from Xenopus oocyte germinal vesicles for electron microscopy. J Microsc 1988; 151(Pt 2):115-26.

74. Jarnik M, Aebi U. Toward a more complete 3-D structure of the nuclear pore complex. J Struct Biol 1991; 107:291-308.

75. Oberleithner H, Brinckmann E, Schwab A et al. Imaging nuclear pores of aldosterone-sensitive kidney cells by atomic force microscopy. Proc Natl Acad Sci USA 1994; 91:9784-8.

76. Wang H, Clapham DE. Conformational changes of the in situ nuclear pore complex. Biophys J 1999; 77:241-7.

77. Nevo R, Markiewicz P, Kapon R et al. High-resolution imaging of the nuclear pore complex by AC scanning force microscopy. Single Mol 2000; 1(1):109-114.

78. Jaggi RD, Franco-Obregon A, Muhlhausser P et al. Modulation of nuclear pore topology by transport modifiers. Biophys J 2003; 84:665-70.

79. Unwin PN, Milligan RA. A large particle associated with the perimeter of the nuclear pore complex. J Cell Biol 1982; 93:63-75.

80. Akey CW. Interactions and structure of the nuclear pore complex revealed by cryo-electron microscopy. J Cell Biol 1989; 109:955-70.

81. Akey CW, Goldfarb DS. Protein import through the nuclear pore complex is a multistep process. J Cell Biol 1989; 109:971-82.

82. Akey CW. Visualization of transport-related configurations of the nuclear pore transporter. Biophys J 1990; 58:341-55.

83. Hinshaw JE, Carragher BO, Milligan RA. Architecture and design of the nuclear pore complex. Cell 1992; 69:1133-41.

84. Akey CW, Radermacher M. Architecture of the Xenopus nuclear pore complex revealed by three-dimensional cryo-electron microscopy. J Cell Biol 1993; 122:1-19.

85. Akey CW. Structural plasticity of the nuclear pore complex. J Mol Biol 1995; 248:273-93.

86. Stoffler D, Feja B, Fahrenkrog B et al. Cryo-electron tomography provides novel insights into nuclear pore architecture: Implications for nucleocytoplasmic transport. J Mol Biol 2003; 328:119-30.

87. Fahrenkrog B, Aebi U. The nuclear pore complex: Nucleocytoplasmic transport and beyond. Nat Rev Mol Cell Biol 2003; 4:757-66.

88. Goldberg MW, Solovei I, Allen TD. Nuclear pore complex structure in birds. J Struct Biol 1997; 119:284-94.

89. Feldherr CM, Akin D. The location of the transport gate in the nuclear pore complex. J Cell Sci 1997; 110(Pt 24):3065-70.

90. Keminer O, Peters R. Permeability of single nuclear pores. Biophys J 1999; 77:217-28.

91. Rout MP, Blobel G. Isolation of the yeast nuclear pore complex. J Cell Biol 1993; 123:771-83.

92. Yang Q, Rout MP, Akey CW. Three-dimensional architecture of the isolated yeast nuclear pore complex: Functional and evolutionary implications. Mol Cell 1998; 1:223-34.

93. Kiseleva E, Allen TD, Rutherford S et al. Yeast nuclear pore complexes have a cytoplasmic ring and internal filaments. J Struct Biol 2004; 145:272-88.

94. Roberts K, Northcote DH. Structure of the nuclear pore in higher plants. Nature 1970; 228:385-6.

95. Franke WW, Scheer U, Fritsch H. Intranuclear and cytoplasmic annulate lamellae in plant cells. J Cell Biol 1972; 53:823-7.

96. Hicks GR, Raikhel NV. Specific binding of nuclear localization sequences to plant nuclei. Plant Cell 1993; 5:983-94.

97. Reichelt R, Holzenburg A, Buhle Jr EL et al. Correlation between structure and mass distribution of the nuclear pore complex and of distinct pore complex components. J Cell Biol 1990; 110:883-94.

98. Krohne G, Franke WW, Scheer U. The major polypeptides of the nuclear pore complex. Exp Cell Res 1978; 116:85-102.

99. Goldberg MW, Allen TD. The nuclear pore complex and lamina: Three-dimensional structures and interactions determined by field emission in-lens scanning electron microscopy. J Mol Biol 1996; 257:848-65.

100. Goldberg MW, Allen TD. The nuclear pore complex: Three-dimensional surface structure revealed by field emission, in-lens scanning electron microscopy, with underlying structure uncovered by proteolysis. J Cell Sci 1993; 106:261-74.

101. Hinshaw JE, Milligan RA. Nuclear pore complexes exceeding eightfold rotational symmetry. J Struct Biol 2003; 141:259-68.

102. Wischnitzer S. The annulate lamellae. Int Rev Cytol 1970; 27:65-100.

103. Stafstrom JP, Staehelin LA. Dynamics of the nuclear envelope and of nuclear pore complexes during mitosis in the Drosophila embryo. Eur J Cell Biol 1984; 34:179-89.

104. Dabauvalle MC, Loos K, Merkert H et al. Spontaneous assembly of pore complex-containing membranes ("annulate lamellae") in Xenopus egg extract in the absence of chromatin. J Cell Biol 1991; 112:1073-82.
105. Kessel RG. Annulate lamellae: A last frontier in cellular organelles. Int Rev Cytol 1992; 133:43-120.
106. Meier E, Miller BR, Forbes DJ. Nuclear pore complex assembly studied with a biochemical assay for annulate lamellae formation. J Cell Biol 1995; 129:1459-72.
107. Cordes VC, Reidenbach S, Franke WW. Cytoplasmic annulate lamellae in cultured cells: Composition, distribution, and mitotic behavior. Cell Tissue Res 1996; 284:177-91.
108. Ewald A, Hofbauer S, Dabauvalle MC et al. Preassembly of annulate lamellae in egg extracts inhibits nuclear pore complex formation, but not nuclear membrane assembly. Eur J Cell Biol 1997; 73:259-69.
109. Ewald A, Kossner U, Scheer U et al. A biochemical and immunological comparison of nuclear and cytoplasmic pore complexes. J Cell Sci 1996; 109(Pt 7):1813-24.
110. Miller BR, Forbes DJ. Purification of the vertebrate nuclear pore complex by biochemical criteria. Traffic 2000; 1:941-51.
111. Onischenko EA, Gubanova NV, Kieselbach T et al. Annulate lamellae play only a minor role in the storage of excess nucleoporins in Drosophila embryos. Traffic 2004; 5:152-64.
112. Maul GG. The nuclear and the cytoplasmic pore complex: Structure, dynamics, distribution, and evolution. Int Rev Cytol Suppl 1977; 75-186.
113. Stafstrom JP, Staehelin LA. Are annulate lamellae in the Drosophila embryo the result of overproduction of nuclear pore components? J Cell Biol 1984; 98:699-708.
114. Harel A, Zlotkin E, Nainudel-Epszteyn S et al. Persistence of major nuclear envelope antigens in an envelope-like structure during mitosis in Drosophila melanogaster embryos. J Cell Sci 1989; 94(Pt 3):463-70.
115. Kiseleva E, Rutherford S, Cotter LM et al. Steps of nuclear pore complex disassembly and reassembly during mitosis in early Drosophila embryos. J Cell Sci 2001; 114:3607-18.
116. Lohka MJ, Masui Y. Formation in vitro of sperm pronuclei and mitotic chromosomes induced by amphibian ooplasmic components. Science 1983; 220:719-21.
117. Zhang H, Ruderman JV. Differential replication capacities of G1 and S-phase extracts from sea urchin eggs. J Cell Sci 1993; 104(Pt 2):565-72.
118. Cameron LA, Poccia DL. In vitro development of the sea urchin male pronucleus. Dev Biol 1994; 162:568-78.
119. Li CJ, Gui JF. Comparative studies on in vitro sperm decondensation and pronucleus formation in egg extracts between gynogenetic and bisexual fish. Cell Res 2003; 13:159-69.
120. Ulitzur N, Gruenbaum Y. Nuclear envelope assembly around sperm chromatin in cell-free preparations from Drosophila embryos. FEBS Lett 1989; 259:113-6.
121. Berrios M, Avilion AA. Nuclear formation in a Drosophila cell-free system. Exp Cell Res 1990; 191:64-70.
122. Burke B, Gerace L. A cell free system to study reassembly of the nuclear envelope at the end of mitosis. Cell 1986; 44:639-52.
123. Suprynowicz FA, Gerace L. A fractionated cell-free system for analysis of prophase nuclear disassembly. J Cell Biol 1986; 103:2073-81.
124. Lohka MJ. The reconstitution of nuclear envelopes in cell-free extracts. Cell Biol Int Rep 1988; 12:833-48.
125. Laskey RA, Leno GH. Assembly of the cell nucleus. Trends Genet 1990; 6:406-10.
126. Zhao Y, Liu X, Wu M et al. In vitro nuclear reconstitution could be induced in a plant cell-free system. FEBS Lett 2000; 480:208-12.
127. Lu P, Zhai ZH. Nuclear assembly of demembranated Xenopus sperm in plant cell-free extracts from Nicotiana ovules. Exp Cell Res 2001; 270:96-101.
128. Newmeyer DD, Wilson KL. Egg extracts for nuclear import and nuclear assembly reactions. Methods Cell Biol 1991; 36:607-34.
129. Murray AW. Cell cycle extracts. Methods Cell Biol 1991; 36:581-605.
130. Smythe C, Newport JW. Systems for the study of nuclear assembly, DNA replication, and nuclear breakdown in Xenopus laevis egg extracts. Methods Cell Biol 1991; 35:449-68.

131. Lohka MJ. Analysis of nuclear envelope assembly using extracts of Xenopus eggs. Methods Cell Biol 1998; 53:367-95.
132. Allan VJ. Organelle motility and membrane network formation in metaphase and interphase cell-free extracts. Methods Enzymol 1998; 298:339-53.
133. Desai A, Murray A, Mitchison TJ et al. The use of Xenopus egg extracts to study mitotic spindle assembly and function in vitro. Methods Cell Biol 1999; 61:385-412.
134. Newmeyer DD, Finlay DR, Forbes DJ. In vitro transport of a fluorescent nuclear protein and exclusion of nonnuclear proteins. J Cell Biol 1986; 103:2091-102.
135. Macaulay C, Forbes DJ. Assembly of the nuclear pore: Biochemically distinct steps revealed with NEM, GTP gamma S, and BAPTA. J Cell Biol 1996; 132:5-20.
136. Goldberg MW, Wiese C, Allen TD et al. Dimples, pores, star-rings, and thin rings on growing nuclear envelopes: Evidence for structural intermediates in nuclear pore complex assembly. J Cell Sci 1997; 110(Pt 4):409-20.
137. Hetzer M, Bilbao-Cortes D, Walther TC et al. GTP hydrolysis by Ran is required for nuclear envelope assembly. Mol Cell 2000; 5:1013-24.
138. Walther TC, Askjaer P, Gentzel M et al. RanGTP mediates nuclear pore complex assembly. Nature 2003; 424:689-94.
139. Harel A, Chan RC, Lachish-Zalait A et al. Importin beta negatively regulates nuclear membrane fusion and nuclear pore complex assembly. Mol Biol Cell 2003; 14:4387-96.
140. Dasso M, Seki T, Azuma Y et al. A mutant form of the Ran/TC4 protein disrupts nuclear function in Xenopus laevis egg extracts by inhibiting the RCC1 protein, a regulator of chromosome condensation. EMBO J 1994; 13:5732-44.
141. Klebe C, Bischoff FR, Ponstingl H et al. Interaction of the nuclear GTP-binding protein Ran with its regulatory proteins RCC1 and RanGAP1. Biochemistry. 1995; 34:639-47.
142. Kutay U, Izaurralde E, Bischoff FR et al. Dominant-negative mutants of importin-beta block multiple pathways of import and export through the nuclear pore complex. EMBO J 1997; 16:1153-63.
143. Cronshaw JM, Krutchinsky AN, Zhang W et al. Proteomic analysis of the mammalian nuclear pore complex. J Cell Biol 2002; 158:915-27.
144. Dabauvalle MC, Loos K, Scheer U. Identification of a soluble precursor complex essential for nuclear pore assembly in vitro. Chromosoma 1990; 100:56-66.
145. Macaulay C, Meier E, Forbes DJ. Differential mitotic phosphorylation of proteins of the nuclear pore complex. J Biol Chem 1995; 270:254-62.
146. Bodoor K, Shaikh S, Salina Det et al. Sequential recruitment of NPC proteins to the nuclear periphery at the end of mitosis. J Cell Sci. 1999; 112(Pt 13):2253-64.
147. Matsuoka Y, Takagi M, Ban T et al. Identification and characterization of nuclear pore subcomplexes in mitotic extract of human somatic cells. Biochem Biophys Res Commun 1999; 254:417-23.
148. Vasu S, Shah S, Orjalo A et al. Novel vertebrate nucleoporins Nup133 and Nup160 play a role in mRNA export. J Cell Biol 2001; 155:339-54.
149. Belgareh N, Rabut G, Bai SW et al. An evolutionarily conserved NPC subcomplex, which redistributes in part to kinetochores in mammalian cells. J Cell Biol 2001; 154:1147-60.
150. Drummond SP, Wilson KL. Interference with the cytoplasmic tail of gp210 disrupts "close apposition" of nuclear membranes and blocks nuclear pore dilation. J Cell Biol 2002; 158:53-62.
151. Daigle N, Beaudouin J, Hartnell L et al. Nuclear pore complexes form immobile networks and have a very low turnover in live mammalian cells. J Cell Biol 2001; 154:71-84.
152. Bucci M, Wente SR. In vivo dynamics of nuclear pore complexes in yeast. J Cell Biol 1997; 136:1185-99.
153. Yokoyama N, Hayashi N, Seki T et al. A giant nucleopore protein that binds Ran/TC4. Nature 1995; 376:184-8.
154. Wilken N, Senecal JL, Scheer U et al. Localization of the Ran-GTP binding protein RanBP2 at the cytoplasmic side of the nuclear pore complex. Eur J Cell Biol 1995; 68:211-9.
155. Bischoff FR, Krebber H, Kempf T et al. Human RanGTPase-activating protein RanGAP1 is a homologue of yeast Rna1p involved in mRNA processing and transport. Proc Natl Acad Sci USA 1995; 92:1749-53.
156. Walther TC, Pickersgill HS, Cordes VC et al. The cytoplasmic filaments of the nuclear pore complex are dispensable for selective nuclear protein import. J Cell Biol 2002; 158:63-77.

157. Sukegawa J, Blobel G. A nuclear pore complex protein that contains zinc finger motifs, binds DNA, and faces the nucleoplasm. Cell 1993; 72:29-38.
158. Shah S, Tugendreich S, Forbes D. Major binding sites for the nuclear import receptor are the internal nucleoporin Nup153 and the adjacent nuclear filament protein Tpr. J Cell Biol 1998; 141:31-49.
159. Fahrenkrog B, Maco B, Fager AM et al. Domain-specific antibodies reveal multiple-site topology of Nup153 within the nuclear pore complex. J Struct Biol 2002; 140:254-67.
160. Cordes VC, Reidenbach S, Rackwitz HR et al. Identification of protein p270/Tpr as a constitutive component of the nuclear pore complex-attached intranuclear filaments. J Cell Biol 1997; 136:515-29.
161. Zimowska G, Aris JP, Paddy MR. A Drosophila Tpr protein homolog is localized both in the extrachromosomal channel network and to nuclear pore complexes. J Cell Sci 1997; 110(Pt 8):927-44.
162. Frosst P, Guan T, Subauste C et al. Tpr is localized within the nuclear basket of the pore complex and has a role in nuclear protein export. J Cell Biol 2002; 156:617-30.
163. Krull S, Thyberg J, Bjorkroth B et al. Nucleoporins as components of the nuclear pore complex core structure and tpr as the architectural element of the nuclear basket. Mol Biol Cell 2004; 15:4261-77.
164. Ullman KS, Shah S, Powers MA. The nucleoporin nup153 plays a critical role in multiple types of nuclear export. Mol Biol Cell 1999; 10:649-64.
165. Walther TC, Fornerod M, Pickersgill H et al. The nucleoporin Nup153 is required for nuclear pore basket formation, nuclear pore complex anchoring and import of a subset of nuclear proteins. EMBO J 2001; 20:5703-14.
166. Kolling R, Nguyen T, Chen EY et al. A new yeast gene with a myosin-like heptad repeat structure. Mol Gen Genet 1993; 237:359-69.
167. Strambio-de-Castillia C, Blobel G, Rout MP. Proteins connecting the nuclear pore complex with the nuclear interior. J Cell Biol 1999; 144:839-55.
168. Kosova B, Pante N, Rollenhagen C et al. Mlp2p, a component of nuclear pore attached intra-nuclear filaments, associates with nic96p. J Biol Chem 2000; 275:343-50.
169. Galy V, Olivo-Marin JC, Scherthan H et al. Nuclear pore complexes in the organization of silent telomeric chromatin. Nature 2000; 403:108-12.
170. Feuerbach F, Galy V, Trelles-Sticken E et al. Nuclear architecture and spatial positioning help establish transcriptional states of telomeres in yeast. Nat Cell Biol 2002; 4:214-21.
171. Hediger F, Dubrana K, Gasser SM. Myosin-like proteins 1 and 2 are not required for silencing or telomere anchoring, but act in the Tel1 pathway of telomere length control. J Struct Biol 2002; 140:79-91.
172. Galy V, Gadal O, Fromont-Racine M et al. Nuclear retention of unspliced mRNAs in yeast is mediated by perinuclear Mlp1. Cell 2004; 116:63-73.
173. Harel A, Orjalo AV, Vincent T et al. Removal of a single pore subcomplex results in vertebrate nuclei devoid of nuclear pores. Mol Cell 2003; 11:853-64.
174. Walther TC, Alves A, Pickersgill H et al. The conserved Nup107-160 complex is critical for nuclear pore complex assembly. Cell 2003; 113:195-206.
175. Loiodice I, Alves A, Rabut G et al. The entire Nup107-160 complex, including three new members, is targeted as one entity to kinetochores in mitosis. Mol Biol Cell 2004; 15:3333-44.
176. Siniossoglou S, Wimmer C, Rieger M et al. A novel complex of nucleoporins, which includes Sec13p and a Sec13p homolog, is essential for normal nuclear pores. Cell 1996; 84:265-75.
177. Boehmer T, Enninga J, Dales S et al. Depletion of a single nucleoporin, Nup107, prevents the assembly of a subset of nucleoporins into the nuclear pore complex. Proc Natl Acad Sci USA 2003; 100:981-5.
178. Grandi P, Dang T, Pane N et al. Nup93, a vertebrate homologue of yeast Nic96p, forms a complex with a novel 205-kDa protein and is required for correct nuclear pore assembly. Mol Biol Cell 1997; 8:2017-38.
179. Zabel U, Doye V, Tekotte H et al. Nic96p is required for nuclear pore formation and functionally interacts with a novel nucleoporin, Nup188p. J Cell Biol 1996; 133:1141-52.

180. Finlay DR, Meier E, Bradley P et al. A complex of nuclear pore proteins required for pore function. J Cell Biol 1991; 114:169-83.

181. Finlay DR, Newmeyer DD, Price TM et al. Inhibition of in vitro nuclear transport by a lectin that binds to nuclear pores. J Cell Biol 1987; 104:189-200.

182. Hanover JA, Cohen CK, Willingham MC et al. O-linked N-acetylglucosamine is attached to proteins of the nuclear pore. Evidence for cytoplasmic and nucleoplasmic glycoproteins. J Biol Chem 1987; 262:9887-94.

183. Schlaich NL, Haner M, Lustig A et al. In vitro reconstitution of a heterotrimeric nucleoporin complex consisting of recombinant Nsp1p, Nup49p, and Nup57p. Mol Biol Cell 1997; 8:33-46.

184. Dabauvalle MC, Schulz B, Scheer U et al. Inhibition of nuclear accumulation of karyophilic proteins in living cells by microinjection of the lectin wheat germ agglutinin. Exp Cell Res 1988; 174:291-6.

185. Wolff B, Willingham MC, Hanover JA. Nuclear protein import: Specificity for transport across the nuclear pore. Exp Cell Res 1988; 178:318-34.

186. Miller MW, Hanover JA. Functional nuclear pores reconstituted with beta 1-4 galactose modified O-linked N-acetylglucosamine glycoproteins. J Biol Chem 1994; 269:9289-97.

187. Miller MW, Caracciolo MR, Berlin WK et al. Phosphorylation and glycosylation of nucleoporins. Arch Biochem Biophys 1999; 367:51-60.

188. Heese-Peck A, Cole RN, Borkhsenious ON et al. Plant nuclear pore complex proteins are modified by novel oligosaccharides with terminal N-acetylglucosamine. Plant Cell 1995; 7:1459-71.

189. Kraemer D, Wozniak RW, Blobel G et al. The human CAN protein, a putative oncogene product associated with myeloid leukemogenesis, is a nuclear pore complex protein that faces the cytoplasm. Proc Natl Acad Sci USA 1994; 91:1519-23.

190. Fornerod M, van Deursen J, van Baal S et al. The human homologue of yeast CRM1 is in a dynamic subcomplex with CAN/Nup214 and a novel nuclear pore component Nup88. EMBO J 1997; 16:807-16.

191. Bastos R, Ribas de Pouplana L, Enarson M et al. Nup84, a novel nucleoporin that is associated with CAN/Nup214 on the cytoplasmic face of the nuclear pore complex. J Cell Biol 1997; 137:989-1000.

192. von Lindern M, van Baal S, Wiegant J et al. Can, a putative oncogene associated with myeloid leukemogenesis, may be activated by fusion of its 3' half to different genes: Characterization of the set gene. Mol Cell Biol 1992; 12:3346-55.

193. von Lindern M, Fornerod M, van Baal S et al. The translocation (6;9), associated with a specific subtype of acute myeloid leukemia, results in the fusion of two genes, dek and can, and the expression of a chimeric, leukemia-specific dek-can mRNA. Mol Cell Biol 1992; 12:1687-97.

194. van Deursen J, Boer J, Kasper L et al. G2 arrest and impaired nucleocytoplasmic transport in mouse embryos lacking the proto-oncogene CAN/Nup214. EMBO J 1996; 15:5574-83.

195. Katahira J, Strasser K, Podtelejnikov A et al. The Mex67p-mediated nuclear mRNA export pathway is conserved from yeast to human. EMBO J 1999; 18:2593-609.

196. Segref A, Sharma K, Doye V et al. Mex67p, a novel factor for nuclear mRNA export, binds to both poly(A)+ RNA and nuclear pores. EMBO J 1997; 16:3256-71.

197. Gruter P, Tabernero C, von Kobbe C et al. TAP, the human homolog of Mex67p, mediates CTE-dependent RNA export from the nucleus. Mol Cell 1998; 1:649-59.

198. Trotman LC, Mosberger N, Fornerod M et al. Import of adenovirus DNA involves the nuclear pore complex receptor CAN/Nup214 and histone H1. Nat Cell Biol 2001; 3:1092-100.

199. Fontoura BM, Blobel G, Matunis MJ. A conserved biogenesis pathway for nucleoporins: Proteolytic processing of a 186-kilodalton precursor generates Nup98 and the novel nucleoporin, Nup96. J Cell Biol 1999; 144:1097-112.

200. Hodel AE, Hodel MR, Griffis ER et al. The three-dimensional structure of the autoproteolytic, nuclear poretargeting domain of the human nucleoporin Nup98. Mol Cell 2002; 10:347-58.

201. Powers MA, Forbes DJ, Dahlberg JE et al. The vertebrate GLFG nucleoporin, Nup98, is an essential component of multiple RNA export pathways. J Cell Biol 1997; 136:241-50.

202. Griffis ER, Altan N, Lippincott-Schwartz J et al. Nup98 is a mobile nucleoporin with transcription-dependent dynamics. Mol Biol Cell 2002; 13:1282-97.

203. Griffis ER, Xu S, Powers MA. Nup98 localizes to both nuclear and cytoplasmic sides of the nuclear pore and binds to two distinct nucleoporin subcomplexes. Mol Biol Cell 2003; 14:600-10.
204. Gilchrist D, Mykytka B, Rexach M. Accelerating the rate of disassembly of karyopherin cargo complexes. J Biol Chem 2002; 277:18161-72.
205. Ben-Efraim I, Gerace L. Gradient of increasing affinity of importin beta for nucleoporins along the pathway of nuclear import. J Cell Biol 2001; 152:411-7.
206. Pyhtila B, Rexach M. A gradient of affinity for the karyopherin Kap95p along the yeast nuclear pore complex. J Biol Chem 2003; 278:42699-709.
207. Denning DP, Uversky V, Patel SS et al. The Saccharomyces cerevisiae nucleoporin Nup2p is a natively unfolded protein. J Biol Chem 2002; 277:33447-55.
208. Denning DP, Patel SS, Uversky V et al. Disorder in the nuclear pore complex: The FG repeat regions of nucleoporins are natively unfolded. Proc Natl Acad Sci USA 2003; 100:2450-5.
209. Chook YM, Blobel G. Structure of the nuclear transport complex karyopherin-beta2-Ran x GppNHp. Nature 1999; 399:230-7.
210. Cingolani G, Bednenko J, Gillespie MT et al. Molecular basis for the recognition of a nonclassical nuclear localization signal by importin beta. Mol Cell 2002; 10:1345-53.
211. Vetter IR, Arndt A, Kutay U et al. Structural view of the Ran-importin beta interaction at 2.3 A resolution. Cell 1999; 97:635-46.
212. Bayliss R, Kent HM, Corbett AH et al. Crystallization and initial X-ray diffraction characterization of complexes of FxFG nucleoporin repeats with nuclear transport factors. J Struct Biol 2000; 131:240-7.
213. Bayliss R, Littlewood T, Strawn LA et al. GLFG and FxFG nucleoporins bind to overlapping sites on importin-beta. J Biol Chem 2002; 277:50597-606.
214. Bednenko J, Cingolani G, Gerace L. Importin beta contains a COOH-terminal nucleoporin binding region important for nuclear transport. J Cell Biol 2003; 162:391-401.
215. Bednenko J, Cingolani G, Gerace L. Nucleocytoplasmic transport: Navigating the channel. Traffic 2003; 4:127-35.
216. Schwoebel ED, Talcott B, Cushman I et al. Ran-dependent signal-mediated nuclear import does not require GTP hydrolysis by Ran. J Biol Chem 1998; 273:35170-5.
217. Schwoebel ED, Ho TH, Moore MS. The mechanism of inhibition of Ran-dependent nuclear transport by cellular ATP depletion. J Cell Biol 2002; 157:963-74.
218. Lyman SK, Guan T, Bednenko J et al. Influence of cargo size on Ran and energy requirements for nuclear protein import. J Cell Biol 2002; 159:55-67.
219. Feldherr CM, Kallenbach E, Schultz N. Movement of a karyophilic protein through the nuclear pores of oocytes. J Cell Biol 1984; 99:2216-22.
220. Dworetzky SI, Feldherr CM. Translocation of RNA-coated gold particles through the nuclear pores of oocytes. J Cell Biol 1988; 106:575-84.
221. Peters R. Nuclear envelope permeability measured by fluorescence microphotolysis of single liver cell nuclei. J Biol Chem 1983; 258:11427-9.
222. Peters R. Nucleo-cytoplasmic flux and intracellular mobility in single hepatocytes measured by fluorescence microphotolysis. EMBO J 1984; 3:1831-6.
223. Peters R. Fluorescence microphotolysis to measure nucleocytoplasmic transport and intracellular mobility. Biochim Biophys Acta 1986; 864:305-59.
224. Peters R, Lang I, Scholz M et al. Fluorescence microphotolysis to measure nucleocytoplasmic transport in vivo and in vitro. Biochem Soc Trans 1986; 14:821-2.
225. Lang I, Scholz M, Peters R. Molecular mobility and nucleocytoplasmic flux in hepatoma cells. J Cell Biol 1986; 102:1183-90.
226. Schulz B, Peters R. Nucleocytoplasmic protein traffic in single mammalian cells studied by fluorescence microphotolysis. Biochim Biophys Acta 1987; 930:419-31.
227. Astumian RD, Derenyi I. Fluctuation driven transport and models of molecular motors and pumps. Eur Biophys J 1998; 27:474-89.
228. Ribbeck K, Gorlich D. Kinetic analysis of translocation through nuclear pore complexes. EMBO J 2001; 20:1320-30.
229. Bickel T, Bruinsma R. The nuclear pore complex mystery and anomalous diffusion in reversible gels. Biophys J 2002; 83:3079-87.

230. Kustanovich T, Rabin Y. Metastable network model of protein transport through nuclear pores. Biophys J 2004; 86:2008-16.
231. Strawn LA, Shen T, Shulga N et al. Minimal nuclear pore complexes define FG repeat domains essential for transport. Nat Cell Biol 2004; 6:197-206.
232. Belanger KD, Kenna MA, Wei S et al. Genetic and physical interactions between Srp1p and nuclear pore complex proteins Nup1p and Nup2p. J Cell Biol 1994; 126:619-30.
233. Hood JK, Casolari JM, Silver PA. Nup2p is located on the nuclear side of the nuclear pore complex and coordinates Srp1p/importin-alpha export. J Cell Sci 2000; 113(Pt 8):1471-80.
234. Dilworth DJ, Suprapto A, Padovan JC et al. Nup2p dynamically associates with the distal regions of the yeast nuclear pore complex. J Cell Biol 2001; 153:1465-78.
235. Haraguchi T, Koujin T, Hayakawa T et al. Live fluorescence imaging reveals early recruitment of emerin, LBR, RanBP2, and Nup153 to reforming functional nuclear envelopes. J Cell Sci 2000; 113(Pt 5):779-94.
236. Lam DH, Aplan PD. NUP98 gene fusions in hematologic malignancies. Leukemia 2001; 15:1689-1695.
237. Cronshaw JM, Matunis MJ. The nuclear pore complex protein ALADIN is mislocalized in triple A syndrome. Proc Natl Acad Sci USA 2003; 100:5823-7.
238. Cronshaw JM, Matunis MJ. The nuclear pore complex: Disease associations and functional correlations. Trends Endocrinol Metab 2004; 15:34-9.
239. Enarson P, Rattner JB, Ou Y et al. Autoantigens of the nuclear pore complex. J Mol Med 2004; 82:423-33.
240. Kau TR, Way JC, Silver PA. Nuclear transport and cancer: From mechanism to intervention. Nat Rev Cancer 2004; 4:106-17.
241. Cohen M, Wilson KL, Gruenbaum Y. Membrane proteins of the nuclear pore complex: Gp210 is conserved in Drosophila, C. elegans, and A. thaliana. Gen Ther Mol Biol 2001; 4:47-55.
242. Rose A, Patel S, Meier I. The plant nuclear envelope. Planta 2004; 218:327-336.
243. Hubner S, Smith HM, Hu W et al. Plant importin alpha binds nuclear localization sequences with high affinity and can mediate nuclear import independent of importin beta. J Biol Chem 1999; 274:22610-7.
244. Merkle T, Leclerc D, Marshallsay C et al. A plant in vitro system for the nuclear import of proteins. Plant J 1996; 10:1177-86.
245. Hicks GR, Smith HM, Lobreaux S et al. Nuclear import in permeabilized protoplasts from higher plants has unique features. Plant Cell 1996; 8:1337-52.
246. Allen TD, Cronshaw JM, Bagley S et al. The nuclear pore complex: Mediator of translocation between nucleus and cytoplasm. J Cell Sci 2000; 113(Pt 10):1651-9.
247. Suntharalingam M, Wente SR. Peering through the Pore: Nuclear pore complex structure, assembly, and function. Developmental Cell 2003; 4:775-789.

CHAPTER 2

Integral Proteins of the Nuclear Pore Membrane

Merav Cohen, Katherine L. Wilson and Yosef Gruenbaum

The nuclear envelope contains three distinct membrane domains. The outer nuclear membrane faces the cytoplasm and is continuous with the rough endoplasmic reticulum (ER). Like the rough ER, the nuclear outer membrane is covered with ribosomes engaged in translating secreted and integral membrane proteins. The inner nuclear membrane faces the nucleoplasm, has its own unique protein composition and interacts with the fibrous meshwork of the nuclear lamina (reviewed in ref. 6). The inner and outer nuclear membranes fuse to form the third membrane domain, termed the pore membrane domain. Nuclear pore complexes (NPCs) are anchored at the pore membrane domain and mediate both passive diffusion and active nucleocytoplasmic transport. Active transport requires signals on the imported or exported macromolecules, termed nuclear localization signals (NLS) and nuclear export signals (NES), respectively. Transport is mediated by soluble NLS and NES receptors (termed importins/exportins/karyopherins/transportins), whose direction of movement is determined by Ran, a small GTP-binding protein (reviewed in refs. 22, 48 and 50). NPC structure includes soluble proteins, termed nucleoporins (nups) and integral membrane proteins, termed POMs. The NPC is anchored to the pore membrane by binding to POMs.[37] POMs are also proposed to have roles in nuclear pore assembly, nucleocytoplasmic transport and NPC organization (see below).

The protein composition of the yeast NPC has been determined.[38] Yeast NPCs consist of multiple copies of at least thirty distinct proteins, with a total estimated mass of 50 MDa. The size and complexity of the NPC appears to have increased during evolution. For example, the vertebrate NPC has an estimated maximum mass of 120 MDa, with an estimated forty different proteins;[35] (reviewed in ref. 48). Many vertebrate nucleoporins have orthologs or functional homologs in yeast and plants. Overall, NPCs are significantly conserved in both structure and protein composition between yeast and humans. One possible exception to this trend are the POMs, which have no obvious similarity between yeast and vertebrates.

Yeast POMs

Five integral membrane proteins have been localized to the pore membrane domain in the yeast *Saccharomyces cerevisiae* (reviewed in ref. 7). These five POMs are named Snl1,[29] Pom152,[55] Ndc1,[5] Pom34[38] and Brr6.[11] Yeast POMs are discussed below.

SNL1 was identified in a genetic screen for high copy suppressors of the lethal phenotype caused by over-expression of the carboxy-terminal 200 residues of Nup116 (*NUP116-C*), in the *nup116* null background. Loss of *NUP116* function causes the nuclear membranes to

Nuclear Import and Export in Plants and Animals, edited by Tzvi Tzfira and Vitaly Citovsky.
©2005 Eurekah.com and Kluwer Academic / Plenum Publishers.

herniate and cover the NPCs. Over-expression of *SNL1* also suppresses the temperature-sensitive phenotypes of mutations in two other genes:[29] *gle2*, which is essential for NPC assembly,[36] and *nic96*, which is involved in the transport of polyadenylated RNA and possibly in protein transport.[21] Cells that lack *SNL1* expression are viable, and so are double-null mutants for *snl1* plus a second POM named *pom152*. Lack of synthetic lethality between *SNL1* and *POM152* suggests a possible functional redundancy with other, possibly unidentified, POMs.[29]

A fraction of Snl1 is localized to the ER,[29] which suggests that Snl1 might shuttle between the pore membrane domain and ER. It would be interesting to compare the diffusional mobility of Snl1 at the NPC versus the ER, to determine if Snl1 proteins in the ER are actively exchanging with pore-localized Snl1.

Pom152 was identified as a glycoprotein with N-linked high mannose oligosaccharide modifications,[55] which localized to NPCs.[55] Pom152 is an integral membrane protein that spans the pore membrane once, using only one of its two predicted transmembrane domains. Its short amino tail (175 residues) faces the NPC, and its long carboxy-tail (1141 residues) is localized in the lumen space between the inner and outer nuclear membranes.[46] *S. cerevisiae* strains that lack the *POM152* gene are viable. However, mutations in *pom152* are lethal in combination with mutations in other genes, including *NUP188* and *NUP170*,[1] which are both involved in establishing the functional diameter of the NPC.[39] It is not clear why mutations in *POM152* are 'synthetically lethal' in combination with mutations in *NUP188* or *NUP170*, but these findings suggest that Pom152 might also be involved in determining the diameter of the pore or NPC. Over-expression of Pom152 reduces the growth rate of cells, for reasons not yet understood.[55]

When *POM152* is ectopically expressed in mammalian cells, it is correctly localized to the pore membrane domain.[55] This result indicates a functional conservation of the pore membrane domain between yeast and mammals. Though there is very little homology between the yeast *POM152* and the vertebrate *GP210* (see below), several characteristics are shared between these pore membrane proteins. Notably, both proteins have a predicted hydrophobic region that is not embedded in the membrane. It would be interesting to test whether *GP210* can functionally complement *pom152* mutations.

NDC1 was originally discovered as an essential gene, which is required at a late stage of spindle pole body (microtubule organizing center) duplication.[53] Yeast spindle pole bodies are embedded in the nuclear envelope, like NPCs, and Ndc1 localizes at both types of structure. This may indicate a functional or assembly-related link between these two nuclear membrane-embedded organelles.[5] In *ndc1*-null cells, the spindle pole body fails to be inserted into the nuclear envelope, but NPCs are positioned and function normally. The lack of a nuclear transport phenotype in cells with mutations in *pom152*, *ndc1* or both, suggests functional redundancy between these POMs and other nucleoporins. Interestingly, in *pom152*-null cells with defective Ndc1 protein, the spindle pole bodies are again inserted into the nuclear envelope, suggesting that *pom152* mutations suppress *ndc1* mutations. Though it is not known if Ndc1 and Pom152 interact directly, it was proposed that the lack of Pom152 releases 'defective' Ndc1 molecules from the NPC, allowing them to function (weakly, but in higher numbers) at spindle pole bodies.[5]

Pom34 was recently identified by mass spectrometry as a component of biochemically-purified yeast NPCs, and localized to the pore membrane domain.[38] Biochemical extraction also revealed that *POM34* encodes an integral membrane protein. Pom34 has a predicted leucine zipper motif and two putative transmembrane domains.[38] Neither the topology of Pom34 in the pore membrane domain, nor its requirement for cell growth or viability, have been determined.

BRR6 was identified by complementation of the *brr6* cold sensitive nuclear transport mutant.[11] *BRR6* encodes a 22.8 kDa integral membrane protein located at the nuclear rim in a punctate pattern characteristic of NPC proteins. Genetic analysis revealed that Brr6 interacts with several soluble nucleoporins including Nup1, Nic96 and Nup188. Depletion of Brr6 from cells causes all NPCs to aggregate in one region, suggesting that Brr6 is required for the normal distribution of NPCs. *BRR6* is also required for normal morphology of the nuclear envelope, since its depletion causes nuclear envelope herniations. Both phenotypes resemble those produced by *gle2* and *nup116* mutations,[36,51] indicating that these proteins may function in the same genetic pathway. Brr6 spans the nuclear envelope once, through a transmembrane domain located at its C-terminus, and its longer N-terminal domain faces the NPC. Brr6 may also be involved in nuclear transport, since mRNA molecules and NLS/NES-GFP fusion proteins all accumulate at the nuclear periphery in *brr6-1* mutants.

Vertebrate POMs

Only two genes encoding integral proteins of the pore membrane have been identified so far in vertebrates, named *POM121* and *GP210*.

POM121 encodes a wheat germ agglutinin (WGA) binding protein in mammals, and is localized at the pore membrane domain.[25] Pom121 contains six XFXFG repeats characteristic of certain nucleoporins that are posttranslationally modified by *O*-linked GlcNAc. Pom121 is an integral membrane protein that spans the pore membrane once, with its short N-terminal head domain in the lumen. The large C-terminal domain faces the NPC, and is localized to the central spoke ring of the NPC.[43] When this large exposed region of Pom121 was over-expressed in green monkey COS cells, it accumulated in cylindrical intranuclear bodies, most of which localized near the inner nuclear membrane.[45] The first 128 amino acids of POM121, which are not predicted to contain a signal peptide, target the protein to ER membranes, but targeting of Pom121 to the NPC requires the region between residues 129 and 618.[44] Pom121 is a highly stable and immobile component of the NPC during interphase;[8] and is therefore likely to anchor NPCs at the pore membrane in vertebrates.

GP210 is the only evolutionarily conserved pore membrane gene among multicellular organisms, including *Arabidopsis, C. elegans, Drosophila* and humans (see ref. 7 and Fig. 1), indicating that it has fundamental roles in NPC formation or function, or both. Rat Gp210 was originally identified as an integral glycoprotein of the NPC.[19] Gp210 spans the pore membrane only once, through a hydrophobic region positioned close to its carboxy terminus. Only a small region of gp210 (its short carboxy-tail) faces the NPC, whereas 95% of its mass is positioned in the perinuclear space.[24] Rat Gp210 has two predicted transmembrane domains, but only one actually spans the membrane. It was hypothesized that the second hydrophobic domain plays a role in membrane fusion during pore formation (see below). The lumenal domain of Gp210 is glycosylated by N-linked high mannose oligosaccharides, and it therefore binds the lectin concanavalin A.[54] Gp210 is organized mostly as dimers, plus higher order multimers.[14] In contrast to Pom121, which is quite stable at the NPC, Gp210 is mobile; at any given time about 20% of all Gp210 molecules are dissociated from NPCs and can diffuse freely throughout the nuclear/ER membrane network (Brian Burke, unpublished observations). During mitosis, Gp210 is phosphorylated on Ser^{1880} by cyclin B-p34^{cdc2} or a related kinase.[15]

Gp210 appears to have a major role in NPC structure, since the lumenal expression of antibodies against the lumenal domain of Gp210, in mammalian cells, inhibited both active transport and passive diffusion through the NPC.[23]

Cell Cycle Dynamics of the NPC

The closed mitosis in yeast does not allow mitotic breakdown and reassembly of the NPCs. Thus in yeast, there are no cell cycle 'dynamics' of nuclear pores. In contrast, higher eukaryotes undergo an open mitosis, in which the nuclear envelope and NPCs disassemble, and nuclear

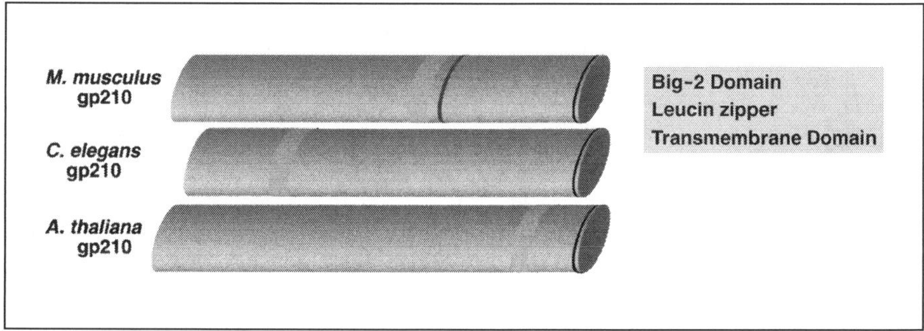

Figure 1. Schematic diagram of the conserved protein domains in Gp210, between vertebrates (*M. musculus*), invertebrates (*C. elegans*) and plants (*A. thaliana*). Big-2 domain is a bacterial Ig like domain found in many bacterial and phage surface proteins (reviewed in ref. 7)

membranes merge into the ER.[13] The timing of NPC disassembly varies among different metazoans. In *C. elegans*, disassembly of the NPCs begins after prometaphase,[31] whereas in *Drosophila* and vertebrates NPC disassembly begins as early as prophase.[18,27] Disassembly of the NPCs is proposed to be a key triggering event for nuclear envelope disassembly.[47] Nuclear envelope reassembly begins at the same time, during late anaphase/early telophase, in all metazoans analyzed. The assembly of NPCs does not require de novo protein synthesis, suggesting that NPC components are both stable and recycled.[34] NPC breakdown and assembly are regulated by cell-cycle dependent phosphorylation of several nucleoporins, including Nup358, CAN/Nup214, Nup153 and Gp210.[15,33]

NPC assembly is being studied in cell-free extracts of *Xenopus* eggs. Assembly initiates in patches of flattened nuclear membranes, which are usually attached to chromatin but do not need to be attached to chromatin. NPC formation begins before the chromatin is fully enclosed by membranes, and proceeds in an asynchronous and rapid manner.[32,33,52] A pathway for NPC assembly has been proposed based on the discovery of structures termed dimples, holes and star-rings, in nuclei assembled in vitro in *Xenopus* cell-free extracts.[20]

Reassembly of NPCs involves the ordered recruitment of NPC components, many of which are sub-complexes of nucleoporins that remain associated during mitosis.[15] However, the order in which different nucleoporins arrive to the site of the NPC is not fully understood, and somewhat controversial. In mammalian cells NPC formation begins during late anaphase, and one of the earliest proteins to accumulate is Nup153,[3,26] a constituent of the nuclear basket.[37] In contrast, in *Xenopus* assembly extracts Nup153 accumulates at NPCs rather late during mitosis, and is reported to bind lamins, which also assemble rather late.[42] As telophase progresses there is the sequential accumulation of Pom121, p62 (a constituent of the central channel),[10] Can/Nup214,[16] and later Gp210 and Tpr (found at NPC baskets and intranuclear structures).[28] However, it is likely that a fraction of Gp210 proteins are present continuously during nuclear envelope formation, even though most gp210 does not accumulate at the nuclear envelope until later in G1.[3,4]

Membrane Fusion and Nuclear Pore Formation

Electrostatic repulsion between phospholipid headgroups in an aqueous environment prevents spontaneous membrane fusion. To fuse, biological membranes must overcome this repulsion. The stalk formation hypothesis of membrane fusion[40,41] predicts several fusion intermediates. Fusion begins with the formation of a stalk between facing phospholipid bilayers. This stalk then forms a dimple. Further horizontal pulling of the dimple produces a hemifusion diaphragm, followed by opening of a fusion pore and its dilation by unknown mechanisms (reviewed by ref. 20).

Membrane fusion plays key roles in many cellular pathways including secretion, synaptic release, endocytosis and ER dynamics, and also in several viral infections.[2] Fusion is also crucial for nuclear pore formation, which occurs within the nuclear lumen, and may therefore resemble the membrane fusion events that take place during viral infection.

Viral fusion proteins (fusogens) are glycoproteins that span the viral membrane once, and contain a relatively large tail exposed on the viral surface.[20] There is very little, if any, sequence similarity between different viral fusogens, yet they all share a key structural feature: a short helical amphiphilic domain with alternating hydrophobic and charged residues. This 'fusion peptide' interacts directly with the target membrane.

Fusion of the influenza virus membrane is triggered by the low pH (pH 5-6) within endosomes, whence the virus fuses to enter the cytoplasm. Low pH induces conformational changes in hemagglutinin (HA), the influenza fusogen, exposing the N-terminal fusion peptide. Once exposed, the fusion peptide inserts into the target membrane,[12] and possibly also into the viral membrane.[49] This insertion is proposed to produce a stalk structure that connects the viral and target membranes. At this stage, expansion of this structure produces a hemifusion diaphragm between the inner leaflets, where a small fusion pore subsequently opens. Dilation of this small pore completes the fusion event.[17] Fusion mediated by influenza HA is cooperative, involving the aggregation of at least three, and probably four, HA trimers.[9]

The fusion mechanism involved in nuclear pore formation would also require a fusogenic protein to overcome surface charge repulsion between the facing leaflets of the inner and outer nuclear membranes. Gp210 has been proposed as a possible fusogen in vertebrates,[54] because it has a large domain in the nuclear lumen that includes a possible amphiphilic helix, similar to viral fusogens, and because it forms multimers (see ref. 14). The evolutionary conservation of gp210 is also consistent with such a fundamental function. These data make Gp210 an excellent candidate fusogen for pore formation, but direct studies will be required to test this model.

In structural terms, NPC formation in yeast is even less well characterized than in vertebrates. Yeast POMs have no obvious metazoan orthologs, though the mechanisms of pore membrane fusion are likely to be conserved. The yeast Pom152 protein might be involved in pore membrane fusion events, since Pom152 resembles Gp210 in having a large lumenal domain with a possible amphiphilic helix. Assuming that the mechanisms of pore formation are conserved in all eukaryotes, the yeast findings might predict that pore membrane fusion involves more than one kind of protein, since yeast with null mutations in *pom152* are viable. Further study of the null phenotypes for vertebrate gp210 and Pom121, and their comparisons with yeast POMs, should help reveal the mechanisms of pore formation.

References

1. Aitchison JD, Rout MP, Marelli M et al. Two novel related yeast nucleoporins Nup170p and Nup157p: Complementation with the vertebrate homologue Nup155p and functional interactions with the yeast nuclear pore-membrane protein Pom152p. J Cell Biol 1995; 131:1133-1148.
2. Bentz J, Mittal A. Deployment of membrane fusion protein domains during fusion. Cell Biol Int 2000; 24:819-838.
3. Bodoor K, Shaikh S, Salina D et al. Sequential recruitment of NPC proteins to the nuclear periphery at the end of mitosis. J Cell Sci 1999; 112:2253-2264.
4. Chaudhary N, Courvalin JC. Stepwise reassembly of the nuclear envelope at the end of mitosis. J Cell Biol 1993; 122:295-306.
5. Chial HJ, Rout MP, Giddings TH et al. Saccharomyces cerevisiae Ndc1p is a shared component of nuclear pore complexes and spindle pole bodies. J Cell Biol 1998; 143:1789-1800.
6. Cohen M, Lee KK, Wilson KL et al. Transcriptional repression, apoptosis, human disease and the functional evolution of the nuclear lamina. Trends Bioc Sci 2001; 26:41-47.

7. Cohen M, Wilson KL, Gruenbaum Y. Membrane proteins of the nuclear pore complex: Gp210 is conserved in vertebrates, Drosophila melanogaster, C. elegans and Arabidopsis. In: Boulikas T, ed. Gene Therapy and Molecular Biology. Vol. 6. Palo Alto: Gene Therapy Press, 2001.

8. Daigle N, Beaudouin J, Hartnell L et al. Nuclear pore complexes form immobile networks and have a very low turnover in live mammalian cells. J Cell Biol 2001; 154:71-84.

9. Danieli T, Pelletier SL, Henis YI et al. Membrane fusion mediated by the influenza virus hemagglutinin requires the concerted action of at least three hemagglutinin trimers. J Cell Biol 1996; 133:559-569.

10. Davis LI, Blobel G. Identification and characterization of a nuclear pore complex protein. Cell 1986; 45:699-709.

11. de Bruyn Kops A, Guthrie C. An essential nuclear envelope integral membrane protein, Brr6p, required for nuclear transport. EMBO J 2001; 20:4183-4193.

12. Durrer P, Gaudin Y, Ruigrok RW et al. Photolabeling identifies a putative fusion domain in the envelope glycoprotein of rabies and vesicular stomatitis viruses. J Biol Chem 1995; 270:17575-17581.

13. Ellenberg J, Siggia ED, Moreira JE et al. Nuclear membrane dynamics and reassembly in living cells: targeting of an inner nuclear membrane protein in interphase and mitosis. J Cell Biol 1997; 138:1193-1206.

14. Favreau C, Bastos R, Cartaud J et al. Biochemical characterization of nuclear pore complex protein gp210 oligomers. Eur J Biochem 2001; 268:3883-3889.

15. Favreau C, Worman HJ, Wozniak RW et al. Cell cycle-dependent phosphorylation of nucleoporins and nuclear pore membrane protein Gp210. Biochemistry 1996; 35:8035-8044.

16. Fornerod M, van Deursen J, van Baal S et al. The human homologue of yeast CRM1 is in a dynamic subcomplex with CAN/Nup214 and a novel nuclear pore component Nup88. EMBO J 1997; 16:807-816.

17. Gaudin Y. Rabies virus-induced membrane fusion pathway. J Cell Biol 2000; 150:601-612.

18. Georgatos SD, Pyrpasopoulou A, Theodoropoulos PA. Nuclear envelope breakdown in mammalian cells involves stepwise lamina disassembly and microtubule-drive deformation of the nuclear membrane. J Cell Sci 1997; 110:2129-2140.

19. Gerace L, Ottaviano Y, Kondor-Koch C. Identification of a major polypeptide of the nuclear pore complex. J Cell Biol 1982; 95:826-837.

20. Goldberg MW, Wiese C, Allen TD et al. Dimples, pores, star-rings, and thin rings on growing nuclear envelopes: Evidence for structural intermediates in nuclear pore complex assembly. J Cell Sci 1997; 110:409-420.

21. Gomez-Ospina N, Morgan G, Giddings TH et al. Yeast nuclear pore complex assembly defects determined by nuclear envelope reconstruction. J Struc Biol 2000; 132:1-5.

22. Gorlich D, Kutay U. Transport between the cell nucleus and the cytoplasm. Annu Rev Cell Dev Biol 1999; 15:607-660.

23. Greber UF, Gerace L. Nuclear protein import is inhibited by an antibody to a lumenal epitope of a nuclear pore complex glycoprotein. J Cell Biol 1992; 116:15-30.

24. Greber UF, Senior A, Gerace L. A major glycoprotein of the nuclear pore complex is a membrane-spanning polypeptide with a large lumenal domain and a small cytoplasmic tail. EMBO J 1990; 9:1495-1502.

25. Hallberg E, Wozniak RW, Blobel G. An integral membrane protein of the pore membrane domain of the nuclear envelope contains a nucleoporin-like region. J Cell Biol 1993; 122:513-521.

26. Haraguchi T, Koujin T, Hayakawa T et al. Live fluorescence imaging reveals early recruitment of emerin, LBR, RanBP2, and Nup153 to reforming functional nuclear envelopes. J Cell Sci 2000; 113:779-794.

27. Harel A, Zlotkin E, Nainudel ES et al. Persistence of major nuclear envelope antigens in an envelope-like structure during mitosis in Drosophila melanogaster embryos. J Cell Sci 1989; 94:463-470.

28. Frosst P, Guan T, Subauste C et al. Tpr is localized within the nuclear basket of the pore complex and has a role in nuclear protein export. J Cell Biol 2002; 156:617-630.

29. Ho AK, Raczniak GA, Ives EB et al. The integral membrane protein snl1p is genetically linked to yeast nuclear pore complex function. Mol Biol Cell 1998; 9:355-373.

30. Jahn R, Sudhof TC. Membrane fusion and exocytosis. Annu Rev Biochem 1999; 68:863-911.

31. Lee KK, Gruenbaum Y, Spann P et al. C. elegans nuclear envelope proteins emerin, MAN1, lamin, and nucleoporins reveal unique timing of nuclear envelope breakdown during mitosis. Mol Biol Cell 2000; 11:3089-3099.

32. Lohka MJ, Masui Y. Roles of cytosol and cytoplasmic particles in nuclear envelope assembly and sperm pronuclear formation in cell-free preparations from amphibian eggs. J Cell Biol 1984; 98:1222-1230.

33. Macaulay C, Meier E, Forbes DJ. Differential mitotic phosphorylation of proteins of the nuclear pore complex. J Biol Chem 1995; 270:254-262.

34. Maul GG. The nuclear and the cytoplasmic pore complex: structure, dynamics, distribution, and evolution. Int Rev Cytol Suppl 1977; 6:75-186.

35. Miller BR, Forbes DJ. Purification of the vertebrate nuclear pore complex by biochemical criteria. Traffic 2000; 1:941-951.

36. Murphy R, Watkins JL, Wente SR. GLE2, a Saccharomyces cerevisiae homologue of the Schizosaccharomyces pombe export factor RAE1, is required for nuclear pore complex structure and function. Mol Biol Cell 1996; 7:1921-1937.

37. Panté N, Aebi U. Molecular dissection of the nuclear pore complex. Crit Rev Biochem Mol Biol 1996; 31:153-199.

38. Rout MP, Aitchison JD, Suprapto A et al. The yeast nuclear pore complex: composition, architecture, and transport mechanism. J Cell Biol 2000; 148:635-651.

39. Shulga N, Mosammaparast N, Wozniak R et al. Yeast nucleoporins involved in passive nuclear envelope permeability. J Cell Biol 2000; 149:1027-1038.

40. Siegel DP. Energetics of intermediates in membrane fusion: Comparison of stalk and inverted micellar intermediate mechanisms. Biophys J 1993; 65:124-140.

41. Siegel DP. The modified stalk mechanism of lamellar/inverted phase transitions and its implications for membrane fusion. Biophys J 1999; 76:291-313.

42. Smythe C, Jenkins HE, Hutchison CJ. Incorporation of the nuclear pore basket protein Nup153 into nuclear pore structures is dependent upon lamina assembly: Evidence from cell-free extracts of Xenopus eggs. EMBO J 2000; 19:3918-3931.

43. Soderqvist H, Hallberg E. The large C-terminal region of the integral pore membrane protein, POM121, is facing the nuclear pore complex. Eur J Cell Biol 1994; 64:186-191.

44. Soderqvist H, Imreh G, Kihlmark M et al. Intracellular distribution of an integral nuclear pore membrane protein fused to green fluorescent protein—Localization of a targeting domain. Eur J Biochem 1997; 250:808-813.

45. Soderqvist H, Jiang WQ, Ringertz N et al. Formation of nuclear bodies in cells overexpressing the nuclear pore protein POM121. Exp Cell Res 1996; 225:75-84.

46. Tcheperegine SE, Marelli M, Wozniak RW. Topology and functional domains of the yeast pore membrane protein Pom152p. J Biol Chem 1999; 274:5252-5258.

47. Terasaki M, Campagnola P, Rolls M et al. A new model for nuclear envelope breakdown. Mol Biol Cell 2001; 12:503-510.

48. Vasu SK, Forbes DJ. Nuclear pores and nuclear assembly. Curr Opin Cell Biol 2001; 13:363-375.

49. Weber T, Paesold G, Galli C et al. Evidence for H(+)-induced insertion of influenza hemagglutinin HA2 N-terminal segment into viral membrane. J Biol Chem 1994; 269:18353-18258.

50. Wente SR. Gatekeepers of the nucleus. Science 2000; 288:1374-1347.

51. Wente SR, Blobel G. A temperature-sensitive NUP116 null mutant forms a nuclear envelope seal over the yeast nuclear pore complex thereby blocking nucleocytoplasmic traffic. J Cell Biol 1993; 123:275-284.

52. Wiese C, Goldberg MW, Allen TD et al. Nuclear envelope assembly in Xenopus extracts visualized by scanning EM reveals a transport dependent envelope 'smoothing' event. J Cell Sci 1997; 13:1489-1502.

53. Winey M, Hoyt MA, Chan C et al. NDC1: A nuclear periphery component required for yeast spindle pole body duplication. J Cell Biol 1993; 122:743-751.

54. Wozniak RW, Blobel G. The single transmembrane segment of gp210 is sufficient for sorting to the pore membrane domain of the nuclear envelope. J Cell Biol 1992; 119:1441-1449.

55. Wozniak RW, Blobel G, Rout MP. POM152 is an integral protein of the pore membrane domain of the yeast nuclear envelope. J Cell Biol 1994; 125:31-42.

CHAPTER 3

Subnuclear Trafficking and the Nuclear Matrix

Iris Meier

The nuclear matrix is the nuclear substructure that remains after the majority of DNA and soluble and chromatin-bound proteins have been removed from the nucleus.[1-3] Electron micrographs show that the animal nuclear matrix consists of the nuclear pore complexes embedded in the nuclear lamina, and a network of internal 10 nm filaments, into which granular structures and the nucleoli are embedded.[4,5] In two-dimensional protein gels of nuclear matrix preparations, more than 200 polypeptides can be distinguished, but only few components of the nuclear matrix have been cloned.[6,7] Best studied from animal systems are the nuclear lamins, a group of intermediate filament proteins that form the lamina, a filamentous protein meshwork that lines the nuclear envelope and is connected to the nuclear pore complexes. The nuclear lamins are attached to the inner envelope membrane by farnesylation and interactions with membrane-associated proteins.[8-10]

The nuclear matrix specifically binds to DNA fragments called matrix attachment regions (MARs). MARs are large, AT-rich DNA fragments with little sequence similarity, and are predicted to form the bases of chromatin loops attached to the nuclear matrix during interphase.[11] MARs bind to nuclear matrix preparations across species borders, and have been implicated in reducing position effects and increasing expression of transgenes in animals and plants.[12-14] Several MAR-binding proteins have been identified, which are components of the nuclear matrix.[15-22] In addition, proteins involved in transcription, splicing and RNA processing have been found to be associated with the nuclear matrix,[23-25] and significantly the respective processes have been shown to take place at specific sites of an isolated nuclear matrix fraction.[26-28] Together, the available data suggest that the nuclear matrix represents a core nuclear structure that is involved in chromatin organization and in different aspects of nucleic acid metabolism.

However, the nuclear matrix as a static, cytoskeleton-like structure is still an issue of intense debate (see for example, refs. 29, 30). The major objections are (1) that the procedures used to isolate the matrix might cause precipitation artifacts that we view as nuclear matrix fibers and (2) that proteins forming the interior matrix (as opposed to the lamins that form the outer shell) remain to be identified. It is probably equally possible that the observed specific subnuclear organization and the spatial "addresses" of chromatin domains, proteins, and protein complexes are caused by dynamic soluble interactions or by the association with a (either dynamic or static) solid-state structure. In any case, the information for specific subnuclear positioning exists, and it will be well worth addressing to what degree this information contributes to the biological functions of the respective molecules.

Nuclear Import and Export in Plants and Animals, edited by Tzvi Tzfira and Vitaly Citovsky.
©2005 Eurekah.com and Kluwer Academic / Plenum Publishers.

This Chapter does not focus on the association of DNA with the nuclear matrix, the function of nuclear matrix attachment regions, or the proteins binding to MARs. Instead, it investigates the information presently available about signals involved in the intranuclear targeting of proteins, either to the nuclear matrix or to specific subnuclear domains.

Of the several arguments that can be made for the functional importance of specific subnuclear protein targeting—and a role for the nuclear matrix in that targeting—three points will be discussed here. First, several proteins have been found to contain specific signals for their association with the nuclear matrix or for their targeting to specific subnuclear sites. These nuclear matrix-targeting signals (NMTSs) sometimes overlap with other functional domains, like DNA-binding domains or nuclear localization signals (NLSs). However, at least in some cases they could be separated from these functions, showing that nuclear matrix association is independent from DNA binding and that targeting to the nuclear matrix requires a signal in addition to the nuclear import signal. In some cases, the NMTS can confer nuclear matrix targeting to a heterologous protein, and in at least one case it aids to the activity of a heterologous transcription factor, thereby suggesting a functional significance associated with the subnuclear targeting of the protein.

Second, the association of proteins with the nuclear matrix and with specific subnuclear sites has been found to be a regulated process in at least some cases, indicating that it is likely of biological significance beyond a simple "sticking together" of cellular components. And finally, the disruption of the specific subnuclear targeting of some nuclear proteins has been found associated with human diseases caused by chromosome translocations. These are strong indications that spatial information is required for the proper functioning of the respective proteins and that subnuclear mislocalization can have a severe impact on their function.

Nuclear Matrix Targeting Signals

For a number of nuclear proteins that have been found either associated with the nuclear matrix, or localized in specific subnuclear domains, amino acid sequences have been identified that are necessary and sufficient for this localization. They range from small peptide motifs capable to confer nuclear matrix localization to a heterologous protein to larger portions of the protein, which in an additive or synergistic way contribute to nuclear matrix association. The following gives an overview over the motifs mapped in different nuclear proteins, including transcription factors, DNA- and RNA-binding proteins, viral proteins, kinases, and kinase adapters. Figure 1 shows a compilation of the locations and sequences of these motifs. As discussed below, there is presently no consensus sequence that can be derived from their comparison.

Steroid Receptors

Steroid receptor binding to the nuclear matrix was first described in the 1980s and was an early realization of a potential functional association of a regulatory protein with a structural component of the nucleus.[29,30] The domain necessary for association with the nuclear matrix has been narrowed down in the human glucocorticoid receptor (hGR) and the human androgen receptor (hAR).[31] Both proteins consist of an N-terminal domain involved in activation, a central DNA-binding domain (DBD) followed by a tau2 transactivation domain, and a C-terminal steroid-binding domain. While in these early studies the domain required for nuclear matrix attachment was localized to the C-terminal steroid binding domain in hAR, both the DNA-binding domain and the C-terminal domain of hGR were found to be required.

A more detailed mapping of the NMTS of the hGR was performed by Tang et al,[32] who showed that the DBD in combination with the C-terminal tau2 transactivation domain constitute an NMTS of the hGR. Neither the DBD nor the tau2 domain alone was sufficient for nuclear matrix binding and the tau2 domain alone could not confer nuclear matrix binding to the heterologous GAL4 DNA-binding domain. Transactivation and nuclear matrix binding

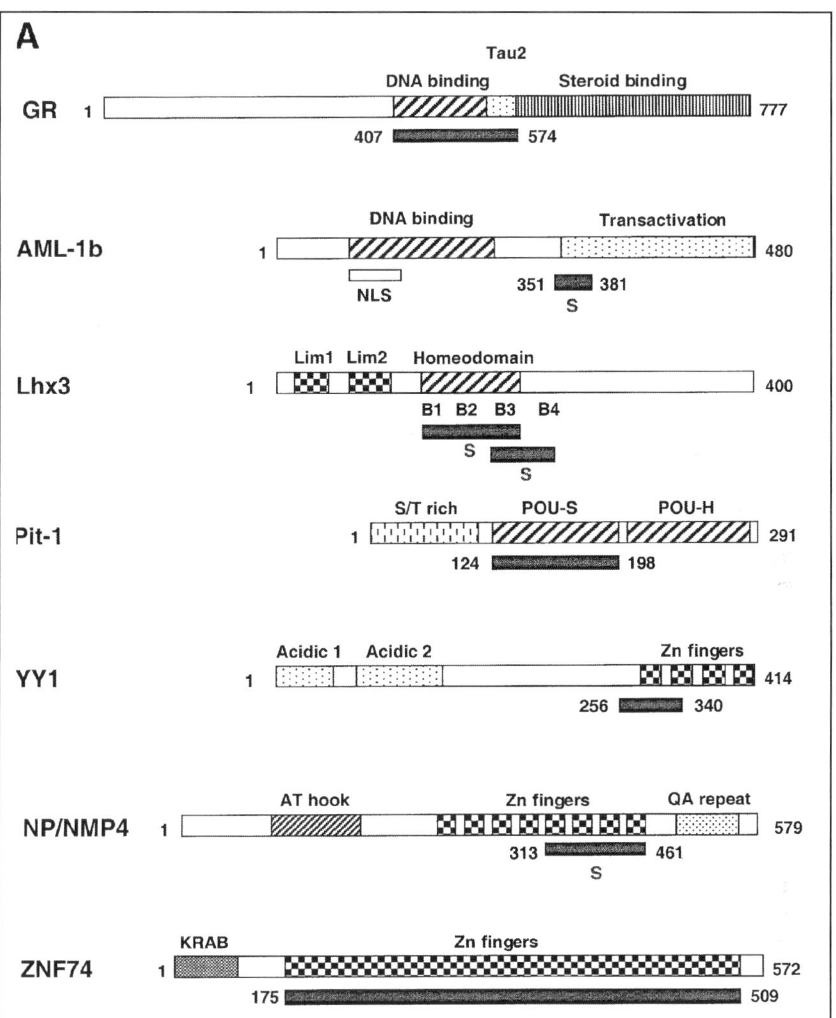

Figure 1. A) Comparison of the location and size of nuclear matrix targeting signals, in transcription factors. NMTSs are indicated as bars underneath the schematic representation of the proteins. "S" indicates that the NMTS has been shown to be sufficient for nuclear matrix targeting. Tau2, tau2 transactivation domain; NLS, nuclear localization signal; Lim1, Lim 2, Lim domains; POU-S, POU-specific domain; POU-H, POU-type homeodomain; Zn fingers, zinc fingers; QA repeat, glutamine alanine repeat; KRAB, Kruppel-associated box. For detailed information, see text. Proteins and domains are only approximately drawn to scale. Figure continued on next page.

could be uncoupled by point mutagenesis, for example in the S573A mutant, which still binds to the nuclear matrix but shows much reduced transcriptional activation. However, the two functions overlap, as is demonstrated e.g., with a L553G/L554G double mutant, which abolishes both nuclear matrix targeting and transactivation. These data show that the 29 amino acid (aa) tau2 domain probably contains two interaction surfaces—one for transcriptional activation and one for the binding of a nuclear matrix acceptor protein—and that nuclear matrix binding is not sufficient for transactivation by hGR.[33]

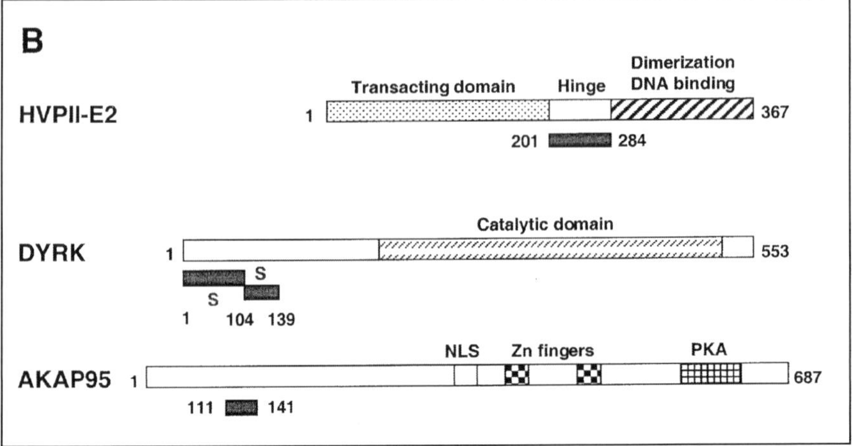

Figure 1, continued. B) Comparison of the location and size of nuclear matrix targeting signals in other nuclear proteins. NMTSs are indicated as bars underneath the schematic representation of the proteins. "S" indicates that the NMTS has been shown to be sufficient for nuclear matrix targeting. NLS, nuclear localization signal; Zn fingers, zinc fingers; PKA, PKA-binding domain. For detailed information, see text. Proteins and domains are only approximately drawn to scale.

In contrast to the tau2 domain, the hGR DBD appears to function as an NMTS in a heterologous context insofar as fusion of the DBD with the Vp16 activation domain reconstitutes a functional NMTS.[32] However, the Vp16 tau2 activation domain shares some structural relatedness to the GR tau2 domain, indicating that their combination might be required in the fusion protein too.

Candidates for Nuclear Matrix Acceptor Proteins

One intriguing—and presently unanswered—question is by which interactions an NMTS targets a protein to the nuclear matrix. Acceptor molecules on the nuclear matrix have been postulated which could be proteins, nucleic acids, or both. In the case of the steroid receptors, several candidates for interaction partners have been identified. However, in no case has a specific interaction been shown to be required for the subnuclear targeting of a protein.

GRIP120, a protein identified as a factor that interacts with GR, was found to be identical to hnRNP U, which belongs to the heterogeneous nuclear ribonucleoproteins, abundant proteins in the eukaryotic nucleus.[34] hnRNP U has in turn been found identical to SAF-A, a protein originally identified by its ability to bind to matrix attachment region (MAR) DNA, and a component of the nuclear matrix.[35] Indirect immunofluorescence microscopy demonstrated that GR and hnRNP U co-localize in nuclear speckles and large clusters. The C-terminal domain (aa 517-806) of hnRNP U and the C-terminal half of GR are sufficient for functional interaction of the two proteins.[34] Because the DBD-tau2 region has been identified as the NMTS of GR, it is therefore possible that hnRNP U constitutes an adapter that associates GR with the nuclear matrix. Overexpression of hnRNP U interferes with glucocorticoid induction and the C-terminal domains of both proteins are sufficient for mediating this effect.

Interestingly, the estrogen receptor (ER) has been shown to interact with another nuclear matrix protein, SAF-B/HAP (Scaffold attachment factor B/hnRNP A1 associated protein).[36] In in vitro binding assays, SAF-B/HAP binds to both the DBD and the hinge domain of ER. Co-immunoprecipitation showed that the binding occurs in vivo in cell lines. SAF-B/HAP was like SAF-A originally isolated as a protein that binds to matrix attachment region DNA and is localized in the nuclear matrix.[19] Subsequently, it was shown that it is identical to hnRNP A1 associated protein, which is itself a bona fide hnRNP protein.[37] SAF-B has recently been shown to interact with RNA polymerase II (Pol II) and with SR proteins in vitro and in vivo and forms a ternary complex with SR proteins and MAR-DNA in vitro. Overexpression of SAF-B represses the expression of a MAR-flanked reporter gene. Oesterreich et al[36] have shown that SAT-B overexpression also decreases the ability of ER to activate transcription and that the ER-DBD is necessary for the repressive effects of SAT-B.[36]

Together, the data connect both GR and ER to proteins that bind to MARs and are components of ribonucleoprotein complexes. While the picture is far from complete, it suggests a connection between the targeting of steroid receptors to the nuclear matrix, the recruitment of Pol II, and a modulation of receptor-induced gene expression by nuclear matrix proteins.

Yang et al[38] have identified by yeast two hybrid screening another acceptor candidate, the protein Hic-5, which binds specifically to the hinge-tau2 domain of GR. Hic-5 (hydrogen peroxide-inducible clone 5) is a previously identified protein that localizes both to focal adhesion points and to the nuclear matrix. Hic-5 stimulates transactivation by several nuclear receptors in combination with the coactivator GRIP1. In addition, it acts as a coactivator for the isolated DBD-tau2 domain of GR, but not the DBD alone. The C-terminus of Hic-5, containing seven zinc fingers, is required for both binding of GR and association to the nuclear matrix. However, in contrast to GR, Hic-5 binding to the nuclear matrix was not altered by ATP depletion, indicating that its NM binding is not simply caused by association with nuclear matrix-bound GR.[38] Hic-5 could therefore be primarily associated with a structural component of the nuclear matrix, where it would bind GR through an interaction of the zinc finger domain with the GR tau2 domain, thereby promoting transactivation.

AML/CBF-α Transcription Factors

The probably best characterized NMTS is that of the bone-specific transcription factor AML1b. The AML/CBF-α *runt* transcription factors are key regulators of hematopoetic and bone tissue-specific gene expression. AML1 and AML1b are two splicing variants, of which only AML1b is transcriptionally active. AML1b, but not AML1 binds to the nuclear matrix. Nuclear matrix association is independent of DNA-binding, shown by point mutations in the *runt* DNA binding-domain, which no longer bind DNA, but still associate with the nuclear matrix. The region necessary for nuclear matrix localization was narrowed down to a 31 aa segment localized near the C-terminus, which is different from the NLS and resides in a region absent in AML1.[39] The 31 aa NMTS is sufficient to target the heterologous protein GAL4 to the nuclear matrix. The NMTS is closely associated with the AML1b activation domain and fusing it to the GAL4 DNA-binding leads to transactivation of a genomically integrated GAL4-responsive reporter gene. Epitope tagged AML1b colocalizes with a subset of hyperphosphorylated Pol II in specific nuclear foci which are linked to the nuclear matrix.[40] This colocalization depends on active transcription and requires the DNA-binding function of the *runt* DNA-binding domain of AML1b.

The crystal structure of the AML1b NMTS fused to glutathione S-transferase has been solved at 2.7 Å resolution[41] and interpreted with respect to its predicted function in nuclear matrix targeting. It consists of two loop domains, which are connected by a flexible hinge region. The two loops of the NMTS could interact with a putative protein or nucleic acid acceptor in the nuclear matrix, a hypothesis supported by mutagenesis studies.[41,42] The 31 aa NMTS is conserved among AML1b homologs in human, mouse, rat and chicken. A similar motif is also present in the related transcription factors AML2 and AML3 from human and mouse. Both proteins were also found associated with the nuclear matrix.[41] A C-terminal deletion mutant of AML3, lacking 150 aa including the NMTS-homologous sequence is not retained in the nuclear matrix, indicating that the NMTS is functional in AML3 too.

Homeodomain Proteins

Different homeodomain-containing transcription factors have been found associated with the nuclear matrix. Lhx3 is a LIM homeodomain transcription factor that is essential for pituitary organogenesis and motor neuron specification. Lhx3 is found both in the nucleoplasm and associated with the nuclear matrix, as measured by in situ nuclear matrix extraction. The protein contains three nuclear localization signals within the homeodomain and one additional one at the C-terminus (B1, B2, B3, and B4). The 60 aa homeodomain (containing regions B1, B2, and B3) is sufficient for nuclear matrix targeting. A second, overlapping domain that also confers association with the nuclear matrix was mapped to a fragment of 28 aa which contains the B3 and B4 domains.[43] This domain is sufficient to target Green Fluorescent Protein to the nuclear matrix, which makes it an interesting tool for nuclear matrix studies due to its small size.

Interestingly, Pit-1, a homeodomain factor that cooperates with Lhx3 in the activation of several pituitary-specific genes, is also a nuclear matrix-associated protein.[44] Pit-1 has a bipartite DNA-binding domain known as the POU domain. It consists of a POU-specific domain and a POU-type homeodomain. The 66 aa POU-specific domain constitutes the domain necessary and sufficient for nuclear matrix targeting.[44] Nuclear matrix-association and DNA binding can be functionally uncoupled, because pointmutants that no longer bind DNA have been shown to remain fully nuclear matrix associated. Both the homeodomain of Lhx3 and the POU-specific domain of Pit-1 adopt a helical structure with basic sequences located at both ends of the domain.[43] Whether these structural features are related to the recognition of acceptor molecules on the nuclear matrix is presently not known. It is however interesting to speculate that the nuclear matrix-association of two functionally cooperating transcription factors might be involved in their biological activity. Another POU homeodomain protein, Oct-1 is also present in the nuclear matrix.[45] Oct-1 partitions, like Pit-1, between soluble and insoluble nuclear fractions. The domain necessary for nuclear matrix association of Oct-1 has not yet been identified. Interestingly, Oct-1 has also been shown to co-localize with lamin B, a component of the nuclear lamina (reviewed in ref. 48). In aging cells, Oct-1 has been found to depart from its location at the nuclear periphery, and this departure correlates with reduced repression of a collagenase gene by Oct-1.[46] These results suggest that Oct-1 is active as a repressor only when located at the nuclear periphery, and imply a connection between association with subnuclear structural elements and regulated function in gene expression.

Zinc Finger Transcription Factors

Yin-Yang 1 (YY1) is a 414 aa zinc finger-containing transcription factor, which acts both as activator and repressor. It is identical to NMP1, a protein found originally as a nuclear–matrix associated protein, which partitions between the nuclear matrix and a 0.4 M salt extract.[47] Its activation domain has been mapped to the N-terminus, while the repression domain overlaps with the zinc-finger DNA-binding domain at the C-terminus. McNeil et al[48] show that the C-terminal domain (aa 201-414) is necessary for high-affinity interaction with the nuclear matrix, while the N-terminus (aa 1-256) shows only weak retention.[48] Bushmeyer et al[49] have narrowed down the NMTS further to the region of aa 256-340. It is at present not known whether this relatively well defined NMTS is sufficient to target a heterologous protein to the nuclear matrix.

The NP/NMP4 transcription factors are nuclear matrix-associated proteins that contain from five to eight C-terminal Cys_2-His_2 Zinc fingers and an N-terminal AT-hook domain.[50] Some NP/NMP4 in-frame splice variants have been proposed to be architectural transcription factors which increase the basal activity of the rat type I collagen $\alpha 1(I)$ polypeptide chain promoter in osteoblast-like cells. They bind to the minor groove of a poly(dT) consensus sequence and bend DNA. The isoform 11H is predominantly located in two non-nucleolar foci inside the nucleus, with smaller amounts diffusely distributed in the non-nucleolar part of the nucleus. The zinc finger domain is necessary and sufficient for this localization. Extracting cells transiently transfected with GFP-NP/NMP4 fusions, it has been shown that the amino terminus plus the AT-hook domain could be extracted, while the full-length GFP fusion protein and the zinc finger fusion protein were retained and diffusely distributed in the nuclear matrix. Because the nuclear matrix fraction showed no evidence for residual DNA by DAPI staining, it was concluded that the zinc-finger mediated nuclear matrix targeting of NP/NMP4 does not require DNA binding. The minimal domain sufficient for nuclear matrix targeting is a 148 aa fragment containing zinc fingers four to eight.

ZNF74 is a developmentally expressed zinc finger gene of the Kruppel-associated box (KRAB) multifinger subfamily and is encoded by a candidate gene for DiGeorge syndrome.[51,52] Grondin et al[53] have shown that the zinc finger nucleic acid binding domain is a multifunctional domain which also acts as nuclear matrix targeting sequence and is involved in protein-protein interactions. ZNF74 interacts with its zinc finger domain with the hyperphosphorylated form of the large subunit of RNA polymerase II (pol IIo), but not with the hypophosphorylated form.[53] In immunofluorescence experiments, ZNF74 co-localized with pol IIo and with the SC35 splicing factor in subnuclear domains.

The smallest region sufficient for association with the nuclear matrix was narrowed down to the zinc finger domain between aa 175 and 509. It is presently not known if a smaller region of ZNF74 still tightly binds to the nuclear matrix, or if the entire zinc finger domain is required. Since the binding occurs after extensive DNase and RNase treatment of the nuclear matrix fraction, the authors concluded that it is independent of the nucleic acid-binding affinity of the zinc finger domain, and therefore most likely mediated by protein-protein interactions. In a search for protein-protein interaction partners, and therefore possible nuclear matrix adapters, only binding to pol IIo was discovered. The two proteins interact in vivo as well as in the absence of nucleic acids and the binding depends on the hyperphosphorylation of pol IIo. This result suggests that ZNF74 is not present in preinitiation complexes but rather associates with elongating RNA polymerase II. Whether and how this complex is associated with the nuclear matrix awaits further investigation. Interestingly, a nuclear matrix protein that interacts with the phosphorylated C-terminal domain of RNA polymerase II has been identified by Patturajan et al[54] and several reports have demonstrated that Pol II itself is associated with the nuclear matrix.[55-57]

Viral Proteins

The Epstein-Barr virus (EBV) nuclear antigen leader protein (EBNA-LP) is the first viral gene product together with EBNA-2 to be expressed after infection of B-cells. EBNA-LP is a nuclear matrix-associated protein and has been suggested to play an important role in EBV-induced transformation.[58] The protein has been shown to bind to p53 and to Rb in vitro and co-localizes with ND10 nuclear domains, indicating a specific spatial sequestering inside the nucleus. To investigate the biological significance of EBNA-LP nuclear matrix binding, Yokoyama et al[58] have attempted to map the nuclear matrix targeting domain and correlate it with the ability of EBNA-LP to co-activate EBNA-2 dependent transactivation. EBNA-LP consists of four W1W2 repeat domains flanked by a Y1Y2 domain. The affinity of nuclear matrix binding is reduced as W1W2 repeats are deleted, indicating that their copy number is involved in high-affinity binding. Fine mapping indicated that the 10 amino acid segment EPRRVRRRVL in the W2 domain is involved in NM binding. However, as substitution of this motif also destroys the NLS of EBNA-LP, leading to accumulation of the protein in the cytoplasm, no clear NLS-independent NMTS could be identified in this study. The mutant protein fails to act as a co-activator, which might be explained by its inability to enter the nucleus.

The papillomavirus E2 protein is a site-specific DNA-binding protein, which functions as the primary origin of replication-recognition protein. In addition, it is involved in regulating transcription from the native viral promoter. The protein is localized in distinct subnuclear foci, which correlate with the replication compartments. It consists of an N-terminal trans-acting domain, a central hinge-domain, and a C-terminal protein-dimerization and DNA-binding domain.[59] The hinge domain confers strong nuclear localization, while the N-terminus and the C-terminus alone are localized both in the nucleus and in the cytoplasm. Both the hinge and the N-terminus also bind to the nuclear matrix, but with reduced affinity compared to the full-length protein. A cluster of basic amino acids in the hinge domain is required for both nuclear and nuclear matrix localization. As in the case of EBNA-LP, the two functions can not be separated, because the mutated protein no longer enters the nucleus. A truncated hinge region (aa 216-255), which contains the basic motif and flanking proline and glycine residues leads to a diffuse nuclear localization, but can not be recovered from a nuclear matrix fraction, indicating that the fragment is not sufficient for the nuclear matrix association observed with the full-length hinge domain. The 82 amino acid hinge domain is therefore the smallest fragment shown to be both necessary and sufficient for nuclear matrix targeting of the HVP-11 protein.

Kinases and Kinase Adapters

DYRK1 is a member of a family of dual-specificity protein kinases involved in brain development. Mutants of the Drosophila homolog *minibrain* have reduced numbers of neurons in some brain areas and show specific behavioral defects.[60] The human homolog maps to the "Down's syndrome critical region" on chromosome 21, but its specific function in neuronal development is not yet known. The mammalian isoform DYRK1a is a nuclear localized protein and the NLS has been mapped to positions 105 to 139. Becker et al[61] have shown that the region between position 1 and 104 is necessary for the specific, punctate nuclear localization pattern of DYRK1a, and that a deletion fragment containing this region only is as tightly nuclear matrix-bound as the full-length protein. A deletion fragment spanning aa 105 to 139 is also partially localized in the nuclear matrix fraction, indicating that the position of potential NMTS sequences in DYRK1a is more complex.[61]

A-kinase anchoring protein (AKAP) 95 is a zinc-finger protein, which binds and anchors cAMP-dependent protein kinase (PKA). The protein is found in the nuclear matrix of a wide variety of mammalian cells.[62,63] The zinc finger domain required for DNA binding and a C-terminal amphipathic helix that serves as an anchoring domain for PKA are both located in the C-terminal half of the 687 amino acid protein. The N-terminal 386 amino acids contain both the NLS and a domain required for nuclear matrix targeting.[64] The NMTS was further narrowed down to a domain between amino acid 111 and 141, which is highly conserved in rat, mouse and human AKAP95 and similar to a sequence found in the nuclear matrix protein ZAN75, and in the genetic neighbor of AKAP95, NAKAP95.[64] The AKAP95 NMTS is independent from both the DNA-binding and the PKA-binding domains, indicating that an additional interaction partner is required for the association with the nuclear matrix.

A candidate for such an acceptor is p68 RNA helicase (p68 RH), which was identified as an AKAP95-binding protein in a yeast two-hybrid screen. The AKAP95 fragment from aa 109 to 201 is sufficient for binding the C-terminal domain of p68 RH, indicating that neither the DNA-binding nor the PKA-binding domain is involved. There is a close correlation between the AKAP95 NMTS (aa 111-141) and the p68 RH-binding domain (aa 109-201), p68 RH has been found to be a nuclear matrix-localized protein, and AKAP95 and p68 RH could be co-immunoprecipitated from nuclear fractions.[64] Together, these findings make p68 RH a potential nuclear matrix-localized acceptor of AKAP95. p68 RH has been previously shown to be associated with cAMP-responsive element binding protein (CREB)-binding protein CBP. Association of AKAP95 with a nuclear matrix-bound complex of p68 RH/CREB/CBP/DNA would therefore be a plausible mechanism to position PKA for its establishes function in CBP/CREB-mediated gene regulation.[64]

Comparison of NMTSs in Different Nuclear Proteins

Figure 1 shows an overview over the position and size of the NMTSs discussed in this Chapter. The regions identified to be either necessary or sufficient for nuclear matrix targeting reach from small, ca. 30 aa motifs to large portions of a protein (e.g., in ZNF74). In the case of transcription factors, there are several cases where they overlap or are identical with the DNA-binding domains. However, as nuclear matrix preparations are usually extensively DNase treated, several authors conclude that nuclear matrix binding is a feature of these domains that is separable from DNA-binding, and therefore probably caused by protein-protein interactions.

The smallest identified peptides sufficient for nuclear matrix targeting are those in AML1b (31 aa), the B3/B4 domain of Lhx3 (28 aa), and the second NMTS mapped in DYRK (34 aa), which confers partial nuclear matrix localization and overlaps with the NLS. Another well-defined region necessary for nuclear matrix localization is the 32 aa fragment from aa 111 to aa 141 in AKAP95. This fragment has so far not been shown to be sufficient for nuclear matrix targeting.

The sequences of the AML1b and AKAP95 NMTS are both well conserved within a family of related proteins, including both their homologs from different species such as human, mouse, rat and chicken, and functionally closely related proteins within the same species.[39,64] However, no consensus can be derived by comparing NMTS sequences of functionally unrelated proteins. This feature of NMTSs might indicate that different proteins use different interaction surfaces on the nuclear matrix, either by binding to different acceptor proteins, or by interacting with different domains of the same ubiquitous nuclear matrix proteins.

To advance our understanding of the nature of NMTSs, interaction partners for the best defined NMTS sequences need to be identified, and experiments need to be designed to show that specific protein-protein interactions are required for the observed targeting events. In addition, more structural information of the respective protein domains—such as the crystal structure of the AML1b NMTS—will allow comparisons beyond the alignment of primary sequences and might uncover presently unknown structural similarities between these domains.

Regulated Nuclear Matrix Interaction

Only a very small number of regulated subnuclear targeting events have been described. Interestingly, one involves the light-regulated subnuclear trafficking of two plant photoreceptors, which have been shown to interact in vivo at their targeting sites. These examples provide first evidence that specific subnuclear targeting might serve as a previously unappreciated mechanism regulating the function of nuclear proteins. This is an area of functional cell biology that clearly awaits further investigation.

It has been shown early that the association of steroid receptors with the nuclear matrix requires the presence of the hormone.[29,30] In addition, the nuclear matrix of ATP depleted cells binds a significantly higher fraction of GRs, while binding equal amounts of SV40 large tumor antigen, indicating that the observation is not caused by an unspecific "collapsing" effect of nuclear material.[65] This retention is reversible by addition of ATP. The data implicate that ATP is required for a step that actively dissociates GR from its acceptor on the nuclear matrix.

A GR-GFP fusion protein has been tested for nuclear import in response to hormone binding. While GR-GFP translocated to the nucleus in the presence of dexamethasone, progesterone, and the glucocorticoid antagonist RU486, a striking difference in the subnuclear distribution of the fusion protein was observed. In dexamethasone-treated cells GR-GFP was predominantly located at bright small foci within the nucleus. In contrast, treatment with progesterone lead to a diffuse nuclear localization, while RU486-treated cells showed a diffuse pattern with regions of condensation in a reticular pattern. These data indicate that while all three hormones were sufficient for the activation of nuclear import of GR-GFP, the specific subnuclear trafficking of the three complexes was regulated differentially.[66] Similar results were obtained with a human estrogen receptor-GFP fusion protein.[67]

Protein kinase CKII is a highly conserved ubiquitous messenger-independent serine/threonine protein kinase, which is localized in the nucleus and the cytoplasm and which is involved in growth and differentiation events. It has been shown to be located in the nuclear matrix and phosphorylates substrates such as topoisomerase II and RNA polymerases.[68] The association of CKII with the nuclear matrix has been found to be influenced by androgen and growth factor stimuli in rat prostate cells.[69] Androgen deprivation leads to a progressive decline in NM-bound CKII while androgen treatment leads to an increase. In addition, three growth factors stimulated the association of CKII with the nuclear matrix.

An interesting case of regulated association with the nuclear matrix is that of the retinoblastoma gene product Rb. Rb is a 110 kD tumor suppressor protein that interacts with several viral oncoproteins that are associated with the nuclear matrix. Rb itself is nuclear matrix-bound, but only specifically during G1 phase, while no nuclear matrix association was detected during S phase. The form bound during G1 is hypophosphorylated and was found at the nuclear periphery as well as in dense fibrogranular masses in immunogold labeling experiments with isolated nuclear matrix fractions.[70] Interestingly, a novel nuclear matrix protein (NRM/B) has been identified that specifically binds to the hypophosphorylated form of Rb.[71] NRB/P is a kelch-domain protein and its expression is limited to neuronal tissue. While it might therefore be an exciting candidate for a nuclear matrix acceptor of Rb, it will also be interesting to find nuclear matrix proteins with affinity for Rb which are less limited to specific cell types.

Subnuclear targeting events and the association of functional proteins with the nuclear matrix are at present a practically uninvestigated field in plant molecular biology. A small number of nuclear matrix-associated and MAR-binding proteins have been identified,[20-22,72-74] but the association of transcription factors and other gene expression-modulating proteins with the nuclear matrix has not yet been studied. Strikingly, and coming from an entirely different angle of investigation, groups studying the biological activity of plant photoreceptors have recently provided first evidence that subnuclear partitioning might both happen and be of functional consequence in plants too. When a fusion of the red-light photoreceptor phytochrome B with GFP was monitored in dark-grown and light-grown plants, the fusion protein was found to be localized in the cytoplasm in the dark, but in the nucleus in the light.[75] Strikingly, the protein was not diffusely distributed within the nucleus, but accumulated in "speckles" of comparable size and distribution to some of the nuclear bodies well characterized in animals, but so far barely investigated in plants.[76-81]

Cryptochrome 2 is a plant blue-light photoreceptor that has also been shown to be located in the nucleus.[82,83] Mas et al[84] have demonstrated that cryptochrome 2 changes its subnuclear localization in response to blue light irradiation. While the protein has a diffuse localization in the dark, it accumulates in speckles after brief blue-light irradiation. Moreover, after treatment with both blue and red light, phytochrome B and cryptochrome 2 co-localize in some nuclear speckles and FRET experiments indicate that the two proteins directly interact. Both photoreceptors are implied in the activation of gene expression. It will be of great interest to see if their interaction takes place on the plant nuclear matrix, if the observed speckles can be identified with respect to known nuclear bodies, and what protein domains and interaction partners of cryptochrome 2 are involved in its light-regulated subnuclear trafficking.

Compromised Subnuclear Localization and Disease

Numerous cytogenetic abnormalities that involve the bone-specific transcription factor AML1 have been identified in acute myelogenous leukemia. In the frequent 8;21 translocation, a fusion protein between AML1b and ETO is created that lacks the C-terminus of AML1b, including the NMTS. The AML1/ETO fusion is still targeted to subnuclear sites, but interestingly they differ from the binding sites of AML1b. Instead, the AML1/ETO protein is redirected by the ETO component to alternative nuclear-matrix associated foci.[85] These findings indicate that modifications in the subnuclear trafficking of transcription factors can disrupt their gene regulatory function and that within the nuclear matrix specific functional subdomains for anchored transcription factors can be defined.

Similarly, the putative transcription factor ALL1 (also called MLL and HRX) normally shows a punctate subnuclear distribution, which is conferred by distinct elements in the N-terminus of the protein.[86] ALL1 is a 430 kD polypeptide, which contains two putative DNA-binding domains, an amino terminal AT-hook motif and two zinc finger regions near the middle of the protein. In the t(1;11) (p32-q23) translocation, which has been described in rare cases of acute myelogenous leukemia,[86] the amino-terminal domain of ALL1 is fused to the carboxy-terminal domain of eps15, a ubiquitously expressed epidermal growth factor receptor substrate, which is localized in the cytoplasm. The ALL1-eps15 fusion protein is localized in the nucleus, but is targeted to different, smaller domains than wildtype ALL1, indicating that the protein fusion created a novel targeting address different from both wildtype proteins.[86]

Interestingly, yet another translocation fusing ALL1 to a heterologous protein and causing acute leukemia creates a fusion between ALL1 and a histone acetyltransferase.[87] Although no localization data exist for this fusion protein, it is tempting to speculate that mistargeting of a functional histone acetyltransferase through fusion with ALL1 might deregulate chromatin structure and gene expression patterns, thereby promoting leukemia.

Besides acute leukemia, subnuclear mislocalization has also been implied in a neurodegenerative disorder, spinocerebellar ataxia type 1 (SCA1). It is caused by expansion of a polyglutamine tract in the *SCA1* gene, coding for ataxin-1. The subnuclear localization patterns of wildtype and mutant ataxin-1 have been compared.[88] Wildtype ataxin-1 localizes to several nuclear structures of about 0.5 μm diameter, while mutant ataxin-1 was found in a single, large 2 μm structure. PML bodies are specific nuclear structures associated with the nuclear matrix, which contain the marker PML protein. Colocalization experiments showed that mutant ataxin-1 sequestered the PML protein to the 2 μm body and altered the distribution and appearance of PML bodies.[88] Both wildtype and mutant ataxin-1 were found bound to the nuclear matrix in cerebral tissue, emphasizing that the interactions between ataxin-1 and PML, which might lead to the observed mislocalizations, are associated with the nuclear matrix.

Concluding Remarks

The case has been made for the presence of sequences different from nuclear localization signals that determine the fate of a protein once inside the nucleus. The number of examples where such signals have been studied in detail is still small, and it is too early to draw conclusions about their similarity or multiplicity. If the described signals act like other targeting domains, they will most likely function by providing a surface for protein-protein interactions. If transcription factors come with signals for specific subnuclear "addresses", and if their disruption can compromise transcription factor function, then one ought to think about what those addresses are, and how they relate to the position and/or compartmentalization of the promoters regulated by these factors. The prevailing evidence for the association of specific chromatin regions (MARs) with the nuclear matrix, and the positive effect MARs have on the transcription of flanking genes has led to the model that the association of genes with the nuclear matrix increases their ability to be expressed, possibly by providing a more "open" chromatin environment. It might be equally attractive to think about another mechanism, by which association with the nuclear matrix of both promoters and transcription factors might increase the probability of productive assembly of transcription initiation complexes. The fact that "transcriptosomes" appear to have specific locations in the nucleus, and that they can function on the isolated nuclear matrix encourages to think about how such complexes might assemble in specific places. It will be highly informative to investigate whether three-way interactions between genes, sequence-specific transcription factors, and nuclear matrix components play a role in their assembly.

References

1. Berezney R, Coffey DS. Identification of a nuclear protein matrix. Biochem Biophys Res Commun 1974; 60:1410-1417.
2. Nickerson JA, Krockmalnic G, Wan KM et al. The nuclear matrix revealed by eluting chromatin from a cross-linked nucleus. Proc Natl Acad Sci USA 1997; 94:4446-4450.
3. Wan KM, Nickerson JA, Krockmalnic G et al. The nuclear matrix prepared by amine modification. Proc Natl Acad Sci USA 1999; 96:933-938.
4. Verheijen R, van Venrooij W, Ramaekers F. The nuclear matrix: structure and composition. J Cell Sci 1988; 90:11-36.
5. Penman S. Rethinking cell structure. Proc Natl Acad Sci USA 1995; 92:5251-5257.
6. Fey EG, Penman S. Nuclear matrix proteins reflect cell type of origin in cultured human cells. Proc Natl Acad Sci USA 1988; 85:121-125.
7. Stuurman N, Meijne AM, van der Pol AJ et al. The nuclear matrix from cells of different origin. Evidence for a common set of matrix proteins. J Biol Chem 1990; 265:5460-5465.
8. McKeon FD, Kirschner MW, Caput D. Homologies in both primary and secondary structure between nuclear envelope and intermediate filament proteins. Nature 1986; 319:463-468.
9. Furukawa K, Pante N, Aebi U et al. Cloning of a cDNA for lamina-associated polypeptide 2 (LAP2) and identification of regions that specify targeting to the nuclear envelope. EMBO J 1995; 14:1626-1636.
10. Wilson KL. The nuclear envelope, muscular dystrophy and gene expression. Trends Cell Biol 2000; 10:125-129.
11. Gasser SM, Laemmli UK. A glimpse at chromosomal order. Trends Genet 1987; 3:16-22.
12. Phi-Van L, von Kries JP, Ostertag W et al. The chicken lysozyme 5' matrix attachment region increases transcription from a heterologous promoter in heterologous cells and dampens position effects on the expression of transfected genes. Mol Cell Biol 1990; 10:2302-2307.
13. Allen GC, Hall GE, Jr., Childs LC et al. Scaffold attachment regions increase reporter gene expression in stably transformed plant cells. Plant Cell 1993; 5:603-613.
14. Allen GC, Hall G, Jr., Michalowski S et al. High-level transgene expression in plant cells: Effects of a strong scaffold attachment region from tobacco. Plant Cell 1996; 8:899-913.
15. von Kries JP, Buhrmester H, Stratling WH. A matrix/scaffold attachment region binding protein: Identification, purification, and mode of binding. Cell 1991; 64:123-135.
16. Dickinson LA, Joh T, Kohwi Y et al. A tissue-specific MAR/SAR DNA-binding protein with unusual binding site recognition. Cell 1992; 70:631-645.
17. Dickinson LA and Kohwi-Shigematsu T. Nucleolin is a matrix attachment region DNA-binding protein that specifically recognizes a region with high base-unpairing potential. Mol Cell Biol 1995; 15:456-465.
18. Gohring F, Schwab BL, Nicotera P et al. The novel SAR-binding domain of scaffold attachment factor A (SAF-A) is a target in apoptotic nuclear breakdown. EMBO J 1997; 16:7361-7371.
19. Renz A, Fackelmayer FO. Purification and molecular cloning of the scaffold attachment factor B (SAF-B), a novel human nuclear protein that specifically binds to S/MAR-DNA. Nucleic Acids Res 1996; 24:843-849.
20. Meier I, Phelan T, Gruissem W et al. MFP1, a novel plant filament-like protein with affinity for matrix attachment region DNA. Plant Cell 1996; 8:2105-2115.
21. Morisawa G, Han-Yama A, Moda I et al. AHM1, a novel type of nuclear matrix-localized, MAR binding protein with a single AT hook and a J domain-homologous region. Plant Cell 2000; 12:1903-1916.
22. Hatton D, Gray JC. Two MAR DNA-binding proteins of the pea nuclear matrix identify a new class of DNA-binding proteins. Plant J 1999; 18:417-429.

23. van Wijnen AJ, Bidwell JP, Fey EG et al. Nuclear matrix association of multiple sequence-specific DNA binding activities related to SP-1, ATF, CCAAT, C/EBP, OCT-1, and AP-1. Biochemistry 1993; 32:8397-8402.
24. Zeng C, McNeil S, Pockwinse S et al. Intranuclear targeting of AML/CBFalpha regulatory factors to nuclear matrix-associated transcriptional domains. Proc Natl Acad Sci USA 1998; 95:1585-1589.
25. Guo B, Odgren PR, van Wijnen AJ et al. The nuclear matrix protein NMP-1 is the transcription factor YY1. Proc Natl Acad Sci USA 1995; 92:10526-10530.
26. Blencowe BJ, Nickerson JA, Issner R et al. Association of nuclear matrix antigens with exon-containing splicing complexes. J Cell Biol 1994; 127:593-607.
27. Spector DL, Schrier WH, Busch H. Immunoelectron microscopic localization of snRNPs. Biol Cell 1983; 49:1-10.
28. Wei X, Somanathan S, Samarabandu J et al. Three-dimensional visualization of transcription sites and their association with splicing factor-rich nuclear speckles. J Cell Biol 1999; 146:543-558.
29. Barrack ER and Coffey DS. The specific binding of estrogens and androgens to the nuclear matrix of sex hormone responsive tissues. J Biol Chem 1980; 255:7265 7275.
30. Barrack E. Specific association of androgen receptors and estrogen receptors with the nuclear matrix: Summary and perspectives. In: Moudgil V, ed. Recent Advances in Steroid Hormone Action. Berlin: Walther de Gruiter, 1987:85-107.
31. van Steensel B, Jenster G, Damm K et al. Domains of the human androgen receptor and glucocorticoid receptor involved in binding to the nuclear matrix. J Cell Biochem 1995; 57:465-478.
32. Tang Y, Getzenberg RH, Vietmeier BN et al. The DNA-binding and tau2 transactivation domains of the rat glucocorticoid receptor constitute a nuclear matrix-targeting signal. Mol Endocrinol 1998; 12:1420-1431.
33. DeFranco DB, Guerrero J. Nuclear matrix targeting of steroid receptors: specific signal sequences and acceptor proteins. Crit Rev Eukaryot Gene Expr 2000; 10:39-44.
34. Eggert M, Michel J, Schneider S et al. The glucocorticoid receptor is associated with the RNA-binding nuclear matrix protein hnRNP U. J Biol Chem 1997; 272:28471-28478.
35. Romig H, Fackelmayer FO, Renz A et al. Characterization of SAF-A, a novel nuclear DNA binding protein from HeLa cells with high affinity for nuclear matrix/scaffold attachment DNA elements. Embo J 1992; 11:3431-3440.
36. Oesterreich S, Zhang Q, Hopp T et al. Tamoxifen-bound estrogen receptor (ER) strongly interacts with the nuclear matrix protein HET/SAF-B, a novel inhibitor of ER-mediated transactivation. Mol Endocrinol 2000; 14:369-381.
37. Weighardt F, Cobianchi F, Cartegni L et al. A novel hnRNP protein (HAP/SAF-B) enters a subset of hnRNP complexes and relocates in nuclear granules in response to heat shock. J Cell Sci 1999; 112:1465-1476.
38. Yang L, Guerrero J, Hong H et al. Interaction of the tau2 transcriptional activation domain of glucocorticoid receptor with a novel steroid receptor coactivator, Hic- 5, which localizes to both focal adhesions and the nuclear matrix. Mol Biol Cell 2000; 11:2007-2018.
39. Zeng C, van Wijnen AJ, Stein JL et al. Identification of a nuclear matrix targeting signal in the leukemia and bone-related AML/CBF-alpha transcription factors. Proc Natl Acad Sci USA 1997; 94:6746-6751.
40. Zeng C, McNeil S, Pockwinse S et al. Intranuclear targeting of AML/CBFalpha regulatory factors to nuclear matrix-associated transcriptional domains. Proc Natl Acad Sci USA 1998; 95:1585-1589.
41. Tang L, Guo B, Javed A et al. Crystal structure of the nuclear matrix targeting signal of the transcription factor acute myelogenous leukemia-1/polyoma enhancer-binding protein 2alphaB/core binding factor alpha2. J Biol Chem 1999; 274:33580-33586.
42. Chen LF, Ito K, Murakami Y et al. The capacity of polyomavirus enhancer binding protein 2alphaB (AML1/Cbfa2) to stimulate polyomavirus DNA replication is related to its affinity for the nuclear matrix. Mol Cell Biol 1998; 18:4165-4176.
43. Parker GE, Sandoval RM, Feister HA et al. The homeodomain coordinates nuclear entry of the Lhx3 neuroendocrine transcription factor and association with the nuclear matrix. J Biol Chem 2000; 275:23891-23898.

44. Mancini MG, Liu B, Sharp ZD et al. Subnuclear partitioning and functional regulation of the Pit-1 transcription factor. J Cell Biochem 1999; 72:322-338.

45. Kim MK, Lesoon-Wood LA, Weintraub BD et al. A soluble transcription factor, Oct-1, is also found in the insoluble nuclear matrix and possesses silencing activity in its alanine-rich domain. Mol Cell Biol 1996; 16:4366-4377.

46. Imai S, Nishibayashi S, Takao K et al. Dissociation of Oct-1 from the nuclear peripheral structure induces the cellular aging-associated collagenase gene expression. Mol Biol Cell 1997; 8:2407-2419.

47. Guo B, Odgren PR, van Wijnen AJ et al. The nuclear matrix protein NMP-1 is the transcription factor YY1. Proc Natl Acad Sci USA 1995; 92:10526-10530.

48. McNeil S, Guo B, Stein JL et al. Targeting of the YY1 transcription factor to the nucleolus and the nuclear matrix in situ: the C-terminus is a principal determinant for nuclear trafficking. J Cell Biochem 1998; 68:500-510.

49. Bushmeyer SM, Atchison ML. Identification of YY1 sequences necessary for association with the nuclear matrix and for transcriptional repression functions. J Cell Biochem 1998; 68:484-499.

50. Feister HA, Torrungruang K, Thunyakitpisal P et al. NP/NMP4 transcription factors have distinct osteoblast nuclear matrix subdomains. J Cell Biochem 2000; 79:506-517.

51. Ravassard P, Cote F, Grondin B et al. ZNF74, a gene deleted in DiGeorge syndrome, is expressed in human neural crest-derived tissues and foregut endoderm epithelia. Genomics 1999; 62:82-85.

52. Cote F, Boisvert FM, Grondin B et al. Alternative promoter usage and splicing of ZNF74 multifinger gene produce protein isoforms with a different repressor activity and nuclear partitioning. DNA Cell Biol 2001; 20:159-173.

53. Grondin B, Cote F, Bazinet M et al. Direct interaction of the KRAB/Cys2-His2 zinc finger protein ZNF74 with a hyperphosphorylated form of the RNA polymerase II largest subunit. J Biol Chem 1997; 272:27877-27885.

54. Patturajan M, Wei X, Berezney R et al. A nuclear matrix protein interacts with the phosphorylated C-terminal domain of RNA polymerase II. Mol Cell Biol 1998; 18:2406-2415.

55. Vincent M, Lauriault P, Dubois MF et al. The nuclear matrix protein p255 is a highly phosphorylated form of RNA polymerase II largest subunit which associates with spliceosomes. Nucleic Acids Res 1996; 24:4649-4652.

56. Mortillaro MJ, Blencowe BJ, Wei X et al. A hyperphosphorylated form of the large subunit of RNA polymerase II is associated with splicing complexes and the nuclear matrix. Proc Natl Acad Sci USA 1996; 93:8253-8257.

57. Chabot B, Bisotto S, Vincent M. The nuclear matrix phosphoprotein p255 associates with splicing complexes as part of the [U4/U6.U5] tri-snRNP particle. Nucleic Acids Res 1995; 23:3206-3213.

58. Yokoyama A, Kawaguchi Y, Kitabayashi I et al. The conserved domain CR2 of Epstein-Barr virus nuclear antigen leader protein is responsible not only for nuclear matrix association but also for nuclear localization. Virology 2001; 279:401-413.

59. Zou N, Lin BY, Duan F et al. The hinge of the human papillomavirus type 11 E2 protein contains major determinants for nuclear localization and nuclear matrix association. J Virol 2000; 74:3761-3770.

60. Tejedor F, Zhu XR, Kaltenbach E et al. minibrain: A new protein kinase family involved in postembryonic neurogenesis in Drosophila. Neuron 1995; 14:287-301.

61. Becker W, Weber Y, Wetzel K et al. Sequence characteristics, subcellular localization, and substrate specificity of DYRK-related kinases, a novel family of dual specificity protein kinases. J Biol Chem 1998; 273:25893-25902.

62. Eide T, Coghlan V, Orstavik S et al. Molecular cloning, chromosomal localization, and cell cycle-dependent subcellular distribution of the A-kinase anchoring protein, AKAP95. Exp Cell Res 1998; 238:305-316.

63. Coghlan VM, Langeberg LK, Fernandez A et al. Cloning and characterization of AKAP 95, a nuclear protein that associates with the regulatory subunit of type II cAMP-dependent protein kinase. J Biol Chem 1994; 269:7658-7665.

64. Akileswaran L, Taraska JW, Sayer JA et al. A-kinase-anchoring Protein AKAP95 Is Targeted to the Nuclear Matrix and Associates with p68 RNA Helicase. J Biol Chem 2001; 276:17448-17454.

65. Tang Y, DeFranco DB. ATP-dependent release of glucocorticoid receptors from the nuclear matrix. Mol Cell Biol 1996; 16:1989-2001.

66. Htun H, Barsony J, Renyi I et al. Visualization of glucocorticoid receptor translocation and intranuclear organization in living cells with a green fluorescent protein chimera. Proc Natl Acad Sci USA 1996; 93:4845-4850.

67. Htun H, Holth LT, Walker D et al. Direct visualization of the human estrogen receptor alpha reveals a role for ligand in the nuclear distribution of the receptor. Mol Biol Cell 1999; 10:471-486.

68. Ahmed K. Nuclear matrix and protein kinase CK2 signaling. Crit Rev Eukaryot Gene Expr 1999; 9:329-336.

69. Guo C, Yu S, Davis AT et al. Nuclear matrix targeting of the protein kinase CK2 signal as a common downstream response to androgen or growth factor stimulation of prostate cancer cells. Cancer Res 1999; 59:1146-1151.

70. Mancini MA, Shan B, Nickerson JA et al. The retinoblastoma gene product is a cell cycle-dependent, nuclear matrix-associated protein. Proc Natl Acad Sci USA 1994; 91:418-422.

71. Kim TA, Lim J, Ota S et al. NRP/B, a novel nuclear matrix protein, associates with p110(RB) and is involved in neuronal differentiation. J Cell Biol 1998; 141:553-566.

72. Masuda K, Xu ZJ, Takahashi S et al. Peripheral framework of carrot cell nucleus contains a novel protein predicted to exhibit a long alpha-helical domain. Exp Cell Res 1997; 232:173-181.

73. Gindullis F, Meier I. Matrix attachment region binding protein MFP1 is localized in discrete domains at the nuclear envelope. Plant Cell 1999; 11:1117-1128.

74. Gindullis F, Peffer NJ, Meier I. MAF1, a novel plant protein interacting with matrix attachment region binding protein MFP1, is located at the nuclear envelope. Plant Cell 1999; 11:1755-1768.

75. Yamaguchi R, Nakamura M, Mochizuki N et al. Light-dependent translocation of a phytochrome B-GFP fusion protein to the nucleus in transgenic Arabidopsis. J Cell Biol 1999; 145:437-445.

76. Boudonck K, Dolan L, Shaw PJ. The movement of coiled bodies visualized in living plant cells by the green fluorescent protein. Mol Biol Cell 1999; 10:2297-2307.

77. Gall JG, Bellini M, Wu Z et al. Assembly of the nuclear transcription and processing machinery: Cajal bodies (coiled bodies) and transcriptosomes. Mol Biol Cell 1999; 10:4385-4402.

78. Gall JG. A role for Cajal bodies in assembly of the nuclear transcription machinery. FEBS Lett 2001; 498:164-167.

79. Matera AG. Nuclear bodies: Multifaceted subdomains of the interchromatin space. Trends Cell Biol 1999; 9:302-309.

80. Maul GG, Negorev D, Bell P et al. Review: properties and assembly mechanisms of ND10, PML bodies, or PODs. J Struct Biol 2000; 129:278-287.

81. Seeler JS, Dejean A. The PML nuclear bodies: actors or extras? Curr Opin Genet Dev 1999; 9:362-367.

82. Kleiner O, Kircher S, Harter K et al. Nuclear localization of the Arabidopsis blue light receptor cryptochrome 2. Plant J 1999; 19:289-296.

83. Guo H, Duong H, Ma N et al. The Arabidopsis blue light receptor cryptochrome 2 is a nuclear protein regulated by a blue light-dependent post-transcriptional mechanism. Plant J 1999; 19:279-287.

84. Mas P, Devlin PF, Panda S et al. Functional interaction of phytochrome B and cryptochrome 2. Nature 2000; 408:207-211.

85. McNeil S, Zeng C, Harrington KS et al. The t(8;21) chromosomal translocation in acute myelogenous leukemia modifies intranuclear targeting of the AML1/CBFalpha2 transcription factor. Proc Natl Acad Sci USA 1999; 96:14882-14887.

86. Yano T, Nakamura T, Blechman J et al. Nuclear punctate distribution of ALL-1 is conferred by distinct elements at the N terminus of the protein. Proc Natl Acad Sci USA 1997; 94:7286-7291.

87. Sobulo OM, Borrow J, Tomek R et al. MLL is fused to CBP, a histone acetyltransferase, in therapy-related acute myeloid leukemia with a t(11;16)(q23;p13.3). Proc Natl Acad Sci USA 1997; 94:8732-8737.

88. Skinner PJ, Koshy BT, Cummings CJ et al. Ataxin-1 with an expanded glutamine tract alters nuclear matrix- associated structures. Nature 1997; 389:971-974.

CHAPTER 4

Nuclear Import and Export Signals

Toshihiro Sekimoto, Jun Katahira and Yoshihiro Yoneda

Eukaryotic cells are separated into two large compartments, namely the nucleus and the cytoplasm, by the nuclear envelope. As a result, macromolecules including RNAs, which are transcribed in the nucleus and nuclear proteins, which are translated in the cytoplasm must cross the double lipid bilayer to reach the intracellular sites where they function. In addition, cumulative evidence suggests that trafficking between the nucleus and the cytoplasm is rather dynamic and some proteins and RNAs cross the nuclear envelope again after being transported to one compartment.

The nuclear pore complex (NPC), a huge proteinaceous channel composed of 50 to 100 different proteins that are collectively termed nucleoporins, is the only known functional path through which soluble molecules are transported.[1,2] The functional diameter of the aqueous channel of the nuclear pore, which is estimated to be 9 to 10 nm, allows the non-selective passive diffusion of small molecules such as ions and low molecular weight metabolites. Small molecules can freely traverse the NPC in both directions in a concentration gradient-dependent manner. On the contrary, molecules with molecular weights in excess of 40 to 60 kDa cannot pass through the NPC by simple diffusion. To accomplish "active transport", which is probably synonymous with transport against a concentration gradient via the consumption of energy, it is generally thought such molecules are directed to the appropriate compartments by specific mechanisms.

Indeed, numerous efforts have revealed that even small proteins, which are smaller than the diffusion limit, are subjected to both the active nuclear import and export. As expected the transport process has been found to be mediated by transferable signals, harbored within the transported substrates themselves. Such signal sequences for nuclear import and export are called nuclear localization signals (NLS) and nuclear export signals (NES), respectively. The aim of this Chapter is to introduce the definition and the mechanism of how such signals function.

Definition of Nuclear Import and Export Signals

During the nuclear import or export process, transport substrates must pass through the NPC by interacting with nucleoporins. Conceptually this may be accomplished by direct interactions of NLS or NES with nucleoporins. The recent identification of various receptors for NLS and NES, however, suggest that this is rather exceptional. Instead, the import and export receptors for various NLS or NES that exhibit intrinsic activities for interacting with the repeat sequences found in some nucleoporins mediate this task.[3-7] To date, at least 21 and 14 members of the importin β (or also called karyopherin β) family proteins have been identified in human and yeast *S. cerevisiae*, respectively, of which at least 9 of the yeast proteins and 5 of the human proteins, their functions in nuclear import have been assigned, whereas 4 of the yeast proteins and 3 of the human proteins have been identified as being involved in the nuclear

Nuclear Import and Export in Plants and Animals, edited by Tzvi Tzfira and Vitaly Citovsky.
©2005 Eurekah.com and Kluwer Academic / Plenum Publishers.

export of various substrates.[8] According to the directions of transport, importin β family proteins are collectively termed importins (for nuclear import) and exportins (for nuclear export). Unidirectional transport by these shuttling receptors is basically maintained by a mechanism whereby substrate-importin complexes are destabilized, whereas substrate-exportin complexes are stabilized in the nucleus. The GTP-bound form of Ran, which can flip between the GDP and the GTP bound state as other small GTPases, plays a pivotal role in the regulation of complex formation. Ran-GTP is enriched in the nucleus due to the biased localization of a specific GTPase activating protein (GAP) and a guanine-nucleotide exchange factor (GEF). For the nuclear import of NLS-bearing substrates, NLS is recognized in the cytoplasm by importins and the complexes are then translocated through the NPC. Once transported into the nucleus, the binding of Ran-GTP to importins destabilizes the import complexes. In contrast, complex formation between NES-bearing substrates and exportins occurs only in the presence of Ran-GTP. In the cytoplasm Ran is rapidly converted from the GTP-bound to the GDP-bound state by the concerted action of RanGAP and RanBP1/BP2, and the cargo molecules are released (for a review see refs. 8 and 9). Thus, NLS or NES recognized by the importin β family proteins basically act only as signals for import or export. In addition to this Ran-dependent regulation, recent publications have established that various mechanisms for regulating the activities of NLS or NES by posttranslational modifications exist. Individual examples of such regulation mechanisms will be described later in this Chapter.

As described above, now we can define the criteria for NLS and NES. One of their functional characteristics is that they could be experimentally identified based on their abilities to stimulate the migration of heterologous reporter proteins, which are otherwise restricted to the cytoplasm or the nucleus, to the opposite compartment. Another important prognosis for NLS or NES, which may also be determined experimentally, is their abilities to be directly recognized by the well-characterized transport receptors in the absence (for NLS) or presence (for NES) of Ran-GTP. In this case, since it has been recently shown that a certain member of the importin β family (Kap142p/Msn5p) has the ability to transport different cargos in an opposite compartment through the NPC,[10] it should be noted that the binding ability of the sequence to importin β family molecules does not necessarily define the direction of transport. The third functional property is that the directionality of transport is exclusively one way. However, various signal sequences that do not fulfill these functional characteristics have emerged. For example a couple of signal sequences enable reporter proteins to be translocated both back and forth through NPC. We will also describe such exceptional sequences below.

Basic Type NLSs

Dingwall et al elegantly showed, using partially digested nucleoplasmin, that a signal required for the nuclear localization of nucleoplasmin exists in a "tail" portion of the molecule.[11] Thereafter, the first NLS was identified in Simian Virus 40 (SV40) large T antigen. This NLS, which is rich in lysine and arginine, consists of only seven amino acids, and has proved to be sufficient for the nuclear localization of the T antigen.[12] Moreover, synthetic peptides containing this sequence act as an NLS when chemically conjugated with a non-nuclear protein such as bovine serum albumin.[13,14] A mutational analysis revealed that lysine and arginine residues in the NLS are essential for nuclear targeting. Although a number of other NLSs had been identified, most have been found to contain one or two basic amino acid clusters composed of several lysine and arginine residues, therefore, these NLSs are referred to as basic type NLS. Thus, it was assumed that most nuclear localizing proteins have a basic type NLS in their primary amino acid sequence. Interestingly, there is no obvious consensus sequence between basic type NLSs. The key word is "basic amino acid cluster". The basic type NLS can be largely classified into two groups: (1) the single basic amino acid cluster type, composed of four to six lysine/arginine resides such as SV40 large T antigen and (2) the bipartite basic amino acid

Table 1. NLSs in various proteins

	Protein	NLS
Basic type	SV40 large T antigen	**PKKKRK**V
	Nucleoplasmin	**KR**PAAIKKAGQA**KKKK**
	CBP80	**RRR**HSDENDGGQPH**KRRK**
(arginine rich)	HIV-1 Rev	RQARRNRRRWE
	HTLV-I Rex	MPKTRRRPRRSQRKRPPT
Non-basic type	hnRNP A1	NQSSNFGPMKGGNFGGRSSGPY-GGGGQYFAKPRNQGGY
	rpL23a	VHSHKKKKIRTSPTFTTPKTLRL-RRQPKYPRKSAPRRNKLDHY

Basic type NLSs from SV40 large T antigen,[12] nucleoplasmin,[15] CBP80,[17] HIV-1 Rev[24] and HTLV-I Rex,[23] and non-basic type NLSs from hnRNP A1[34,35] and rpL23a[39] are shown.

clusters type, composed of a dozen amino acids containing two basic amino acid clusters spaced by about ten amino acids and include nucleoplasmin and cap-binding protein (CBP) 80 (Table 1).[15-17] Since both the deletion of a basic amino acid cluster and the introduction of mutations into the clusters in the bipartite type NLS affects nuclear import activity, both structures are critical in its function. These basic type NLS containing proteins are directly recognized by importin α[3,4,18-20] and form a complex with importin β and are then transported into nucleus with assistance by Ran and p10/NTF2.

The importin β binding domain (IBB domain), an N-terminal region of importin α which is rich in arginine residues, is recognized by the HEAT repeats of importin β.[21,22] It is likely that arginine rich basic type NLSs are directly recognized by importin β because of their similarity to the IBB domain of importin α. In fact, several basic, arginine-rich type NLSs are specifically recognized by importin β in the absence of importin α, although the majority of basic type NLSs are recognized directly by importin α. For example, human T-cell leukemia virus type I (HTLV-I) Rex protein,[23] human immunodeficiency virus (HIV-1) Rev and Tat proteins[24] contain arginine-rich type NLS and are transported into nucleus by importin β without the aid of importin α (Table 1).

Certain basic type NLS containing proteins exist predominantly in the cytoplasm until cells are stimulated by appropriate signals. Regulation of recognition of such NLS containing proteins by transport factors may control the import of nuclear protein. The precursor of the p50 NFκB subunit has a classical basic type NLS, but persists in the cytoplasm because of intramolecular masking by its C-terminal portion.[25] After proteolytic cleavage, the NLS is exposed and is then recognized by importin α.[26] It has been shown that the NLS masking of the NFκB p65 subunit by IκBα is regulated by proteolytic degradation and phosphorylation of IκBα.[27] Extracellular signals cause the nuclear translocation of signal transduction factors, and in most cases, the regulation of phosphorylation is critical for their nuclear import. Phosphorylation-dependent NLS masking and unmasking has also been reported for NF-AT4 by calcium signaling[28] and Smad2 by TGF-β stimulation,[29] respectively.

Another interesting example is the cytokine dependent nuclear localization of STAT1. In response to interferon, STAT1 is tyrosine phosphorylated, forms homodimer and then is translocated into the nucleus. Although a basic type NLS has not been found in the STAT1 primary amino acid sequence, the nuclear import of STAT1 is mediated by the conventional importin/Ran-dependent transport system.[30,31] Recently, based on a crystallographic study, it has been shown that a lysine and arginine rich element within the DNA binding domain of STAT1 is involved in the interferon-dependent nuclear import of STAT1.[32] This element becomes accessible through tyrosine phosphorylation and homodimerization on the surface of the three-dimensional structure of STAT1, and constitutes a novel type of NLS which is different from the conventional basic type NLSs contained in the primary amino acid sequence.

Although the basic type NLS is the only well defined general sequence of an NLS and the basic amino acid clusters in the deduced primary amino acid sequences of various proteins are easily deduced, basic amino acid clusters similar to functional NLS are also found in non-nuclear proteins.[33] In addition, even basic amino acid clusters in nuclear proteins are not able to promote protein nuclear import as the case may be. Therefore, a functional analysis is needed in order to determine whether these sequences function as an NLS in the context of full length proteins.

Non-Basic Type NLSs

Basic type NLSs are general sequences for protein nuclear import. However, an early work showed that only half of the known nuclear proteins have basic amino acid clusters in their primary sequences.[16] Recently, it has been shown that many proteins that are able to be translocated into the nucleus have no sequence similarities to the classical basic type NLSs. It has been documented that, in addition to basic type NLSs, several other sequence motifs are also able to mediate the nuclear import of proteins. In most cases, these sequences are longer than the basic type NLSs. A well-known signal of this type is the NLS of the heterogeneous nuclear ribonucleoprotein (hnRNP) A1. The sequence, referred to as M9, consists of 38 amino acids and shows no similarity to the basic type NLSs.[34,35] M9 mediates not only the nuclear import of hnRNP, but is also responsible for the re-export of hnRNP which shuttles between the cytoplasm and nucleus (see also below).[36] The M9 is recognized by transportin, a member of the importin β family proteins[37] for nuclear import. Several ribosomal proteins in mammals and yeast have 30 to 40 amino acid long nuclear localization signals. Another examples is the helix-loop-helix-leucine zipper domain of SREBP2.[38] These proteins are transported to the nucleus through the direct binding to importin β families.[39,40]

Strikingly, an NLS rich in negatively charged amino acids has been identified in human DNA Topoisomerase I. This enzyme has two functional NLSs, one is the basic type NLS and the other is an acidic type NLS.[41] The latter is rich in acidic amino acids, such as aspartic and glutamic acids and is sufficient to promote the nuclear localization of non-nuclear protein. Because the acidic type NLS is significantly different from classical NLSs and other types of NLSs, it is likely that the mechanism of protein nuclear translocation by this signal may be different from previously identified ones.

NESs Recognized by Importin β Related Proteins

The first example of NESs came from the analysis of retroviral mRNA exporter proteins called Rev[42] and Rex[43] and the inhibitor of cAMP-dependent protein kinase (PKI) protein.[44] These NESs consist of comparably short canonical amino acid sequences rich in hydrophobic amino acids, often leucine or isoleucine residues (Table 2), and therefore they are usually termed "leucine-rich NESs". It has been shown that the leucine-rich NESs are recognized by CRM1 (Xpo1p in yeast) in the presence of Ran-GTP in the nucleus.[45-47] Following the identification of the export receptor, successful and extensive usage of a specific inhibitor leptomycine B (LMB), which binds to a cystein residue in CRM1,[48-51] blocks the formation of trimeric complexes by Ran-GTP, leucine-rich NESs and CRM1, and therefore inhibits CRM1 dependent nuclear export,[45,51] has led to identification of similar leucine-rich NESs in a variety of proteins.[52-56] Although a sequence comparison of leucine-rich NESs derived from different pro-

Table 2. Leucine-rich NESs

HIV-1 Rev	L P P – L E R – L T L
HTLV-I Rex	L S A Q L Y S S L S L
PKI	L A L K L A G – L D I
Gle1p	L P - - L G K – L T L
MAPKK	L Q K K L E E – L E L
cyclin B1	L C Q A F S D V I L A
β-actin	L P H A I L R – L D L
PHAX	V A T E L C I – L C M
Consensus	**L** X_{2-3} (**F, I, L, V, M**) X_{2-3} (**L, I**) X (**L, I**)

Sequence alignment of leucine-rich NESs from HIV-1 Rev,[42] HTLV-I Rex,[43] inhibitor of cAMP-dependent protein kinase (PKI),[44] *S. cerevisiae* Gle1p,[56] MAP kinase kinase,[96] cyclin B1,[61,63,97] β-actin,[52] and the phosphorylated adaptor for RNA export (PHAX)[73] gives a consensus sequence (bottom).

teins gives the consensus sequence as depicted in Table 2, accumulating data indicate that leucine-rich NESs can also be divergent from the consensus,[9] as has been shown for basic-type NLSs.

The leucine-rich NES of cyclin B1 provides an example of an NES, the activity of which is regulated by a post-translational modification (for review see ref. 57). Cyclin B1 is a cell cycle regulator that is localized within the cytoplasm during S to G2 phase of cell-cycle. It then migrates into the nucleus at the onset of metaphase before the nuclear envelope begins to break down.[58] A deletion analysis revealed that the N-terminal 42-amino acid region contains a signal which is necessary for the cytoplasmic localization.[59] The export activity was then shown to be blocked by LMB and the signal sequence exhibits a similarity to leucine-rich NES, although it slightly differs from the consensus.[60-62] To achieve the cell-cycle dependent repression of NES, the leucine-rich NES of cyclin B1 undergoes specific phosphorylation at the onset of mitosis, which prevents its interaction with CRM1, blocking its export to the cytoplasm, and thus cyclin B1 accumulates in the nucleus in a cell-cycle dependent manner.[63]

Major spliceosomal U snRNAs including U1, U2, U4, and U5 are transcribed in the nucleus and acquire a monomethylated m^7G-cap structure. They are transported to the cytoplasm and form complexes with a series of proteins which are collectively termed Sm proteins. The m^7G-cap is then hypermethylated to form the mature 2,2,7-trimethylated m_3G-cap structure, and the mature U snRNPs are transported back into the nucleus.[64,65] The monomethylated cap structure of U snRNAs, which has been shown to be the major determinant for their nuclear export,[66,67] is directly recognized by the nuclear cap binding complex (CBC), a heterodimeric complex composed of CBP20 and CBP80,[68-70] while the m_3G-cap structure is recognized by snurportin 1 for subsequent nuclear import.[71] CRM1 has been shown to be the crucial factor in U snRNA biogenesis.[42,45,72] In the export leg, it was revealed that the CBC-U snRNA complex is recognized by CRM1. However, the interaction is not direct and, instead, a protein called PHAX (phosphorylated adaptor for RNA export) harboring leucine-rich NES acts as an adaptor protein that serves as a bridge between CBC and CRM1.[73] Interestingly it was shown that PHAX undergoes compartment specific phosphorylation; i.e., it is phosphorylated in the nucleus while rapidly dephosphorylated in the cytoplasm, and that the dephosphorylation triggers disassembly of the export complex even in the presence of the Ran mutant which is locked in GTP-bound form. The posttranslational modification dependent release mechanism seems to be integrated in the constitutive nuclear transport cycle. This clearly contrasts with the case of cyclin B1's or Pho4p's (see below) NESs whose interactions with exportins are regulated by a type of exogeneous stimuli. In the import leg, CRM1 acts as an exportin for snurportin 1.[72] Biochemical analysis revealed that CRM1 binds to snurportin 1 with an affinity several orders of magnitude higher than authentic leucine-rich NES and that snurpotin 1 binds to CRM1 not through a short contiguous peptide sequence but through larger domain, suggesting that the mode of interaction differs from that of the leucine-rich NES. It has been suggested that several NES-like sequences that show imperfect homology to the authentic leucine-rich NES interact synergistically with CRM1.[72]

Other NESs in this class can be found in the classical NLS receptor Importin-α and the eukaryotic translation initiation factor eIF-5A. These NESs are not only different in their primary sequences from leucine-rich NES but are also recognized by different exportins CAS (or Cse1p in yeast)[74] and exportin 4,[75] respectively. In contrast to the short canonical leucine-rich NESs, both these NESs are reported to be comparably large in their sizes. In the case of eIF-5A, even an entire molecule of eIF-5A and a unique hypsine [Nϵ-(4-amino-2-hydroxybutyl)-L-lysine] modification at lysine 50 are indispensable for its export as well as for its efficient interaction with exportin 4, indicating that exportin 4 may recognize not a short contiguous sequence but, rather, larger parts of the eIF-5A molecule as NES.[75]

Yeast proteins recognized by exportin Msn5p provide another example of an NES. The NES of Pho4p, a transcriptional activator involved in phosphate metabolism, does not show any sequence similarity to the leucine-rich type NES. The Pho4p protein is recognized by the exportin Msn5p and is exported from the nucleus only when specific serine residues are phosphorylated.[76] The authors have shown that the phosphorylation of, not only serine residues within the minimal NES, but also those at remote sites are required for regulated export.[76] Therefore, the two possibilities that phosphorylation induces conformational changes to make NES accessible or that Msn5p preferentially recognizes the phosphopeptide sequences as NES are both viable at this moment.

Sequences Acting As Both NES and NLS

A series of nuclear proteins that are specifically associated with pre-mRNAs are collectively termed hnRNPs (heterogeneous nuclear ribonucleoproteins).[77] Among these, the hnRNPs A1, A2, D, E, I, and K proteins have been shown to bind to pre-mRNAs in both the nucleus and the cytoplasm, and thus they shuttle continuously between the two compartments.[36,78] A mutational analysis has revealed that the hnRNP A1 protein harbors a modular domain organization and the activities of nuclear export and RNA binding can be assigned to different domains.[34,35,79] Thus the hnRNP A1 protein is not exported from the nucleus by simply being associated with mRNA molecules. The shuttling ability of hnRNP A1 is controlled by a 38-amino acid sequence (aa 265-303 of human hnRNP A1) termed M9 that is rich in glycines and aromatic residues and can confer nuclear export and import abilities to otherwise non-shuttling reporter proteins (Table 1).[34,36] These data indicate that the M9 sequence is recognized by shuttling intracellular receptor for nuclear export and import. Indeed, as discussed in the section on NLS, the nuclear import activity of the M9 sequence is mediated by transportin, an importin β type shuttling transporter.[37] In contrast, the identification of a specific export receptor for M9 is complicated due to the fact that the M9 sequence could not be functionally divided into an NES and an NLS, despite the extensive mutagenesis analysis that has been conducted.[36,80] It is also possible that transportin is implicated in both the nuclear import and export of hnRNP A1. However, this seems unlikely, since an in vitro study revealed that high concentrations of Ran-GTP, which mimics intranuclear conditions, is able to dissociate the transportin-M9 complex.[81] These observations suggest that another non-importin β type export receptor recognizes M9 sequence as NES only in the presence of additional help from other factors present in the nucleus. In this case, an intriguing candidate for such a cofactor is mRNA. Alternatively, posttranslational modifications that occur only in the nucleus may be required for the M9 sequence to be a functional export signal.

The KNS sequence is a 39-amino acid signal sequence that lies between amino acids 323 to 361 of the human hnRNP K protein.[82] Although the sequence is not similar to M9, KNS also confers both nuclear import and export abilities to a heterologous reporter protein as in the case of the M9 sequence. By deletion analysis, import and export activities of the KNS signal were shown to be partially separated. The entire 39-amino acids sequence is required for nuclear import, whereas the C-terminal 23-amino acid subregion is sufficient to activate the nuclear export of heterologous reporter protein. However, a specific export (and also import) receptor for KNS has not yet been identified. That the importin β, as well as transportin-mediated nuclear import, could be inhibited by an excess of KNS and that the nuclear import of KNS does not require exogeneously added soluble factors in an in vitro import assay system suggest that the KNS signal may have an intrinsic activity for interacting with nucleoporins to achieve pore translocation.[83] Further analysis will be required to determine if KNS is also able to inhibit nuclear export pathways and thus indicating a direct interaction with nucleoporins also being involved in nuclear export. It would also be interesting to identify which factor is the interaction counterpart in mediating KNS-dependent nuclear import and export.

Concluding Remarks

This review mainly focused on signal sequences recognized by importin β related proteins for nuclear import and export. However, the possibility also exists that a protein that does not contain its own functional NLS or NES could be imported into or exported from the nucleus by a "piggy-back" mechanism through interactions with other export substrates containing a bona fide NLS or NES, and indeed such examples have been reported.[53,54,73,84] Alternatively, nuclear transport receptors such as the importin β related proteins, NTF2/p10, a specific import receptor for Ran, or Mex67p/Tap, an essential mRNA exporter conserved from yeast to human, contain sequences required for NPC translocation.[5,85-93] These sequences unequivocally exhibit NPC binding ability and can support the translocation of exogeneous reporter proteins both back and forth through NPC without the aid of soluble transport factors. Although binding to nucleoporins has not been established, the ankyrin repeat of IκBα[94] and the un-identified domain of β catenin[95] also act as NLSs and mediate nuclear translocation in a Ran- and importin β type receptor-independent manner. Thus it is also possible to experimentally define such sequences as NLS or NES. However, because of space limitations, these sequences are not addressed in this Chapter. Readers interested in such examples are referred to recent excellent reviews and references therein.[7,83]

Since eukaryotic cells are subdivided into the nucleus and the cytoplasm, nuclear transport is indispensable for gene expression and the maintenance of cellular homeostasis. As depicted here, numerous signal sequences harbored within transported substrates themselves primarily determine the transport pathway and their final destination. In addition, there are a number of mechanisms that regulate the affinity or accessibility of the signal sequences by the transport receptors and thus control the timing of the initiation of a transport reaction. Cells have evolved these elaborate mechanisms to achieve "fine tuning" of the activities of NLS and NES, and this results in a major control mechanism of cellular processes, such as cell-cycle and cell differentiation.

References

1. Goldberg MW, Allen TD. Structural and functional organization of the nuclear envelope. Curr Opin Cell Biol 1995; 7(3):301-309.
2. Doye V, Hurt E. From nucleoporins to nuclear pore complexes. Curr Opin Cell Biol 1997; 9(3):401-411.
3. Görlich D, Vogel F, Mills AD et al. Distinct functions for the two importin subunits in nuclear protein import. Nature 1995; 377(6546):246-248.

4. Moroianu J, Hijikata M, Blobel G et al. Mammalian karyopherin alpha 1 beta and alpha 2 beta heterodimers: Alpha 1 or alpha 2 subunit binds nuclear localization signal and beta subunit interacts with peptide repeat-containing nucleoporins. Proc Natl Acad Sci USA 1995; 92(14):6532-6536.
5. Kose S, Imamoto N, Tachibana T et al. Ran-unassisted nuclear migration of a 97-kD component of nuclear pore- targeting complex. J Cell Biol 1997; 139(4):841-849.
6. Ribbeck K, Görlich D. Kinetic analysis of translocation through nuclear pore complexes. EMBO J 2001; 20(6):1320-1330.
7. Stewart M, Baker RP, Bayliss R et al. Molecular mechanism of translocation through nuclear pore complexes during nuclear protein import. FEBS Lett 2001; 498(2-3):145-149.
8. Görlich D, Kutay U. Transport between the cell nucleus and the cytoplasm. Annu Rev Cell Dev Biol 1999; 15:607-660.
9. Mattaj IW, Englmeier L. Nucleocytoplasmic transport: The soluble phase. Annu Rev Biochem 1998; 67:265-306.
10. Yoshida K, Blobel G. The karyopherin Kap142p/Msn5p mediates nuclear import and nuclear export of different cargo proteins. J Cell Biol 2001; 152(4):729-740.
11. Dingwall C, Sharnick SV, Laskey RA. A polypeptide domain that specifies migration of nucleoplasmin into the nucleus. Cell 1982; 30(2):449-458.
12. Kalderon D, Richardson WD, Markham AF et al. Sequence requirements for nuclear location of simian virus 40 large-T antigen. Nature 1984; 311(5981):33-38.
13. Goldfarb DS, Gariepy J, Schoolnik G et al. Synthetic peptides as nuclear localization signals. Nature 1986; 322(6080):641-644.
14. Yoneda Y, Arioka T, Imamoto-Sonobe N et al. Synthetic peptides containing a region of SV 40 large T-antigen involved in nuclear localization direct the transport of proteins into the nucleus. Exp Cell Res 1987; 170(2):439-452.
15. Dingwall C, Robbins J, Dilworth SM et al. The nucleoplasmin nuclear location sequence is larger and more complex than that of SV-40 large T antigen. J Cell Biol 1988; 107(3):841-849.
16. Dingwall C, Laskey RA. Nuclear targeting sequences—A consensus? Trends Biochem Sci 1991; 16(12):478-481.
17. Weis K, Mattaj IW, Lamond AI. Identification of hSRP1 alpha as a functional receptor for nuclear localization sequences. Science 1995; 268(5213):1049-1053.
18. Imamoto N, Shimamoto T, Takao T et al. In vivo evidence for involvement of a 58 kDa component of nuclear pore-targeting complex in nuclear protein import. EMBO J 1995; 14(15):3617-3626.
19. Adam SA, Gerace L. Cytosolic proteins that specifically bind nuclear location signals are receptors for nuclear import. Cell 1991; 66(5):837-847.
20. Conti E, Uy M, Leighton L et al. Crystallographic analysis of the recognition of a nuclear localization signal by the nuclear import factor karyopherin alpha. Cell 1998; 94(2):193-204.
21. Görlich D, Henklein P, Laskey RA et al. A 41 amino acid motif in importin-alpha confers binding to importin- beta and hence transit into the nucleus. EMBO J 1996; 15(8):1810-1817.
22. Cingolani G, Petosa C, Weis K et al. Structure of importin-beta bound to the IBB domain of importin-alpha. Nature 1999; 399(6733):221-229.
23. Palmeri D, Malim MH. Importin beta can mediate the nuclear import of an arginine-rich nuclear localization signal in the absence of importin alpha. Mol Cell Biol 1999; 19(2):1218-1225.
24. Truant R, Cullen BR. The arginine-rich domains present in human immunodeficiency virus type 1 Tat and Rev function as direct importin beta-dependent nuclear localization signals. Mol Cell Biol 1999; 19(2):1210-1217.
25. Henkel T, Zabel U, van Zee K et al. Intramolecular masking of the nuclear location signal and dimerization domain in the precursor for the p50 NF-kappa B subunit. Cell 1992; 68(6):1121-1133.
26. Nadler SG, Tritschler D, Haffar OK et al. Differential expression and sequence-specific interaction of karyopherin alpha with nuclear localization sequences. J Biol Chem 1997; 272(7):4310-4315.
27. Zabel U, Henkel T, Silva MS et al. Nuclear uptake control of NF-kappa B by MAD-3, an I kappa B protein present in the nucleus. EMBO J 1993; 12(1):201-211.
28. Zhu J, Shibasaki F, Price R et al. Intramolecular masking of nuclear import signal on NF-AT4 by casein kinase I and MEKK1. Cell 1998; 93(5):851-861.
29. Xu L, Chen YG, Massague J. The nuclear import function of Smad2 is masked by SARA and unmasked by TGFbeta-dependent phosphorylation. Nat Cell Biol 2000; 2(8):559-562.

30. Sekimoto T, Nakajima K, Tachibana T et al. Interferon-gamma-dependent nuclear import of Stat1 is mediated by the GTPase activity of Ran/TC4. J Biol Chem 1996; 271(49):31017-31020.
31. Sekimoto T, Imamoto N, Nakajima K et al. Extracellular signal-dependent nuclear import of Stat1 is mediated by nuclear pore-targeting complex formation with NPI-1, but not Rch1. EMBO J 1997; 16(23):7067-7077.
32. Melen K, Kinnunen L, Julkunen I. Arginine/lysine-rich structural element is involved in interferon-induced nuclear import of STATs. J Biol Chem 2001; 276(19):16447-16455.
33. Boulikas T. Putative nuclear localization signals (NLS) in protein transcription factors. J Cell Biochem 1994; 55(1):32-58.
34. Siomi H, Dreyfuss G. A nuclear localization domain in the hnRNP A1 protein. J Cell Biol 1995; 129(3):551-560.
35. Weighardt F, Biamonti G, Riva S. Nucleo-cytoplasmic distribution of human hnRNP proteins: a search for the targeting domains in hnRNP A1. J Cell Sci 1995; 108(Pt 2):545-555.
36. Michael WM, Choi M, Dreyfuss G. A nuclear export signal in hnRNP A1: a signal-mediated, temperature- dependent nuclear protein export pathway. Cell 1995; 83(3):415-422.
37. Pollard VW, Michael WM, Nakielny S et al. A novel receptor-mediated nuclear protein import pathway. Cell 1996; 86(6):985-994.
38. Nagoshi E, Imamoto N, Sato R et al. Nuclear import of sterol regulatory element-binding protein-2, a basic helix-loop-helix-leucine zipper (bHLH-Zip)-containing transcription factor, occurs through the direct interaction of importin beta with HLH-Zip. Mol Biol Cell 1999; 10(7):2221-2233.
39. Jakel S, Görlich D. Importin beta, transportin, RanBP5 and RanBP7 mediate nuclear import of ribosomal proteins in mammalian cells. EMBO J 1998; 17(15):4491-4502.
40. Schlenstedt G, Smirnova E, Deane R et al. Yrb4p, a yeast ran-GTP-binding protein involved in import of ribosomal protein L25 into the nucleus. EMBO J 1997; 16(20):6237-6249.
41. Mo YY, Wang C, Beck WT. A novel nuclear localization signal in human DNA topoisomerase I. J Biol Chem 2000; 275(52):41107-41113.
42. Fischer U, Huber J, Boelens WC et al. The HIV-1 Rev activation domain is a nuclear export signal that accesses an export pathway used by specific cellular RNAs. Cell 1995; 82(3):475-483.
43. Bogerd HP, Fridell RA, Benson RE et al. Protein sequence requirements for function of the human T-cell leukemia virus type 1 Rex nuclear export signal delineated by a novel in vivo randomization-selection assay. Mol Cell Biol 1996; 16(8):4207-4214.
44. Wen W, Meinkoth JL, Tsien RY et al. Identification of a signal for rapid export of proteins from the nucleus. Cell 1995; 82(3):463-473.
45. Fornerod M, Ohno M, Yoshida M et al. CRM1 is an export receptor for leucine-rich nuclear export signals. Cell 1997; 90(6):1051-1060.
46. Fukuda M, Asano S, Nakamura T et al. CRM1 is responsible for intracellular transport mediated by the nuclear export signal. Nature 1997; 390(6657):308-311.
47. Stade K, Ford CS, Guthrie C et al. Exportin 1 (Crm1p) is an essential nuclear export factor. Cell 1997; 90(6):1041-1050.
48. Kudo N, Matsumori N, Taoka H et al. Leptomycin B inactivates CRM1/exportin 1 by covalent modification at a cysteine residue in the central conserved region. Proc Natl Acad Sci USA 1999; 96(16):9112-9117.
49. Kudo N, Wolff B, Sekimoto T et al. Leptomycin B inhibition of signal-mediated nuclear export by direct binding to CRM1. Exp Cell Res 1998; 242(2):540-547.
50. Neville M, Rosbash M. The NES-Crm1p export pathway is not a major mRNA export route in Saccharomyces cerevisiae. EMBO J 1999; 18(13):3746-3756.
51. Wolff B, Sanglier JJ, Wang Y. Leptomycin B is an inhibitor of nuclear export: inhibition of nucleocytoplasmic translocation of the human immunodeficiency virus type 1 (HIV-1) Rev protein and Rev-dependent mRNA. Chem Biol 1997; 4(2):139-147.
52. Wada A, Fukuda M, Mishima M et al. Nuclear export of actin: a novel mechanism regulating the subcellular localization of a major cytoskeletal protein. EMBO J 1998; 17(6):1635-1641.
53. Adachi M, Fukuda M, Nishida E. Nuclear export of MAP kinase (ERK) involves a MAP kinase kinase (MEK)- dependent active transport mechanism. J Cell Biol 2000; 148(5):849-856.

54. Huang TT, Kudo N, Yoshida M et al. A nuclear export signal in the N-terminal regulatory domain of IkappaBalpha controls cytoplasmic localization of inactive NF- kappaB/IkappaBalpha complexes. Proc Natl Acad Sci USA 2000; 97(3):1014-1019.

55. Jans DA, Xiao CY, Lam MH. Nuclear targeting signal recognition: a key control point in nuclear transport? Bioessays 2000; 22(6):532-544.

56. Watkins JL, Murphy R, Emtage JL et al. The human homologue of Saccharomyces cerevisiae Gle1p is required for poly(A)+ RNA export. Proc Natl Acad Sci USA 1998; 95(12):6779-6784.

57. Hood JK, Silver PA. In or out? Regulating nuclear transport. Curr Opin Cell Biol 1999; 11(2):241-247.

58. Pines J, Hunter T. Human cyclins A and B1 are differentially located in the cell and undergo cell cycle-dependent nuclear transport. J Cell Biol 1991; 115(1):1-17.

59. Pines J, Hunter T. The differential localization of human cyclins A and B is due to a cytoplasmic retention signal in cyclin B. EMBO J 1994; 13(16):3772-3781.

60. Hagting A, Jackman M, Simpson K et al. Translocation of cyclin B1 to the nucleus at prophase requires a phosphorylation-dependent nuclear import signal. Curr Biol 1999; 9(13):680-689.

61. Toyoshima F, Moriguchi T, Wada A et al. Nuclear export of cyclin B1 and its possible role in the DNA damage-induced G2 checkpoint. EMBO J 1998; 17(10):2728-2735.

62. Yang J, Bardes ES, Moore JD et al. Control of cyclin B1 localization through regulated binding of the nuclear export factor CRM1. Genes Dev 1998; 12(14):2131-2143.

63. Yang J, Song H, Walsh S et al. Combinatorial control of cyclin B1 nuclear trafficking through phosphorylation at multiple sites. J Biol Chem 2001; 276(5):3604-3609.

64. Mattaj IW. Cap trimethylation of U snRNA is cytoplasmic and dependent on U snRNP protein binding. Cell 1986; 46(6):905-911.

65. Fischer U, Luhrmann R. An essential signaling role for the m3G cap in the transport of U1 snRNP to the nucleus. Science 1990; 249(4970):786-790.

66. Hamm J, Mattaj IW. Monomethylated cap structures facilitate RNA export from the nucleus. Cell 1990; 63(1):109-118.

67. Jarmolowski A, Boelens WC, Izaurralde E et al. Nuclear export of different classes of RNA is mediated by specific factors. J Cell Biol 1994; 124(5):627-635.

68. Izaurralde E, Stepinski J, Darzynkiewicz E et al. A cap binding protein that may mediate nuclear export of RNA polymerase II-transcribed RNAs. J Cell Biol 1992; 118(6):1287-1295.

69. Kataoka N, Ohno M, Moda I et al. Identification of the factors that interact with NCBP, an 80 kDa nuclear cap binding protein. Nucleic Acids Res 1995; 23(18):3638-3641.

70. Izaurralde E, Lewis J, Gamberi C et al. A cap-binding protein complex mediating U snRNA export. Nature 1995; 376(6542):709-712.

71. Huber J, Cronshagen U, Kadokura M et al. Snurportin1, an m3G-cap-specific nuclear import receptor with a novel domain structure. EMBO J 1998; 17(14):4114-4126.

72. Paraskeva E, Izaurralde E, Bischoff FR et al. CRM1-mediated recycling of snurportin 1 to the cytoplasm. J Cell Biol 1999; 145(2):255-264.

73. Ohno M, Segref A, Bachi A et al. PHAX, a mediator of U snRNA nuclear export whose activity is regulated by phosphorylation. Cell 2000; 101(2):187-198.

74. Kutay U, Bischoff FR, Kostka S et al. Export of importin alpha from the nucleus is mediated by a specific nuclear transport factor. Cell 1997; 90(6):1061-1071.

75. Lipowsky G, Bischoff FR, Schwarzmaier P et al. Exportin 4: a mediator of a novel nuclear export pathway in higher eukaryotes. EMBO J 2000; 19(16):4362-4371.

76. Kaffman A, Rank NM, O'Neill EM et al. The receptor Msn5 exports the phosphorylated transcription factor Pho4 out of the nucleus. Nature 1998; 396(6710):482-486.

77. Dreyfuss G, Matunis MJ, Pinol-Roma S et al. hnRNP proteins and the biogenesis of mRNA. Annu Rev Biochem 1993; 62:289-321.

78. Pinol-Roma S, Dreyfuss G. Shuttling of pre-mRNA binding proteins between nucleus and cytoplasm. Nature 1992; 355(6362):730-732.

79. Burd CG, Dreyfuss G. RNA binding specificity of hnRNP A1: significance of hnRNP A1 high-affinity binding sites in pre-mRNA splicing. EMBO J 1994; 13(5):1197-1204.

80. Bogerd HP, Benson RE, Truant R et al. Definition of a consensus transportin-specific nucleocyto-plasmic transport signal. J Biol Chem 1999; 274(14):9771-9777.
81. Izaurralde E, Kutay U, von Kobbe C et al. The asymmetric distribution of the constituents of the Ran system is essential for transport into and out of the nucleus. EMBO J 1997; 16(21):6535-6547.
82. Michael WM, Eder PS, Dreyfuss G. The K nuclear shuttling domain: a novel signal for nuclear import and nuclear export in the hnRNP K protein. EMBO J 1997; 16(12):3587-3598.
83. Michael WM. Nucleocytoplasmic shuttling signals: two for the price of one. Trends Cell Biol 2000; 10(2):46-50.
84. Caceres JF, Screaton GR, Krainer AR. A specific subset of SR proteins shuttles continuously between the nucleus and the cytoplasm. Genes Dev 1998; 12(1):55-66.
85. Kose S, Imamoto N, Tachibana T et al. beta-subunit of nuclear pore-targeting complex (importin-beta) can be exported from the nucleus in a Ran-independent manner. J Biol Chem 1999; 274(7):3946-3952.
86. Kutay U, Izaurralde E, Bischoff FR et al. Dominant-negative mutants of importin-beta block multiple pathways of import and export through the nuclear pore complex. EMBO J 1997; 16(6):1153-1163.
87. Katahira J, Strasser K, Podtelejnikov A et al. The Mex67p-mediated nuclear mRNA export pathway is conserved from yeast to human. EMBO J 1999; 18(9):2593-2609.
88. Strasser K, Basler J, Hurt E. Binding of the Mex67p/Mtr2p Heterodimer to FXFG, GLFG, and FG Repeat Nucleoporins Is Essential for Nuclear mRNA Export. J Cell Biol 2000; 150(4):695-706.
89. Schmitt I, Gerace L. In vitro analysis of nuclear transport mediated by the C-terminal shuttle domain of tap. J Biol Chem 2001; 276:42355-42363.
90. Bear J, Tan W, Zolotukhin AS et al. Identification of novel import and export signals of human TAP, the protein that binds to the constitutive transport element of the type D retrovirus mRNAs. Mol Cell Biol 1999; 19(9):6306-6317.
91. Guzik BW, Levesque L, Prasad S et al. NXT1 (p15) is a crucial cellular cofactor in TAP-dependent export of intron-containing RNA in mammalian cells. Mol Cell Biol 2001; 21(7):2545-2554.
92. Ribbeck K, Lipowsky G, Kent HM et al. NTF2 mediates nuclear import of Ran. EMBO J 1998; 17(22):6587-6598.
93. Bayliss R, Ribbeck K, Akin D et al. Interaction between NTF2 and xFxFG-containing nucleoporins is required to mediate nuclear import of RanGDP. J Mol Biol 1999; 293(3):579-593.
94. Sachdev S, Bagchi S, Zhang DD et al. Nuclear import of IkappaBalpha is accomplished by a ran-independent transport pathway. Mol Cell Biol 2000; 20(5):1571-1582.
95. Yokoya F, Imamoto N, Tachibana T et al. beta-catenin can be transported into the nucleus in a Ran-unassisted manner. Mol Biol Cell 1999; 10(4):1119-1131.
96. Fukuda M, Gotoh I, Gotoh Y et al. Cytoplasmic localization of mitogen-activated protein kinase kinase directed by its NH2-terminal, leucine-rich short amino acid sequence, which acts as a nuclear export signal. J Biol Chem 1996; 271(33):20024-20028.
97. Hagting A, Karlsson C, Clute P et al. MPF localization is controlled by nuclear export. EMBO J 1998; 17(14):4127-4138.

CHAPTER 5

Nuclear Import of Plant Proteins

Glenn R. Hicks

In this Chapter, we will focus primarily on protein import into the nucleus of plants. As in other eukaryotes the partitioning of genetic information into the nucleus necessitates the import and export of macromolecules such as proteins, nucleic acids and protein/nucleic acid complexes across the nuclear envelope. These transport processes are essential and are subject to stringent regulatory controls. In plants, it is clear that in addition to the maintenance of basic cellular processes, the regulated import of proteins plays a vital role in development. As we will discuss, the nuclear import of proteins in a selective manner is essential in responses of plants to light and results in the dramatic morphological changes that occur as plants switch from growth in the dark to growth in the light.[1] Protein translocation across the nuclear envelope is also a process that is utilized by pathogenic viruses[2] and tumor-inducing bacteria in the genus *Agrobacterium*[3] to transport protein/nucleic acid complexes into the nucleus for replication and even incorporation of pathogen DNA into the host genome.

The import pathway for proteins, themselves key factors regulating nuclear transport processes, is the best understood of the nuclear transport processes in plants. Numerous import signals have been characterized, and we are beginning to identify and understand the major components of the import machinery. As expected many of these factors are conserved between plants, animals and fungi, but there are surprising results indicating subtle differences in preferences for nuclear localization signals (NLSs),[3,4] the mechanics of import receptor function[5], and potential plant-specific import factors.[6] As we move forward plants are contributing new knowledge such as potential mechanisms for the targeting of proteins to the nuclear envelope and nuclear pore complex (NPC).[7] Recent efforts have begun to focus on other nuclear transport processes in plants such as the export of proteins and nucleic acids.[8,9]

Protein Import in Animals and Yeast

Our knowledge of nuclear transport in vertebrates and yeast is more advanced than our understanding in plants. Among the reasons for this historically are: (1) the ease of recovering nuclei from *Xenopus* oocytes that are intact for morphological studies, (2) the development of in vitro systems in *Xenopus* for reconstituting the assembly of nuclei and NPCs, (3) the reconstitution of cytosol-dependent nuclear import in permeabilized mammalian cells, (4) the rapid genetics possible in yeast, (5) the large biomedical research community and potential importance of nuclear processes for medicine. However as we are discovering nuclear translocation is a critical aspect of plant growth and development and thus has broad implications for agriculture in terms of disease and stress resistance and other crop improvements, which are the keys to feeding an expanding worldwide population. To place our knowledge of plant nuclear import in perspective, we will overview processes in animals and yeast throughout the

Nuclear Import and Export in Plants and Animals, edited by Tzvi Tzfira and Vitaly Citovsky.
©2005 Eurekah.com and Kluwer Academic / Plenum Publishers.

Chapter highlighting important advances and unique aspects in plants. Import from the perspective of non-plant systems is covered in greater detail in other Chapters. There are also excellent recent reviews that are focused on the nuclear/cytoplasmic transport of proteins, nucleic acids and their complexes[4,10-22] as well as the structure and function of the NPCs in vertebrates and yeast.[23-30]

Nuclear Translocation in Plants

Nuclear Pore Complex

The channels through which all transport substrates must pass are the NPCs which are macromolecular complexes embedded in the double-membrane nuclear envelope. The NPCs are estimated to have a mass of 125 MDa in higher eukaryotes and to be composed of 50 to 100 different proteins collectively known as nucleoporins. Morphologically, an NPC is composed of a nucleoplasmic and a cytoplasmic ring. Eight spokes are found within the rings that extend toward a central channel resulting in eight 9 nm channels thought to function in the diffusion of small molecules across the nuclear envelope. A basket-like structure extends from the nucleoplasmic face and fibrils have been observed extending from the cytoplasmic face. Thus far, fewer than 20 nucleoporins have been purified from vertebrates. The NPCs in yeast are less complex (mass about 66 MDa) and are not dissociated and reassembled during mitosis as in higher eukaryotes; nevertheless the development of methods to purify intact complexes[31] has permitted the identification and sequencing of all 30 of its nucleoporins. Even with such information available understanding the assembly and function of the NPC will be a daunting task.[32]

A number of vertebrate NPC proteins have been implicated in nuclear import.[30] They include Nup358 which is located on the cytoplasmic filaments. Nup 358 has multiple Ran binding sites and binds to the protein transporter importin β (See *Components and Mechanisms of Protein Import*) via FXFG (single amino acid code where X represents an amino acid with a small or polar side chain) repeats[33] which are characteristic of many nucleoporins. Other nucleoporins reported to bind to importin β include Nup153,[34] Nup214,[35] Nup116p, Nup100p,[36] and p62.[37] The p62 protein was one of the first nucleoporins to be purified, and like many nucleoporins from vertebrates it is modified by single *O*-linked *N*-acetylglucosamine (O-GlcNAc) residues. While the O-GlcNAc is probably not essential for import, the binding of the lectin wheat germ agglutinin (WGA) inhibits nuclear import in vertebrates and has been used for the identification and purification of nucleoporins from higher eukaryotes (for review see ref. 38).

From electron microscopy studies beginning in the 1970s we know that nuclear pores in plants are morphologically similar to those of other organisms.[39] However there have been few reports in which plant nucleoporins have been purified (for review see ref. 40). Scofield et al[41] localized a protein at the NPC and identified a 100 kDa polypeptide in a nuclear matrix fraction from carrots using an antibody to the yeast nucleoporin NSP100; however successful purification was not reported. Other studies indicate that nuclear envelope fractions from maize and tobacco nuclei contain a subset of proteins in the NPC fraction that can bind NLSs specifically.[42-44] The binding site is at the NPC indicating a role for plant nucleoporins in protein import.[42] This finding is consistent with the unusually tight association of at least one plant importin α NLS receptor with the nuclear envelope in purified nuclei and intact cells.[45,46]

Using WGA as a probe it is clear that GlcNAc-modifications are present at the periphery of the nucleus,[47-49] and electron microscopy has shown that some of these modifications are present at the NPC.[47] In fact biochemical characterization of tobacco nuclear fractions has shown that the glycans are attached to proteins via an *O*-linkage and the moities are longer than five sugar residues in length ending with a terminal GlcNAc residue. This is a novel modification not found in vertebrates, which contain only a single O-GlcNAc residue.[47] Although the functional significance of O-GlcNAc modifications is not clear the modification can be used to advantage for purifying plant NPC proteins. Using lectin affinity chromatography four O-GlcNAc proteins were purified from nuclear envelope fractions of cultured tobacco cells. Peptide sequence was obtained from a protein of 40 kDa (gp40) that shares about 30% identity with aldose-1-epimerases (also known as mutarotases) which are involved in aldose sugar metabolism in bacteria.[38] Their role in higher eukaryotes is not well defined but they do share structural similarities to the glucose carrier from erythrocytes. Interestingly, it has been reported that glycosylated proteins can be imported in a sugar-specific and NLS-independent manner in mammalian cells in vitro.[50] Thus one speculation is that gp40 may be involved in such an import system in plants by binding to glycosylated proteins destined for import.[40] Regardless of the role eventually assigned to gp40 a procedure to purify NPC proteins from plants should permit additional proteins to be characterized. The availability of *Arabidopsis* genome sequence should also contribute to the identification of plant nucleoporins, although it should be noted that beyond several very short repeat motifs such as FXFG there are few similarities even between vertebrate and yeast nucleoporins. While interesting in itself, it suggests that the identification of plant nucleoporins based on protein identity across kingdoms will be only partially successful.

Import Signals in Plants

Several types of transport can occur through NPCs. Ions and small proteins that are typically less than 20 to 30 kDa can pass by simple diffusion through the 9 nm channel. However even small macromolecules such as histones (14 kDa)[51] or tRNAs[52] cross the NPC via active processes permitting their translocation to be under cellular control. There is evidence that even calcium ions may be subject to selective concentration within the nucleus;[53] this may relate to the suggestion that calcium plays a role in the regulation of active import.[54] Nuclear transport processes are mediated by specific import receptors that recognize signals located within their respective substrates. In the case of protein import, NLSs interact with the import machinery that facilitates translocation through the NCP. Unlike most signals for organelles including the chloroplasts, mitochondria, vacuoles, peroxisomes and endoplasmic reticulum, NLSs are not proteolytically removed following import. This permits NLS-containing proteins to shuttle in and out of the nucleus and to be re-imported following post-mitotic nuclear assembly. Many transcription factors and cell cycle-regulatory proteins are able to exert their activities upon the cell based on their relative abundances in the cytosol compared to the nuceloplasm.[10,54] Most NLSs are classified as either monopartite or bipartite. The classic monopartite NLS is from the SV40 large T-antigen and is composed of a single short region enriched in the basic residues arginine and lysine, whereas the nucleoplasmin NLS, which first defined the bipartite class, is composed of two basic domains separated by a spacer of variable length and composition. There are also NLSs that are less typical. One unusual NLS class is typified by the signal within the Matα2 protein which requires both basic and hydrophobic residues (for review see ref. 44).

As in other organisms nuclear localization signals in plants cannot be defined by a strict consensus sequence and regions of basic amino acids are common within proteins, particularly regions involved in DNA binding. Thus, putative NLSs must be examined for activity in vivo to confirm their function. A number of NLSs have been carefully defined in such a manner in plants. These signals include at least one member for each of the three NLS classes. As examples, the transcriptional activitor R from maize possesses three functional NLSs, two SV40-like (one of them, NLS M, is defined as MSERKRREKL) and one Matα2-like signal (NLS C is defined as MISEALRKAIGKR), each of which are sufficient to target a reporter protein to the nucleus in vivo. Another transcription factor from maize, Opaque-2 possesses two signals, one SV40-like and one bipartite signal (NLS B is defined as RKRKESNRESARRSRYRK), that are sufficient to target a reporter protein to the nucleus. For R and Opaque2, mutations within the intact proteins indicate that multiple NLSs are necessary for efficient targeting in vivo suggesting cooperativity among NLSs for import which is also true in vertebrates.[4]

It is generally accepted that most NLSs can function across kingdoms pointing to a high degree of functional conservation. For example the SV40 large T-antigen NLS functions in plants (see for example refs. 55, 56) and the single bipartite NLS from the VirD2 protein of the plant pathogen *Agrobacterium* functions in plant, *Xenopus, Drosophila,* mammalian, and yeast cells.[57-59] *Agrobacterium* is an opportunistic pathogen that infects a wide variety of plant species.[3] In the coarse of pathogenesis *Agrobacterium* interacts with the host cell and transfers pathogen DNA (T-DNA) into the host cell nucleus through the NPC via the cooperative action of the VirD2 and VirE2 proteins. The T-DNA integrates into the plant genome and utilizes host factors to transcribe pathogen sequences. *Agrobacterium* has been a valuable system for studying import and will be discussed in later sections.

Plant import is not strictly conserved when compared to import in other kingdoms however. The yeast Matα2 NLS targets a β-glucuronidase (GUS) reporter protein to the nucleus in onion epidermal cells which is consistent with the specific association of this class of NLSs with an import receptor from *Arabidopsis.*[44-46] Interestingly, the Matα2 signal does not function in mammals[60,61] although other yeast NLSs are known to function in vertebrates.[62] Another exception is from *Agrobacterium.* As mentioned the NLS from VirD2 is broadly functional across kingdoms. Fascinatingly, VirE2 which contains two functional bipartite signals does not function in any of the non-plant organisms described for VirD2.[57,3,59] However a single amino acid change which alters one NLS to conform to the animal bipartite consensus permits the NLS to function in *Xenopus* and *Drosophila.*[57] Overall these results indicate that there may be subsets of import receptors or other components in plants that are not present in animals and fungi.

Components and Mechanisms of Protein Import

The key components of the classical NLS protein import pathway have been identified within the past decade and are the NLS-receptor importin α, the broad specificity transporter importin β, the GTPase Ran, and the Ran-interacting factor NTF2.[21,30,63] In the first event of the protein import pathway, NLSs within nuclear proteins are recognized and bound by the NLS receptor importin α in the cytoplasm (Fig. 1). Another factor, importin β interacts with importin α (via an importin β binding domain within importin α) completing the trimeric import complex. It is importin β that then interacts with specific proteins of the NPC facilitating translocation through the NPC. Thus for protein import importin α functions as an adapter that recognizes the NLS and associates with importin β, whereas importin β functions as the actual transporter. The directionality of import is determined by the binding of the small *ras*-related GTPase Ran to importin β. Following import of the trimeric complex, the GTP-bound form of Ran (Ran-GTP) binds to importin β in the nucleoplasm resulting in the release of importin α and the NLS-containing cargo from the complex. The importin β/Ran-GTP heterodimer is then exported to the cytoplasm where Ran-GTP is hydrolyzed to Ran-GDP leading to the release of importin β for subsequent rounds of import via reassociation with importin α and NLS-containing cargo. Since monomeric importin α does not typically interact directly with the NPC it is exported back to the cytoplasm via its own export receptor, CAS,[64] for subsequent rounds of import.

The energy and directionality of import are hypothesized to depend on the enrichment of Ran-GTP in the nucleoplasm compared to the cytoplasm which contains mostly Ran-GDP. This is accomplished through the action of the nucleotide exchange factor RCC1 in the nucleus

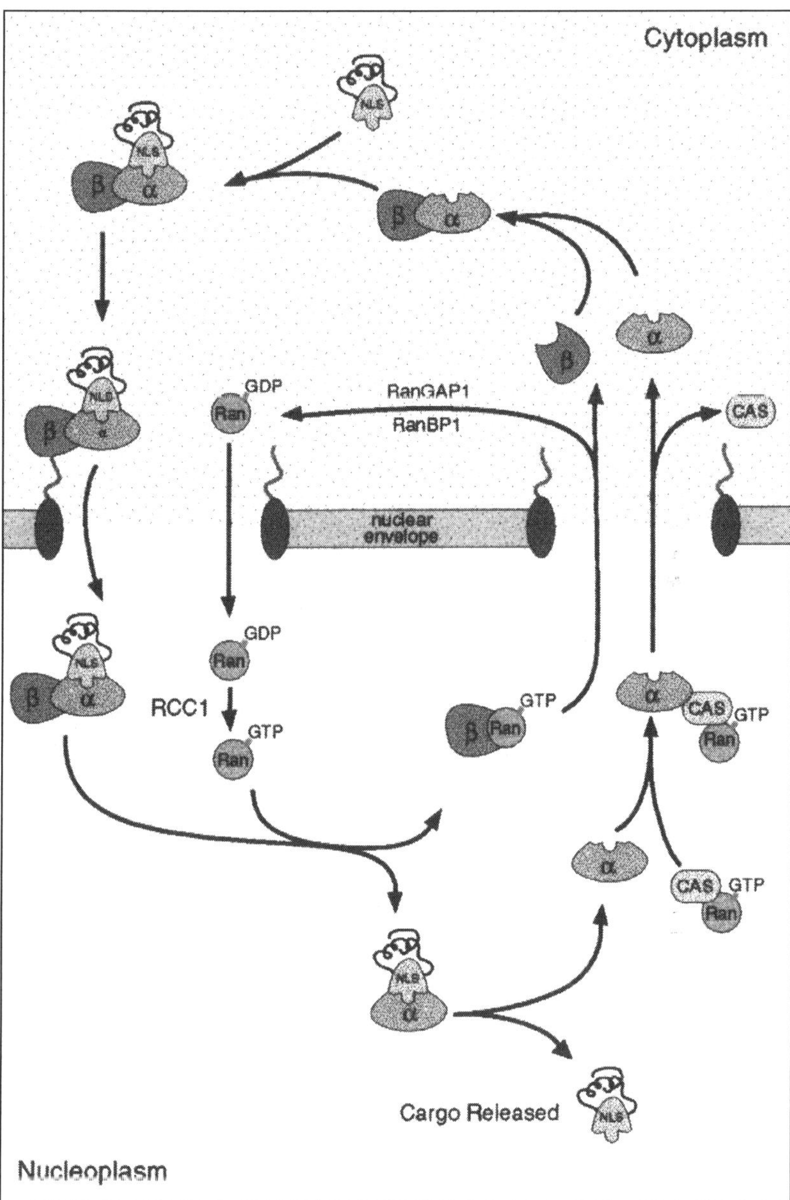

Figure 1. Schematic overview of nuclear protein import. Import of NLS-containing proteins is dependent upon the formation of a trimeric import complex in the cytoplasm. Following translocation, binding of Ran-GTP to importin β releases importin α and the NLS cargo. Importin α and importin β are exported for subsequent rounds of import. Ran is reimported preventing its depletion in the cytoplasm.

that enhances nucleotide exchange favoring Ran-GTP and the GTPase-activating protein RanGAP1 that favors conversion to Ran-GDP in the cytoplasm. The GTPase activity is further stimulated by Ran-BP1. Import is effectively a process of facilitated diffusion utilizing a gradient of Ran-GTP; a loose analogy would be to envision the import apparatus as an antiporter that causes accumulation of protein against a gradient of Ran-GTP. In the cytoplasm, the factor NTF2 interacts with Ran-GDP and NPC proteins and functions as a receptor/transporter for the re-import of Ran preventing its depletion from the nucleoplasm.[65,66] It should be noted that Ran plays essential roles beyond nuclear trafficking such as regulation of cell cycle progression.[63]

The selectivity and range of cargoes shuttled across the NPC is determined by different isoforms of importin α and importin β within the cell. The importin α receptor for NLS-mediated protein import has been found either as a single gene (SRP1) in yeast or as a small gene family in other organisms. In vertebrates at least six genes encoding importin α isoforms has been reported.[21] There is also evidence of distinct but overlapping preferences for different NLSs and possible cell-specific roles for the different importin αs.[57,67-71] Conversely, the range of importin β-like proteins, many of which are only distantly related, is far more diverse. This is due to the fact that members of the importin β family participate not only in the import of NLS-containing proteins but also function directly as import (importins) and export (exportins) receptor/transporters for other essential cargoes via their interaction with the NPC. Some examples include transportin 1 (import of hnRNP proteins), CRM1 (exportin 1; export of proteins containing a nuclear export signal), exportin-t (export of tRNAs), importin 7 (import of ribosomal proteins), and snurportin 1 (import of U snRNPs) (for review see ref. 21). The recent understanding of this receptor/transporter family has provided mechanistic details about the transport of diverse cargoes and highlighted potential ways in which the translocation of essential macromolecules is coordinated during the cell cycle.

In higher plants we have only recently begun to identify components of the import apparatus. Hicks et al[48] identified a homologue of the importin α receptor, At-IMP α from *Arabidopsis*. Immunologically-related proteins are found in all organs examined including roots, stems, leaves, and flowers as expected for an essential factor. There is also evidence that At-IMP α is phosphylated in vitro in the presence of cytosolic extracts suggesting a potential mechanism for controlling NLS binding or interaction with other proteins (Hicks and Raikhel unpublished). At the cellular level, At-IMP α is localized in the cytoplasm and nucleoplasm and at the nuclear envelope as expected for a receptor that shuttles between compartments. One unusual aspect of At-IMP α is its tight association with the nuclear envelope even in plant cells that have been treated to permeabilize the plasma membrane and deplete cytosolic contents;[48] in animal cells endogenous importin α is mostly cytosolic and easily depleted from permeabilized cells. The plant receptor binds specifically in vitro to monopartite, bipartite and Matα2-like NLSs indicating broad NLS selectivity.[5,45] Unlike yeast and mammalian orthologs that require importin β for high-affinity binding to NLSs, At-IMP α binds with high affinity (K_d of 5 to10 nM) in the absence of an importin β subunit, although At-IMP α is capable of binding to mouse importin β. This is consistent with the finding that At-IMP α can mediate association of import substrate at the nuclear envelope in permeabilized animal cells, whereas mouse importin α absolutely requires importin β.[5] In fact, At-IMP α can mediate nuclear protein import in permeabilized animal cells in the absence of exogenous importin β indicating that At-IMP α shares some properties with importin β.[5] This is a surprising result given that At-IMPα shares significant homology with other importin αs which require importin β and indicates the possibility of an importin β-independent pathway in plants.

Although At-IMP α has some unusual properties, other plant importin α homologues are more typical of their animal and yeast counterparts. Using the *Agrobacterium* protein VirD2 as bait in a yeast two-hybrid screen of an *Arabidopsis* library, Ballas and Citovsky[72] identified an importin α homologue (AtKAP α) that has high identity with At-IMP α except for a 64 amino acid extension at the amino-terminus. AtKAP α interacts specifically with the carboxy-terminal NLS of VirD2 both in vitro and in vivo in the two-hybrid assay, and the AtKAP α gene was found to complement a temperature-sensitive yeast mutant *srp1-31* (yeast importin α). Cytosolic extract containing AtKAP α protein was able to restore import to cells of *srp1-31* in an in vitro import system using permeabilized yeast cells in which import is dependent upon exogenous cytosol. Interestingly, VirE2 does not interact with AtKAP α suggesting an alternative pathway for its import (for discussion see ref. 72). A two-hybrid screen has identified a candidate for a VirE2 import protein (VIP1) that is related to basic leucine zipper proteins rather than to importin α.[3] One possibility is that VirE2 is imported via a "piggy-back" mechanism through association with VIP1 which is a nuclear protein likely having functions other than in import. It is likely that importin α in *Arabidopsis* is encoded by a small gene family as at least 4 members have been reported.[73] Another member of this family has been found in a two-hybrid screen using a WD40 type regulatory protein as a bait.[74] PRL1 when disrupted by the insertion of a T-DNA tag results in a pleiotropic phenotype conferring hypersensitivity to glucose, sucrose and hormones. PRL1 was found to interact in vivo and in vitro with ATHKAP2 which has high identity with At-IMP α, but ATHKAP2 has a truncated carboxy terminal end being about 60 amino acids shorter than AtIMP α. As with the other characterized importin αs it possesses motifs required for interaction with importin β as well as the characteristic eight conserved armadillo repeats presumably for protein-protein interactions.[45]

An importin α homologue (importin α1) from rice has been characterized that is 76% identical to At-IMP α. Expression of the gene in rice is suppressed by light in both etiolated seedlings and in leaves but not in roots or calli which display constitutive expression.[75] Binding to NLSs was examined in vitro, and importin α1 was found to bind to the SV40 large T antigen NLS and the bipartite signal from Opaque 2. However no association was observed between importin α1 and either the yeast Matα2 signal or the Matα2-like signal from the R protein.[76] This again points to some selectivity in NLS recognition in plants. Rice importin α binds to importin β from mouse, although with an affinity much lower than that of mouse importin α. Nevertheless in vitro import in permeabilized HeLa cells could be made dependent upon the presence exogenous rice importin α1 and mouse importin β indicating that rice importin α1 can function in a manner similar to that of the animal and yeast homologues.[76]

Importin β in plants has been less studied than importin α, although there should be a significant number of genes for importin βs by analogy with vertebrates. There are a few plant importin βs that are characterized, and the results indicate a high degree of functional conservation between kingdoms. The first importin β homologues to be reported are from rice,[77] and these are designated rice importins β1 and β2. Recombinant importin β1 can interact in vitro with rice importin α1 and a second rice importin α homologue (rice importin α2). This was studied by examining mobility shifts of protein complexes on native polyacrylamide gels. Using this approach it is argued that importin β1 specifically interacts with Ran-GTP and not Ran-GDP which is consistent with the ability of Ran-GTP to dissociate the importin α/importin β import complex in animals and yeast. In permeabilized HeLa cells, exogenous rice importin α1 and importin β1 can support import in the presence of mouse Ran.[78] This result plus the finding that exogenous importin β1 can bind directly to the nuclear envelope in permeabilized tobacco BY2 cells argues that importin α and β functions are conserved. It will be interesting to examine the importin βs and other components for tissue specific or light regulation as has been observed for rice importin α1 because regulation of development by light is an essential and unique feature of plants.

The Ran GTPase has been characterized in plants and homologues have been reported in *Arabidopsis*,[79] tomato[80] and tobacco[81] among other species.[82,83] While a direct role for Ran in nuclear import in plants has not been established, the protein is localized to the nucleus and appears to be encoded by a small gene family of at least 3 members[79] that are expressed throughout the plant including tissues that are not actively growing. Furthermore, it has been demonstrated that homologues from tomato or tobacco when expressed in *Schizosaccharomyces pombe* can suppress the phenotype of the *pim46-1* mutant which is defective in cell cycle progression.[80,81] Ran is known to be involved in cell cycle control suggesting that the tomato and tobacco Ran homologues have functions analogous to those in other organisms.

Both RanGAP1 and RanBP1 can stimulate Ran GTPase activity in the cytoplasm favoring Ran-GDP in that compartment. RanBP1 has been identified functionally from *Arabidopsis* by using the Ran homologue AtRan1 as bait in a two-hybrid screen in yeast. Haizel et al[79] found interaction with two RanBP1-like proteins (At-RanBP1a and At-RanBP1b) having 60% identity with vertebrate Ran BP1 and possessing conserved domains for Ran binding. Further study indicates that both At-RanBP1s are capable of associating in vivo in yeast with the GTP-bound form of each of the three Ran homologues from *Arabidopsis* (AtRan1, AtRan2, AtRan3). Neither the intracellular location nor the ability of At-RanBP1a or At-RanBP1b to stimulate GTPase activity has been reported. Likewise, neither RanGAP1 nor the nuclear exchange factor RCC1 has been characterized functionally in plants. Two putative RanGAPs have been identified from *Arabidopsis*, and they appear to have a unique domain not present in Ran GAPs from animals and yeast.[84] The motif which has been named a WPP domain is shared with MAF1.[85] MAF1 is a recently discovered protein that is composed mostly of WPP repeats and appears to be localized to the nuclear envelope via interaction with another envelope protein, MFP1.[86] The WPP domain appears to be specific to plants and is hypothesized to be involved in protein-protein interaction.[84] If true, RanGAP may associate with the nuclear envelope and NPCs in plants through this interaction. Table 1 describes components of the nuclear import pathway in plants that have been characterized to date.

It is increasingly apparent that there is functional conservation of the basic nuclear import pathway in plants, vertebrates and yeast. However many of the interesting biological questions will no doubt reside in the ways that plants have adapted the basic nuclear import pathways to fit their sessile photosynthetic life style. Exceptions have been noted already (importin β independence of At-IMP α, a potential alternative pathway for VirE2 import, differences in NLS selectivity) and others will be discussed below.

Systems to Study Import in Plants

The essential breakthrough that permitted the biochemical identification and purification of the factors now known to be involved in nuclear translocation of proteins and other substrates was the development of an in vitro import system utilizing permeabilized animal cells.[87,88] The method relies on the fact that plasma membranes from animals can be selectively permeabilized with digitonin, a reagent that aggregates to form pores upon binding to cholesterol. This sterol is abundant in the plasma membranes of animal cells but not in other membranes such as the nuclear envelope. The effect of the reagent is to permit the selective depletion of soluble factors from cultured cells while leaving the integrity of the nuclear envelope intact. Nuclear import thus occurs only by authentic facilitated translocation rather than by simple diffusion into damaged nuclei through tears in the nuclear envelope. Import can be directly visualized by microscopy following the addition of fluorescently labeled NLS-containing proteins. Another accepted method to examine import is microinjection into *Xenopus* oocytes or mammalian cells which is analytical and not amenable to fractionation of components. Other approaches that have been reported include the use of purified nuclei or nuclei mixed with *Xenopus* cytosol extracts (for review see ref. 4); however the ease of the permeabilization assay resulted in broad acceptance.

Although digitonin is not the reagent of choice in yeast and plants due to differences in membrane composition, the principle of selective permeabilization has been utilized successfully. A method in yeast was developed in which the plasma membrane of cell wall-less

Table 1. Components of the nuclear protein import pathway in plants

Component	Species	Characterization	Reference
AtIMPa1 (Y14615)	*Arabidopsis*	no functional characterization	73
AtIMPa2 (Y14616)	*Arabidopsis*	as above	73
AtIMPa3 (Y15224)	*Arabidopsis*	as above	73
AtIMPa4 (Y15225)	*Arabidopsis*	as above	73
AtIMPα	*Arabidopsis*	supports importin β-dependent import in HeLa cells	48
AtKAPα (U695333)	*Arabidopsis*	interacts with VirD2 in vivo	72
AtKAP2 (Y09511)	*Arabidopsis*	interacts with PRL1 in vivo	74
Importin α1	rice	interacts with rice importin β1 and supports import in HeLa cells	75
Importin α2	rice	no functional characterization	75
Import β1	rice	interacts with rice importin α1 and supports importin in HeLa cells	77
Importin β2	rice	no functional characterization	77
AtRan1	*Arabidopsis*	interacts with AtRanBP1b in vivo	79
AtRan2	*Arabidopsis*	as above	79
AtRan3	*Arabidopsis*	as above	79
Ran	tomato	suppresses yeast Ran mutant	80
Ran	tobacco	suppresses yeast Ran mutant	81
AtRanBP1a	*Arabidopsis*	interacts with AtRan1	79
AtRanBP1b	*Arabidopsis*	as above	79
AtRanGAP1 (AF214559)	*Arabidopsis*	unique WPP interaction motif	84
AtRanGAP2 (AF214560)	*Arabidopsis*	as above	84
Ran GAP (AF215731)	*Medicago*	as above	84
Ran GAP (AAD27557)	rice	as above	84
AtXPO1 (exporter)	Arabidopsis	interacts in vivo with AtRan1 and nuclear export signals	140

spheroplasts was selectively permeabilized via a simple freeze-thaw technique.[89] As in animal cells, import is dependent upon ATP (which can be converted to GTP for import), temperature, and the presence of exogenous cytosol (from yeast or mammalian cells). Again, alternative methods have been reported.[90]

In plants the development of in vitro import systems was difficult technically due to the fact that plant cells possess a thick cell wall and are highly vacuolated. The first report of in vitro import was an antibody cotranslocation assay in which antibodies to G-box binding factors (GBFs) are translocated into the nucleus (presumably by association with the GBFs) of Triton X-100 permeabilized parsley cells.[91] The results indicate that GBFs involved in light-regulated gene expression are imported in response to light and that import is ATP and temperature dependent. The assay is indirect however relying of protease protection of antibody associated with the nucleus. Two groups reported the development of direct import assays using evacuolated protoplasts from tobacco.[92] The approaches were similar with Merkle et al[49] using Triton X-100 to permeabilize the plasma membrane, whereas Hicks et al[48] used an osmotic shift to achieve permeabilization without detergents. In both cases, direct visualization of fluorescent import substrates indicates specific import that is dependent upon GTP hydrolysis and is specific for proteins containing functional NLSs. Some interesting differences are apparent between import in plants and import in animals and yeast. Import in plants is only partially inhibited on ice compared to an almost complete block in yeast and animals (perhaps an adaptation), and import was not blocked by WGA as in animals (perhaps due to the unusual NPC modifications). The most fundamental difference, however, is that import in permeabilized plant cells can occur in the absence of exogenous cytosol. This is not to suggest that cytoplasmic factors are not required in plants as in animals and yeast. However, a significant fraction of specific import factors may be tightly associated with cellular structures such as the cytoskeleton that are not disrupted by the permeabilization techniques used. There is support for this notion.

By direct observation in permeabilized protoplasts[48] and by fractionation of purified nuclei[45] it is clear that in the presence of high concentrations of Triton X-100 significant fractions of At-IMP α remain in the cytoplasm and in association with the nuclear envelope in addition to a soluble pool. Whereas these observations provide an opportunity to investigate potential mechanisms for retention of At-IMP α, unfortunately they limit the utility of the assays for the identification of essential import factors in plants. Alternatives have been used most of which are heterologous systems. As noted Ballas and Cytovsky[72] have used the yeast permeabilized system to demonstrate the function of AtKAPα in the import of VirD2. Other heterologous systems have been used to examine *Agrobacterium* Vir proteins including *Xenopus, Drosophila,* mammalian, and yeast cells.[57-59] Other examples previously cited are the functional characterizations of At-IMP α and rice importins α and β in permeabilized HeLa cells. An alternative heterologous approach is to utilize plant cytosol extracts to support import in animal cells. This approach was examined, and it was found that cytosolic extract from petunia could in fact support import in permeabilized HeLa cells. As in animal cells import is temperature dependent, requires GTP hydrolysis and is blocked by WGA.[93] The final approach that has proven useful is microinjection of import substrates into the stamen hairs of *Tradescantia*. This was valuable in characterizing the nuclear import of VirE2-single stranded DNA complexes.[94] VirE2 facilitates *Agrobacterium* infection by associating with pathogen DNA in the plant cell cytoplasm and assisting in its nuclear import. Import of the complexes was found to be dependent upon GTP hydrolysis and was inhibited by WGA. The inconsistency of WGA inhibition upon injection compared to a lack of inhibition in permeabilized cells in unclear. Perhaps the large mass of the VirE2-protein complexes renders them more susceptible to import inhibition than simple protein substrates.

Regulated Protein Import in Plant Development

Plants lead a sessile life style and thus have evolved sensitive mechanisms to control development and withstand environmental challenges. A large number of gene products are induced in response to light, which is an essential developmental stimulus. Seedlings respond to light by undergoing photomorphogenesis, a process that includes altered morphology (shorter stems, leaf development), the development of chloroplasts and induction of the photosynthetic machinery (greening). Even fully differentiated plants must continually respond to quantitative and qualitative differences in light and other environmental challenges such as temperature fluctuations. For the purposes of illustrating the important roles that nuclear protein import play in such responses, we will discuss several interesting examples in which regulation of import potentiates a response to environmental cues.

Photomorphogenesis

Genetic screens have identified components in light signaling that include photoreceptors (for review see ref. 95) and downstream components that couple light signals to gene expression during photomorphogenesis (see for example refs. 96-99) The understanding of photomorphogenesis including the mechanisms of light-modulated gene expression is major area of plant biology beyond the scope of this Chapter but recently reviewed in detail.[100-103] In *Arabidopsis*, screens for defective or inappropriate responses to light have resulted in mutants that are photomorphogenic (ie they possess a light-grown phenotype) in complete darkness and define at least 11 loci known as the COP/DET/FUS loci.[1,104,105] One protein that acts as a negative regulator of photomorphogenesis is COP1. Mutations in COP1 result in plants that develop a light-grown phenotype including chloroplast development in complete darkness. When expressed as a GUS fusion protein in *Arabidopsis*, COP1 localizes primarily to the nucleus in the dark in leaves and shoots but is exclusively nuclear localized in the roots.[104] However in the light COP1 partitions between the nucleus and cytoplasm in leaves and shoots. COP1 possesses, in addition to a bipartite NLS, a zinc-binding domain, a coiled-coil domain, and WD-40 repeats. Recent characterization of COP1 domain structure indicates that the amino terminal region containing the zinc-binding domain is essential for basal function, whereas the carboxyl terminal domain is necessary for the repression of photomorphogenesis in the dark.[105] Although the precise mechanism by which COP1 targeting is regulated by light is not fully understood, a potential route by which photomorphogenic repression is achieved is via interaction of COP1 with the basic leucine-zipper (bZIP) transcription factor Hy5. Hy5 binds to G-box containing promoters for light regulated genes such as ribulose bisphosphate carboxylase/oxygenase small subunit (RBCS) and chalcone synthase (CHS). It has been identified as a suppressor of the *cop1* mutant and has been shown to interact physically with COP1 protein.[106] Several studies indicate that Hy5 is an exclusively nuclear protein that is abundant in the light but degraded in the dark. Furthermore, the degradation is clearly dependent upon interaction with COP1 in the nucleus because a truncated Hy5 protein lacking a COP1-interaction domain is no longer degraded in the dark.[1] Thus, COP1 functions as a repressor of photomorphogenesis by signaling the selective degradation of downstream affectors including Hy5 which participates gene expression essential for development. Recent evidence suggests that degradation of HY5 is mediated by the ubiquitin pathway by interaction with a COP1-containing proteosome[107] and that HY5 degradation can be further modulated by phosphorylation of the COP1 interaction domain.[108] Expression of HY5 is interestingly itself under negative regulation in the light by the action of a recently discovered calcium-binding protein, SUB1[109] The story is more complicated as COP1 appears to interact via its coiled-coiled domain with yet another protein (CIP1) that is cytoplasmic and is capable of interacting with the cytoskeleton.[110] One hypothesis is that CIP1 is involved in the retention of COP1 in the cytoplasm in the light essentially excluding it from the nucleus.

The perception of light involves multiple photoreceptors that detect blue light (cryptochrome), UV-B, and red light (phytochrome).[95,111] The best characterized family of receptors are the phytochromes which are soluble proteins possessing a tetrapyrrole chromophore for activation by red light. After light absorption, the inactive form of phytochrome (Pr) is converted to the active far-red light absorbing form (Pfr) that participates in signal transduction and expression of genes involved in photomorphogenesis. The morphological consequences of red light perception include characteristic hypocotyl shortening and red light dependence of seed germination. There are five genes encoding phytochromes in *Arabidopsis* (phy A through phy E). The best characterized are phy A and phy B each of which detects red light in different ways as developmental cues. Phy A is rapidly degraded in the light and much more abundant in dark-grown plants than phy B. PhyA is responsible for the so-called very low fluence responses (VLFR) and for absorption of continuous far-red light known as the high irradiance responses (HIR). Phy B is stable in the light and is responsible for the detection of red light known as the low fluence responses (LFR) which are reversible by far-red light.[95]

For the induction of gene expression in response to light signals there must be communication between the soluble photoreceptors (that are for the most part cytoplasmic) and the nucleus. Recent experiments indicate that phy A and phy B are transported from the cytoplasm to the nucleus in response to red light. Rice phy A and tobacco phy B were fused to green fluorescent protein (GFP) and overexpressed in tobacco.[112] When adapted to growth in the dark, phy A-GFP and phy B-GFP were not detected. However upon exposure to as little as 5 min of far-red light phy A-GFP (i.e., VLFR) localized to the nucleus. In contrast, upon exposure to red light (i.e., LFR), but not far-red light, Phy B-GFP was found in the nucleus and the localization was reversible upon irradiation by far-red light. Furthermore, the nuclear localization of phy B is dependent upon the presence of the chromophore. The kinetics of the light-dependent relocalization have been examine in detail.[113,114] These results are consistent with the biological activities of phy A and phy B and indicate that their regulated targeting in response to the quality of the red light is an important step in phytochrome signaling. The mechanism of import inhibition in the dark is hypothesized to involve the masking of putative NLSs within the carboxyl terminus via structural changes that are dependent upon the presence of functional chromophore or perhaps a specific retention of P_r in the cytoplasm in the dark (for discussion see ref. 112).

The light-regulated nuclear import of several classes of transcription factors has also been described recently and may provide additional pathways for the control of photomorphogenesis. Nuclear translocation of the common plant regulatory factor (CPRF) proteins in parsley cells has been shown to be under red light control and is far-red reversible.[115] The CPRFs are bZIP transcription factors that bind to G-box elements found adjacent to many light-regulated genes. Of the three CPRFs that have been examined (CPRF1, CPRF2, CPRF4), CPRF2 was found in the cytoplasm in the dark and relocated to the nucleus in the light. Immunolocalization indicates that phy A HIR and phy B LFR responses are involved implicating CPRFs in phytochrome signaling and provides regulation in addition to light-modulated nuclear import of the receptors. A deletion analysis of CPRF2 reveals two potential domains involved in cytoplasmic retention in the dark. Neither domain has homology to the COP1 retention factor CIP1[110] nor to a retention domain in the bZIP factor G-box binding factor 1 (GBF1; discussed below).[116] However, one retention domain of CPRF2 shares 25% identity with an α-helical retention domain from mammalian heat shock factor 2.[115]

Other examples of light-regulated nuclear import are known. Using the previously discussed parsley in vitro antibody cotranslocation assay, Harter et al[91] have found evidence for a cytoplasmic pool of GBF-transcription factors involved in light-regulated gene expression. The GBFs are another class of bZIP transcription factors that bind to G-box elements and participate in light-regulated gene expression. Upon exposure to white light, GBFs were found in the nucleus, presumably due to light stimulated relocalization. More recently, Kircher et al[117] have cloned several CPRFs from parsely and using the antibody cotranslocation assay find that parsley CPRF1, CPRF2 and GBF2 are translocated to the nucleus in response to UV light. GBFs have been examined further using three different *Arabidopsis* GBF genes fused to the reporter GUS and examined by transient expression in soybean protoplasts.[116] About 50% of one fusion protein (GUS:GBF2) was found in the nucleus in the dark, whereas this increased to about 80% upon exposure to blue light. Deletion analysis of a different fusion protein (GUS:GBF1) resulted in an increase in nuclear localization from a maximum of 50% to about 90%. This analysis may have identified a region involved in cytoplasmic retention of GBFs in the dark. One caviat is that GUS:GBF1 itself is not under light control being about 50% nuclear under all conditions tested. Given the importance of light in plant development additional examples of modulated nuclear import will no doubt be identified.

Other Examples of Regulated Import

Plants must continually respond to environmental and pathogen challenges and examples indicating the involvement of regulated import are being discovered. In tomato, the import of several heat shock transcription factors (Hsfs) requires protein-protein interaction. The expression of HsfA1 is constitutive but is accompanied by the expression of several heat shock inducible forms called HsfA2 and HsfB1. HsfA2 has been shown in tomato and tobacco protoplasts to be mostly cytoplasmic upon heat shock, even though related factors such as HsfA1 are translocated to the nucleus under these conditions. If a short region is deleted from the carboxyl terminus HsfA2, the protein is strongly localized to the nucleus.[118] Interestingly, when coexpressed with HsfA1 in tomato protoplasts, HsfA2 is efficiently translocated following heat shock. The cotranslocation is dependent upon the physical interaction of HsfA1 and HsfA2 as demonstrated by coimmunoprecipitation and a two-hybrid assay.[119] The stress induction of HsfA2 and its interaction with constitutively expressed HfA1 to form a transcriptionally active heterodimer provides a mechanism for dynamic changes in the intracellular distribution of HsfA2.

We have already discussed aspects of the *Agrobacterium* system in which the pathogen utilizes endogenous plant import components to assist is the infection process. Nuclear import in plants can also be under viral control. An interesting example occurs in plants infected with the squash leaf curl virus (SqLCV), a geminivirus (for review see refs. 2, 120). The virus encodes two movement proteins, BR1 and BL1, which cooperativey participate in cell-to-cell spread of the virus. BR1 is an NLS-containing protein that shuttles between the nucleus and cytoplasm and binds to single-stranded DNA. BL1 is localized to peripheral regions of cytoplasm and appears to function in the movement of the BR1:viral DNA complexes across the cell wall to adjacent cells. When expressed transiently in tobacco protoplasts BR1 is strongly localized to the nucleus[121,122] However when coexpressed with BL1, BR1 is relocalized to the cell periphery via specific interaction between the proteins[122] providing a mechanism for delivery of viral genomes to the cell periphery for cell-to-cell spread. Other examples of viral protein nuclear import and cytoplasmic retention controlling nuclear import in viruses exist.[123,124]

Besides the specific examples of regulated nuclear import cited above, there is almost certainly broader control of development and environmental responses through modulation of the nuclear import apparatus itself. For example in rice, it is known that light exposure results in the down-regulation of importin α in leaves and dark-grown seedlings.[75] Potentially broad control of import in plants could be imparted through phophorylation, which has been clearly implicated in both overall control of the cell cycle and in the specific regulation of imported proteins in animals and yeast (for reviews see refs. 4, 10, 125).

Recent Advances in Plant Nuclear Translocation

Several recent advances in our understanding of nuclear protein translocation in plants are worthy of mention as they have a potential impact of the field in general.

Targeting to the NPC

Many of the essential components of the import pathway have been identified animals and yeast, and are beginning to be identified in plants. In addition the NPCs have been the focus of intense investigation in animals and yeast. One fundamental question that has received little or no attention is how proteins in the cytoplasm are targeted to the NPCs for import. The notion that proteins are freely soluble in the cytoplasm where they associate with importins and diffuse to the NPCs for translocation is too simplistic. It is known that organelles and mRNAs can be transported along the cytoskeleton to specific sites.[126,127] In fact, animal viruses can be targeted to the nucleus along microtubules,[128] and there are strong indications that plant viral movement proteins can associate with the cytoskeleton.[2,129] Could the cytoskeleton play a role in transporting complexes to the NPC for nuclear import?

Immunolocalization of At-IMP α in tobacco protoplasts is suggestive of cytoskeleton, extending from the nucleus throughout the cytoplasm to the cell periphery.[45] In addition as noted previously, At-IMP α like other importin αs contains hydrophobic armadillo repeats implicated in protein-protein interactions including association with the cytoskeleton[130] Furthermore, At-IMP α cannot be fully depleted from the cytoplasm of permeabilized cells indicating a tight association with intracelluar components.[48] These observations prompted Smith and Raikhel[7] to investigate the role of the cytoskeleton in NPC targeting using double-immunofluorescence and confocal microscopy.[7,46] Importin α was found to colocalize with microfilaments and microtubules in tobacco protoplasts, whereas depolymerization of cytoskeleton results in loss of the cytoskeleton-like staining pattern. Depolymerization of microtubules results in diffuse cytoplasmic staining (Fig. 2). Interestingly, depolymerization of microfilaments results in accumulation of receptor in the nucleus, suggesting that microfilaments may be involved in retention of importin α in the cytoplasm.[7] An examination of At-IMP α association in vitro in a cytoskeleton-binding assay indicates that association with microtubules and microfilaments requires the presence of a functional NLS. The NLS-dependent association of At-IMP α with cytoskeleton may represent a mechanism for the assembly and transport of import complexes to the NPC. Based upon the data, a working model has been proposed[46] in which microfilaments serve as sites for assembly of importin α-NLS protein complexes (Fig. 3). Transport to the NPC would likely require the participation of a microtubule motor protein. It is possible that proteins translated from polysomes associated with the cytoskeleton could be assembled into complexes following synthesis. The model is supported by observations of the movement of NLS-containing substrates along neurons toward the nucleus, which is microtubule dependent.[131] It is unclear at this time what role importin β would play in the formation of cytoskeletal import complexes. Although At-IMP α appears to function in import in an importin β-independent manner, this is probably not true for other importin αs. Other connections between importins and the cytoskeleton are becoming apparent. For example, it is now known that importin β can inhibit microtubule assembly in *Xenopus* egg extracts, and it is suggested this serves to suppress aster assembly until interaction with Ran-GTP releases importin β and assemble can proceed during mitosis.[132]

Nuclear Export

Recently the nuclear shuttling protein BR1 from SqLCV protein has been examined for a nuclear export signal (NES) that would permit its export from the nucleus as has been found in the viral protein HIV Rev[133] and others.[134,135] This signal, like NLSs, is not strictly conserved, but is a leucine-rich hydrophobic sequence of 10 to 13 amino acids. Such as motif was found

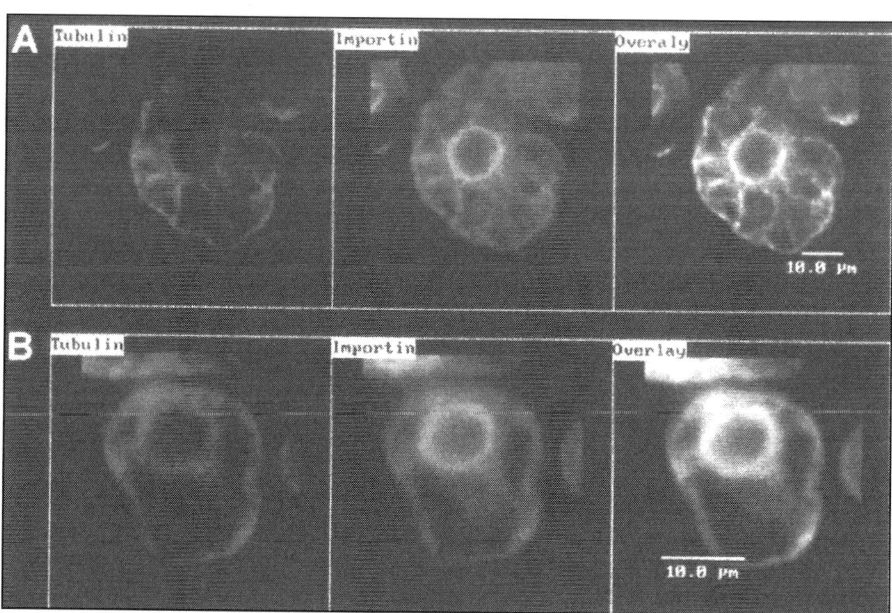

Figure 2. Colocalization of importin α with microtubules in the cytoplasm of tobacco cells. Fixed proto-plasts were double immunolabeled for tubulin (Tubulin panels) and importin α (Importin panels) and examined for coalignment (Overlay panels). A) Importin α (red) displays a pattern of labeling similar to that of cytoplasmic microtubules (green). Superimposition of the images clearly demonstrates coalignment of microtubules and importin α (yellow). B) Depolymerization of microtubules with 10 µM oryzalin results in a loss of importin α cytoplasmic strands (red) and microtubules (green) that is also evident when the images are superimposed (yellow). Note that in (A) and (B) importin α also localizes to the nucleoplasm and labels intensely at the nuclear envelope. Scale bars are as indicated in the panels. Figure reprinted with permission from Smith and Raikhel Plant Cell 1998; 10:1791-1799. © American Society of Plant Biologists. A color version of this figure is available online at http://www.Eurekah.com.

within BR1, and when its three leucine residues were mutated to alanines, viral pathogenicity was lost indicating the essential nature of these residues.[9] It was further reasoned that only export competent BR1 protein could be relocalized to the cell periphery by association with BL1. In fact, GUS-BR1 fusion proteins when coexpressed in tobacco protoplasts are only relocalized to the cell periphery when the NES is present. Fusion proteins without the NES remain in the nucleus. Using this assay it was demonstrated that the NES from the *Xenpous* transcription factor IIIA can functionally replace the endogenous NES of BR1 and even restore viral infectivity. A homologue of the NES transporter CRM1 also known as exportin 1[135-139] was recently cloned from *Arahidopsis* (AtXPO1) and characterized functionally in a yeast two-hybrid assay. The AtXPO1 interacts with the functional NES from AtRanBP1a as well as the NES from HIV Rev.[140] In another protoplast expression system utilizing an NLS and NES fused to GFP, the NES from HIV Rev functions in export, which is inhibited by leptomycine B as in animal cells. The NES from AtRan BP1a, but not a version in which three leucines were mutated to alanines, also functions in this assay. These are the first examples in plants of char-acterized NESs and suggest that along with the protein import pathway, there is functional conservation with animals and yeast. Again, the interesting biology will reside in the details of how such pathways are utilized in plants, and the development of a straightforward assay for nuclear export in plants should encourage progress.

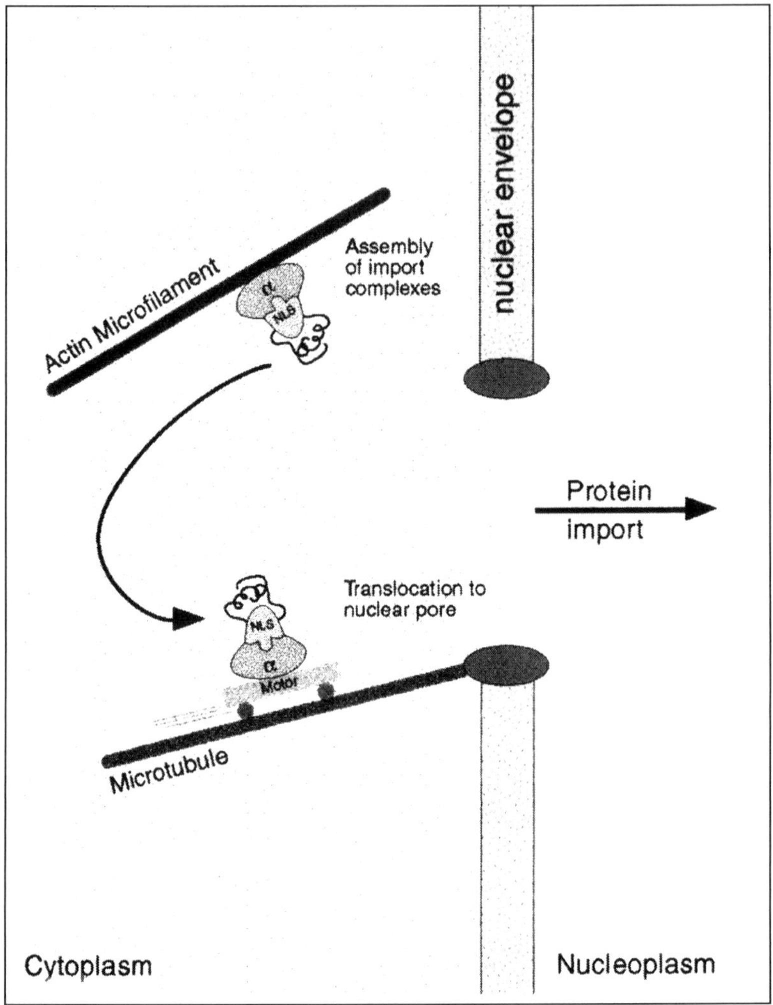

Figure 3. Diagram depicting hypothetical model for intracellular retention and translocation of import complexes prior to nuclear import. Actin microfilaments serve as a site for assembly of import complexes. Importin α probably does not bind to microfilaments directly but through association with an actin-binding protein such as ACT2 which has been shown in yeast to bind to importin α.[46] Assembled complexes can then be loaded onto microtubules for targeting to the NPCs for import. A microtubule motor such as kinesin or dynein would provide directional movement toward the nuclear envelope.

Conclusions

The study of protein import in plants is beginning to yield insight into not only the similarities with other kingdoms, but also the interesting differences that we have described throughout this Chapter. Plants are essential to all life on our planet and are the foundation for our food chain. Protein import processes and their role in development and environmental responses are essential to our understanding of plant biology, an important goal in itself. As our knowledge increases about nuclear protein import in plants, contributions to our general understanding of these processes in all organisms will increase.

Some of the important areas to be addressed in the future include:
1. The complete characterization of import components from higher plants as has been underway
2. Answers to the question, is there an importin β-independent pathway in plants?
3. The further investigation of plant NPCs
4. Taking full advantage of pathogens such as *Agrobacterium* and viruses in understanding pathogenesis and import and export pathways
5. A focus on the regulation of import in essential developmental pathways in plants such as photomorphogensis, stress responses, and perhaps phytohormone signaling
6. The molecular mechanism of import complex targeting to the NPCs for translocation including the development of a system to examine NLS protein movement along microtubules and a search for factors that may mediate complex association with the cytoskeleton

There is much to be learned and the future will surely present opportunities for new discovery.

References

1. Hartke CA, Deng X-W. The cell biology of the CP/DET/FUS proteins. Regulating proteolysis in photomorphogenesis and beyond? Plant Physiol 2000; 124:1548-1557.
2. Lazarowitz SG, Beachy RN. Viral movement proteins as probes for intracellular and intercellular trafficking in plants. Plant Cell 1999; 11:535-548.
3. Tzfira T, Rhee Y, Chen M-H et al. Nucleic acid transport in plant microbe interactions: The molecules that walk through the walls. Annu Rev Microbiol 2000; 54:187-219.
4. Hicks GR, Raikhel NV. Protein import into the nucleus: An integrated view. Annu Rev Cell Dev Biol 1995; 11:155-188.
5. Hubner S, Smith HMS, Hu W et al. Plant importin α binds nuclear localization sequences with high affinity and can mediate nuclear import independent of importin β. J Biol Chem 1999; 274(32):22610-22617.
6. Deng W, Chen L, Wood DW et al. Agrobacterium VirD2 protein interacts with plant host cyclophilins. Proc Natl Acad Sci USA 1998; 95:7040-7045.
7. Smith HMS, Raikhel NV. Nuclear localization signal receptor importin α associates with the cytoskeleton. Plant Cell 1998; 10:1791-1799.
8. Haasen D, Kohler C, Neuhaus G et al. Nuclear export of proteins in plants: AtXPO1 is the export receptor for leucine-rich nuclear export signals in Arabidopsis thaliana. Plant J 1999; 20(6): 695-705.
9. Ward BM, Lazarowitz SG. Nuclear export in plants: Use of geminivirus movement proteins for an in vivo cell based export assay. Plant Cell 1999; 11: 1267-1276.
10. Jans DA, Hubner S. Regulation of protein transport to the nucleus: Central role of phosphorylation. Physiol rev 1996; 76(3):651-685.
11. Boulikas T. Nuclear localization signal peptides for the import of plasmid DNA in gene therapy (Review). Int J Oncol 1996; 10:301-309.
12. Corbett AH, Silver PA. Nucleocytoplasmic transport of macromolecules. Microbiol Mol Biol Rev 1997; 61(2):193-211.
13. Goldfarb DS. Whose finger is on the switch? Science 1997; 276:1814-1816.
14. Lee MS, Silver PA. RNA movement between the nucleus and the cytoplasm. Curr Opin Gen Dev 1997; 7:212-219.
15. Moroianu J. Molecular mechanisms of nuclear protein import. Crit Rev Eukaryot Gene Expr 1997; 7(1-2):61-72.
16. Nakielny S, Fischer U, Michael WM et al. RNA transport. Annu Rev Neurosci 1997; 20:269-301.
17. Yoneda Y. How proteins are transported from the cytoplasm to the nucleus. J Biochem 1997; 121:811-817.
18. Mattaj IW, Englmeier L. Nucleocytoplasmic transport: the soluble phase. Annu Rev Biochem 1998; 67:265-306.
19. Weis K. Importins and exportins: How to get in and out of the nucleus. Trends Biochem Sci 1998; 23(5):185-189.

20. Whittaker GR, Helenius A. Nuclear import and export of viruses and virus genomes. Virology 1998; 246(1):1-23.
21. Gorlich D, Kutay U. Transport between the cell nucleus and the cytoplasm. Annu Rev Cell Dev Biol 1999; 15:607-660.
22. Hood JK, Silver PA. In or out? Regulating nuclear transport. Curr Opin Cell Biol 1999; 11(2):241-247.
23. Davis LI. The nuclear pore complex. Annu Rev Biochem 1995; 64:865-896.
24. Pante N, Aebi U. Molecular dissection of the nuclear pore complex. Crit Rev Mol Biol 1996; 31:153-159
25. Fabre E, Hurt E. Yeast genetics to dissect the nuclear pore complex and nucleocytoplasmic trafficking. Annu Rev Genet 1997; 31:277-313.
26. Nigg EA. Nucleocytoplasmic transport: Signals, mechanisms and regulation. Nature 1997; 386:779-787.
27. Gant TM, Goldberg MW, Allen TD. Nuclear envelope and nuclear pore assembly: Analysis of assembly intermediates by electron microscopy. Curr Opin Cell Biol; 10:409-415.
28. Yang Q, Rout MP, Akey CW. Three-dimensional architecture of the isolated yeast nuclear pore complex: Function and evolutionary implications. Mol Cell 1998; 1:223-234.
29. Badoor K, Shaikh S, Enarson P et al. Function and assembly of nuclear pore complex proteins. Biochem Cell Biol 1999; 77(4):321-329.
30. Allen TD, Cronshaw JM, Bagley S et al. The nuclear pore complex: Mediator of translocation between the nucleus and cytoplasm. J Cell Sci 2000; 113:1651-1659.
31. Rout MP, Aitchison JD, Suprapto A et al. The yeast nuclear pore complex: Composition, architecture and transport mechanism. J Cell Biol 148:635-651.
32. Rout MP, Aitchison JD, Suprapto A et al. The yeast nuclear pore complex: Composition, architecture and transport mechanism. J Cell Biol 2000; 148:635-651.
33. Delphin C, Guan T, Melchior F et al. RanGTP targets p97 to RanBP2, a filamentous protein localized at the cytoplasmic periphery of the nuclear pore complex. Mol Biol Cell 1997; 8:2379-2390.
34. Shah S, Tugendreich S, Forbes DJ. Major binding sites for the nuclear import receptor are the internal nucleoporin Nup153 and the adjacent nuclear filament protein Tpr. J Cell Biol 1998; 141:31-49.
35. Fornerod M, van Deursen J, van Baal S et al. The human homologue of yeast CRM1 is in a dynamic subcomplex with CAN/Nup214 and a novel nuclear pore component Nup88. EMBO J 1997; 16:807-816.
36. Iovine MK, Wente SR. A nuclear export signal in Kap95p is required for both recycling the import factor and interaction with the nucleoporin GLFG repeat regions of Nup116p and Nup100p. J Cell Biol 1997; 137:797-811.
37. Percipalle P, Clarkson WD, Kent HM et al. Molecular interactions between the importin alpha/beta heterodimer and proteins involved in vertebrate nuclear protein import. J Mol Biol 1997; 266:722-732.
38. Heese-Peck A, Raikhel NV. A glycoprotein modified with terminal N-acetylglucosamine and localized at the nuclear rim shows sequence similarity to aldose-1-epimerases. Plant Cell 1998; 10:599-612.
39. Roberts K, Northcoat DH. Structure of the nuclear pore complex in higher plants. Nature 1970; 228: 385-386.
40. Heese-Peck A, Raikhel NV. The nuclear pore complex. Plant Mol Biol 1998a; 38(1-2):145-162.
41. Scofield GN, Beven AF, Shaw PJ et al. Identification and localization of a nucleoporin-like protein component of the plant nuclear matrix. Plant 1992; 187: 414-420.
42. Hicks GR, Raikhel NV. Specific binding of nuclear localization sequences to plant nuclei. Plant Cell 1993; 5:983-994.
43. Hicks GR, Raikhel NV. Nuclear localization signal binding proteins in higher plant nuclei. Proc Natl Acad Sci USA 1995; 92:734-938.
44. Hicks GR, Smith HMS, Shieh M et al. Three classes of nuclear import signals bind to plant nuclei. Plant Physiol 1995; 107:1055-1058.

45. Smith HMS, Hicks GR, Raikhel NV. Importin α from Arabidopsis thaliana is a nuclear import receptor that recognizes three classes of import signals. Plant Physiol 1997; 114:411-417.

46. Smith HMS, Raikhel NV. Protein targeting to the nuclear pore. What can we learn from plants? Plant Physiol 1999; 119:1157-1163.

47. Heese-Peck A, Cole RN, Borkhsenious ON et al. Plant nuclear pore complex proteins are modified by novel oligosaccharides with terminal N-acetylglucosamine. Plant Cell 1995; 7:1459-1471.

48. Hicks GR, Smith HMS, Lobreaux S et al. Nuclear import on permeabilized protoplasts from higher plants has unique features. Plant Cell 1996; 8:1337-1352.

49. Merkle T, Leclerc D, Marshallsay C et al. A plant in vitro system for the nuclear import of proteins. Plant J 1996; 10:1177-1186.

50. Duverger E, Pellerin, Mendes C, Mayer R et al. Nuclear import of glycoconjugates is distinct from the classical NLS pathway. J Cell Sci 1995; 108:1325-1332.

51. Breeuwer M, Goldfarb D. Facilitated nuclear trasnport of histone H1 and other small nucleophilic proteins. Cell 1990; 60:999-1008.

52. Zasloff M. tRNA transport from the nucleus in a eukaryotic cell: Carrier-mediated translocation process. Proc Natl Acad Sci USA 1983; 80:6436-6440.

53. Al-Mohanna FA, Caddy KWT, Bolsover SR. The nucleus is insulated from large cytosolic calcium ion changes. Nature 1994; 367:745-750.

54. Sweitzer TD, Love DC, Hanover JA. Regulation of nuclear import and export. Curr Top Cell Regul 2000; 36:77-94.

55. Lassner MW, Jones A, Daubert S et al. Targeting of T7 RNA polymerase to tobacco nuclei mediated by an SV40 nuclear localization signal. Plant Mol Biol 1991; 17:229-234.

56. van der Krol AR, Chua N-H. The basic domain of plant B-ZIP proteins facilitates import of a reporter protein into plant nuclei. Plant Cell 1991; 3:667-675.

57. Guralnick B, Thomsen G, Citovsky V. Transport of DNA into the nuclei of Xenopus oocytes be a modified VirE2 protein of Agrobacterium. Plant Cell 1996; 8:363-373.

58. Relic B, Andjelkovic M, Rossi L et al. Interaction of the DNA modifying proteins VirD1 and VirD2 of Agrobacterium tumefaciens: Analysis by subcellular localization in mammalian cells. Proc Natl Acad Sci USA 1998; 95:9105-9110.

59. Rhee Y, Gurel F, Gafni Y et al. A genetic system for detection of protein nuclear import and export. Nat Biotechnol 2000; 18:433-437.

60. Chelsky D, Ralph R, Jonak G. Sequence requirements for synthetic peptide-mediated translocation to the nucleus. Mol Cell Biol 1989; 2487-2492.

61. Lanford RE, Feldherr CM, White RG et al. Comparison of diverse transport signals in synthetic peptide-induced nuclear transport. Exp Cell Res 1990; 186:32-38.

62. Wagner P, Hall MN. Nuclear transport is functionally conserved between yeast and higher eukaryotes. FEBS Lett 1993; 321:261-266.

63. Sazer S, Dasso M. The Ran decathlon: Multiple roles of Ran. J Cell Sci 2000; 113:1111-1118.

64. Kutay U, Bischoff FR, Kostka S et al. Export of importin alpha from the nucleus is mediated by a specific nuclear transport factor. Cell 1997; 90:1061-0171.

65. Ribbeck K, Lipowsky G, Kent HM et al. NTF2 mediates nuclear import of Ran. EMBO J 1998; 17:6587-6598.

66. Smith A, Brownawell A, Macara IG. Nuclear import of ran is mediated b the transport factor NTF2. Curr Biol 1998; 8:103-1406.

67. Kussel P, Frasch M. Pendulin, a Drosophila protein with cell cycle-dependent nuclear localization is required for normal cell proliferation. J Cell Biol 1995; 129:1491-1507.

68. Torok I, Strand D, Schmitt R et al. The overgrown hematopoietic organs-31 tumor suppressor gene of Drosophila encodes an importin-like proteins accumulating in the nucleus at the onset of mitosis. J Cell Biol 1995; 129:1473-1489.

69. Kohler M, Ansieau S, Prehn S et al. Cloning of two novel human importin-alpha subunits and analysis of the expression pattern of the importin-alpha protein family. FEBS Lett 1997; 417:104-108.

70. Tsuji L, Takumi T, Imamoto N et al. Identification of novel homologues of mouse importin alpha, the alpha subunit of the nuclear pore-targeting complex, and their tissu-specific expression. FEBS Lett 1997; 416:30-34.

71. Kohler M, Speck C, Christiansen M et al. Evidence for distinct substrate specificities of importin α-family members in nuclear protein import. Mol Cell Biol 1999; 19(11):7782-7791.

72. Ballas N, Citovsky V. Nuclear localization signal binding protein from Arabidopsis mediates nuclear import of Agrobacterium VirD2 protein. Proc Natl Acad Sci USA 1997; 94:10723-10728.

73. Schledz M, Leclerc D, Neuhaus G et al. Characterization of four cDNAs encoding different importin aloha homologs from Arabidopisis thaliana. Plant Physiol 1998; 116:868.

74. Nemeth K, Sakchert K, Putnoky P et al. Pleiotropic control of glucose and hormone resposnes by PRL1, a nuclear WD protein, in Arabidopsis. Genes and Dev 1998; 12:3059-3073.

75. Shoji K, Iwasaki T, Matsuki R et al. Cloning of an importin-a and down-regulation of the gene by light in rice leaves. Gene 1998; 212:279-286.

76. Jiang C-J, Imamoto N, Matsuki R et al. Functional characterization of a plant importin a homologue. J Biol Chem 1998; 273(37):24083-24087.

77. Matsuki R, Iwasaki T, Shoji K et al. Isolation and characterization of two importin-beta genes from rice. Plant Cell Physiol 1998; 39(8):879-884.

78. Jian C-J, Imatoto N, Matsuki R et al. In vitro characterization of rice importin β1: molecular interaction with nuclear transport factors and mediation of nuclear protein import. FEBS Lett 1998; 437:127-130.

79. Haizel T, Merkle T, Pay A et al. Characterization of proteins that interact with the GTP-bound form of the regulatory GTPase Ran in Arabidopsis. Plant J 1997; 11:93-103.

80. Ach RA, Gruissem W. A small GTP-binding protein from tomato suppresses a Schizosaccharomyces pombe cell-cycle mutant. Proc Natl Acad Sci USA 1994; 91:5863-5867.

81. Merkle T, Haizel T, Matsumoto T et al. Phenotype of the fission yeast cell cycle regulatory mutant pim1-46 is suppressed by a tobacco cDNA encoding a small, Ran-like GTP-binding protein. Plant J 1994; 6:555-565.

82. Saalbach G, Christov V. Sequence of a plant cDNA from Vicia faba encoding a novel Ran-related GTP-binding protein. Plant Mol Biol 1994; 24:969-972.

83. Borg S, Brandstrup B, Jenson TJ et al. Identification of a new protein species among 33 different small GTP-binding proteins encoded by cDNAs from Lotus japonicus, and expression of corresponding mRNAs in developing root nodules. Plant J 1997; 11:237-250.

84. Meier I. A Novel link between Ran signal transduction and nuclear envelope proteins in plants. Plant Physiol 2000; 124:1507-1510.

85. Gindullis F, Peffer NJ, Meier I. MAF1, a novel plant protein interacting with matrix attachment region binding protein MFP1, is located at the nuclear envelope. Plant Cell 1999; 11:1755-1767

86. Gindullis F, Meier I. Matrix attachment region binding protein MFP1 is localized in discrete domains at the nuclear envelope. Plant Cell 1999; 11:1117-1128

87. Adam SA, Sterne-Marr R, Gerace L. Nuclear protein import in permeabilized mammalian cells requires soluble cytoplasmic factors. J Cell Biol 1990; 11:807-816.

88. Adam SA, Sterne-Marr R, Gerace L. Nuclear protein import using digitonin-permeabilized cells. Methods Enzymol 1992; 219:97-110.

89. Schlenstedt G, Hurt E, Doye V et al. Reconstitution of nuclear protein transport with semi-intact yeast cells. J Cell Biol 1993; 123:785-798.

90. Kalinich JF, Douglas MG. In vitro translocation through the yeast nuclear envelope. J Biol Chem 1989; 264:17979-17989.

91. Harter K, Kircher S, Frohmeyer H et al. Light regulated modification and nuclear translocation of cytosolic G-box binding factors in parsley. Plant Cell 1994; 6:545-559.

92. Griesbach RJ, Sink KC. Evacuolation of mesophyll protoplasts. Plant Sci Lett 1983; 30:297-301.

93. Broder YC, Stanhill A, Zakai N et al. Translocation of NLS-BSA conjugates into nuclei of permeabilized mammalian cells can be supported by protoplast extract. FEBS Lett 1997; 412:535-539.

94. Zupan JR, Citovsky V, Zambryski P. Agrobacterium VirE2 protein mediates nuclear uptake of single-stranded DNA in plant cells. Proc Natl Acad Sci USA 1996; 93:2392-2397.

95. Batschauer A. Photoreceptors of higher plants. Planta 1998; 206:479-492.

96. Buche C, Poppe C, Schafer E et al. eid1: a new Arabidopsis mutants hypersensitive in phytochrome A-dependent high-irradiance responese. Plant Cell 2000: 12:547-558.

97. Fairchild CD, Schumaker MA, Quail PH. HFR1 encodes an atypical bHLH protein that acts in phytochrome A signal transduction. Genes Dev 2000; 14:2377-2391.

98. Fankhauser C, Chory J. RSF1, an Arabidopsis locus implicated in phytochrome A signaling. Plant Physiol 2000; 124:39-46.

99. Hsieh HL, Okamoto H, Wang A et al. FIN219, an auxin-regulated gene, defines a link between phytochrome A and the downstream regulator COP1 in light control of Arabidopsis development. Genes Dev 2000; 14:1958-1970.

100. Casal JJ. Phytochromes, cryptochromes, phototropin: photoreceptor interactions in plants. Photochem Photobiol 71:1-11.

101. Nagy F, Schafer E. Nuclear and cytosolic events of light-induced, phytochrome-regulated signaling in higher plants. EMBO J 2000; 19:157-163.

102. Neff MM, Fankhauser C, Chory J. Light: An indicator of time and place. Genes Dev 2000; 14:257-271.

103. Wei N, Deng XW. Making sense of the COP9 signalosome: A regulatory protein complex conserved from Arabidopsis to human. Trends Genet 2000; 15:98-103.

104. von Armin AG, Deng XW. Light inactiviation of Arabidopsis photomorphogenic repressor COP1 involves a cell-specific regulation of its nucleocytoplasmic partitioning. Cell 1994; 79:1035-1045.

105. Stacey MG, Kopp OR, Kim T-H et al. Modular domain structure of Arabidopsis COP1. Reconstitution of activity by fragment complementation and mutational analysis of a nuclear localization signal in planta. Plant Physiol 2000; 124:979-989.

106. Ang LH, Chattopadhyay S, Wei N et al. Molecular interaction between COP1 and HY5 defines a regulatory switch for light control of Arabidopsis development. Mol Cell 1998; 1:213-222.

107. Osterlund MT, Hardtke CS, Wei N et al. Targeted destabilization of HY5 during light-regulated development of Arabidopsis. Nature 2000; 405:462-466.

108. Hardtke CS, Gohda K, Osterlund MT et al. HY5 stability and activity in Arabidopsis is regulated by a phosphorylation within COP1 binding domain. EMBO J 2000; 19:4997-5006.

109. Guo H, Mockler T, DuongH et al. SUB1, an Arabidopsis Ca^{2+}-binding protein involved in cryptochrome and phytochrome coaction. Science 2001; 291:487-490.

110. Matsui M, Stoop CD, von Armin AG et al. Arabidopsis COP1 protein specifically interactsin vitro with a cytoskeleton-associated protein, CIP1. Proc Natl Acad Sci USA 1995; 92:4239-4243.

111. Batschauer A. Light perception in higher plants. Cell Mol Sci 1999; 55:163-166.

112. Kircher S, Kozma-Bognar L, Kim L et al. Light quality-dependent import of plant photoreceptors phytochrome A and B. Plan Cell 1999; 11:1445-1456.

113. Gil P, Kircher S, Adam E et al. Photocontrol of subcellular partitioning of phytochrome-B:GFP fusion protein in tobacco seedlings. Plant J 2000: 22(2):1335-145

114. Kim L, Kircher S, Toth R et al. Light-induced nuclear import of phytochrome-A:GFP fusion proteins is differentially regulated in transgenic tobacco and Arabidopsis. Plant J 2000: 22(2):125-133.

115. Kircher S, Wellmer F, Nick P et al. Nuclear import of the parsley bZIP transcription factor CPRF2 is regulated by phytochrome photoreceptors. J Cell Biol 1999; 144(2):201-211.

116. Terzaghi WB, Bertekap RL, Cashmore AR. Intracellular localization of BGF proteins and blue light-induced import of GBF2 fusion proteins into the nucleus of cultured Arabidopsis and soybean cells. Plant J 1997; 11(5):967-982.

117. Kircher S, Ledger S, Hayashi H et al. CPRF4a, a novel plant bZIP protein of the CPRF family: comparative analysis of light-dependent expression, post-transcriptional regulation, nuclear import and heterodimerisation. Mol Gen Gen 1998; 257:595-605.

118. Lyck R, Harmening U, Hohfeld I et al. Intracellular distribution and identification of the nuclear localization signals of two plant heat-stress transcription factors. Planta 1997; 202:117-125.

119. Scharf K-D, Heider H, Hohfeld I et al. The tomato Hsf system: HsfA2 needs interaction with HsfA1 for efficient nuclear import and may be localized in cytoplasmic heat shock granules. Mol Cell Biol 1998; 18(4):2240-2251.

120. Sanderfoot AA, Lazarowitz SG. Getting it together in plant virus movement: Cooperative interactions between bipartite geminivirus movement proteins. Trends Cell Biol 1996; 6:353-358.

121. Pascal E, Sanderfoot AA, Ward BM et al. The geminivirus BR movement protein binds single-stranded DNA and localizes to the cell nucleus. Plant Cell 1994; 6:995-1006.

122. Sanderfoot AA, Ingham DJ, Lazarowitz SG. A viral movement protein as a nuclear shuttle: The geminivirus BR1 movement protein contains domains essential for interaction with BL1 and nuclear localization. Plant Physiol 1996; 110:23-33.

123. Restrepo-Hartwig MA, Carrington JC. The tobacco etch potyvirus 6-kilodalto protein is membrane associated and involved in viral replication. J Virol 1994; 68:2388-2397.

124. Kunik T, Palanichelvam K, Czosnek H et al. Nuclear import of the capsid protein of tomato yellow leaf curl virus (TYLCV) in plant and insect cells. Plant J 1998; 13(3):393-399.

125. Nagatani A. Regulated nuclear targeting. Curr Opin Plant Biol 1998; 1:470-474.

126. Bassel G, Singer RH. MRNA and cytoskeletal filaments. Curr Opin Cell Biol 1997; 9:109-115.

127. Hirokawa N. Kinesin and dynein superfamily proteins and the mechanism of organelle transport. Science 1998; 279:519-526.

128. Sodeik B, Ebersold MW, Helenius A. Microtuble-mediated transport of incoming herpes simplex virus 1 capsids to the nucleus. J Cell Biol 1997; 136:1007-1021.

129. Heinlein M, Epel BL, Padgett et al. Interaction of tobamovirus movement proteins with the plant cytoskeleton. Science 1995; 270:1983-1985.

130. Barth AI, Nathke IS, Nelson WJ. Cadherins, catenins and AC signaling; Interplay between cytoskeletal complexes and signaling pathways. Curr Opin Cell Biol 1997; 9:693-690.

131. Ambron RT, Schmied R, Huang CC. A signal sequence mediates the retrograde transport of proteins from the axon periphery to the cell body and then into the nucleus. J Neurosci 1992; 12:2813-2818.

132. Wiese C, Wilde A, Moore MS et al. Role of importin-β in coupling Ran to downstream targets in microtubule assembly. Science 2001; 291(5504):653-656.

133. Fischer U, Huber J, Boelens WC et al. The HIV-1 Rev activation domain is a nuclear export signal that accesses an export pathway used by specific cellular RNAs. Cell 1995; 82:475-483.

134. Wen W, Meinkoth JL, Tsien RY et al. Identification of a signal for rapid export of proteins from the nucleus. Cell 1995; 82:463-473.

135. Fridell RA, Fischer U, Luhrmann R et al. Amphibian transcription factor IIIA proteins contain a sequence element functionally equivalent to the nuclear export signal of human immunodeficiency virus type 1 Rev. Proc Natl Acad Sci USA 1996; 93:2936-2940.

136. Fornerod M, Ohno M, Yoshida M et al. CRM1 is an export receptor for leucine-rich nuclear export signals. Cell 1997; 90:1051-1060.

137. Fukuda M, Asano S, Nakamura T et al. CRM1 is responsible for intracellular transport mediated by the nuclear export signal. Nature 1997; 390:308-311.

138. Ossareh-Nazari B, Bachlerie F, Dargemont C. Evidence for a role of CRM1 in signal-mediated nuclear protein export. Science 1997; 278:141-144.

139. Stade K, Ford CS, Guthrie C et al. Exportin 1 (CRM1p) is an essential nuclear export factor. Cell 1997; 90:1041-1050.

140. Haasen D, Kohler C, Neuhaus G et al. Nuclear export of proteins in plants: AtXPO1 is the export receptor for leucine-rich nuclear export signals in Arabidopsis thaliana. Plant J 1999; 20(6):695-705.

CHAPTER 6

Nuclear Import of *Agrobacterium* T-DNA

Tzvi Tzfira, Benoit Lacroix and Vitaly Citovsky

Agrobacterium-mediated genetic transformation is a process by which genetic material is transported from the bacterium into the host nucleus, where it stably integrates. The transferred DNA (T-DNA) is escorted, by two bacterial proteins, as a single-stranded DNA-protein complex (a T-complex), which mediate its transport to the host nucleus. The large size and mass of this DNA-protein complex raise questions as to the molecular machinery and mechanism by which the T-complex passes the nuclear pore barrier. Recent studies have revealed the important role of specific host proteins in interacting with and guiding the T-complex through the nuclear pore, and to its point of integration. In this chapter, we summarize our knowledge of the function of T-DNA bacterial and host protein chaperones, and draw a model for their action during the nuclear import and intranuclear transport of *Agrobacterium* T-DNA.

Introduction

Agrobacterium-mediated genetic transformation of plant cells is a unique and complicated process by which genetic material is transported from the bacterium into the host nucleus, where it stably integrates (see Fig. 1 and reviews in refs. 1-5). Although in nature, *Agrobacterium* transforms mainly dicotyledonous plants,[6] under controlled culture conditions it has been shown to possess a broader host range which includes monocot plants, yeast[7] and other fungi,[8,9] and even human cells.[10] In modern plant breeding, *Agrobacterium* is widely used for plant genetic engineering.[11] The molecular basis for the transformation process has therefore been the subject of numerous studies over the past several decades (reviewed in refs. 1-5).

Agrobacterium tumefaciens is a gram-negative bacterium which is the causative agent of crown-gall disease in many dicotyledonous plant species.[6] The disease symptoms result from the transfer, integration and expression of a specific DNA fragment (known as transferred DNA or T-DNA) from the bacterial tumor-inducing (Ti) plasmid into the plant cell genome. The machinery needed for the generation of transferred T-DNA and for its transport into the host cell is encoded by a serie of chromosomal (*chv*) and Ti-plasmid-encoded virulence (*vir*) genes (reviewed in refs. 1-5). Interestingly, although the bacterial proteins possess many of the functions needed for the transformation process, host-plant factors also play a crucial role in the T-DNA nuclear import and integration (reviewed in refs. 2-4). In this chapter, we discuss the role of both *Agrobacterium* and host proteins during *Agrobacterium* T-DNA long voyage in the host cell, from its cytoplasmic point of entry to the nucleus.

Nuclear Import and Export in Plants and Animals, edited by Tzvi Tzfira and Vitaly Citovsky.
©2005 Eurekah.com and Kluwer Academic / Plenum Publishers.

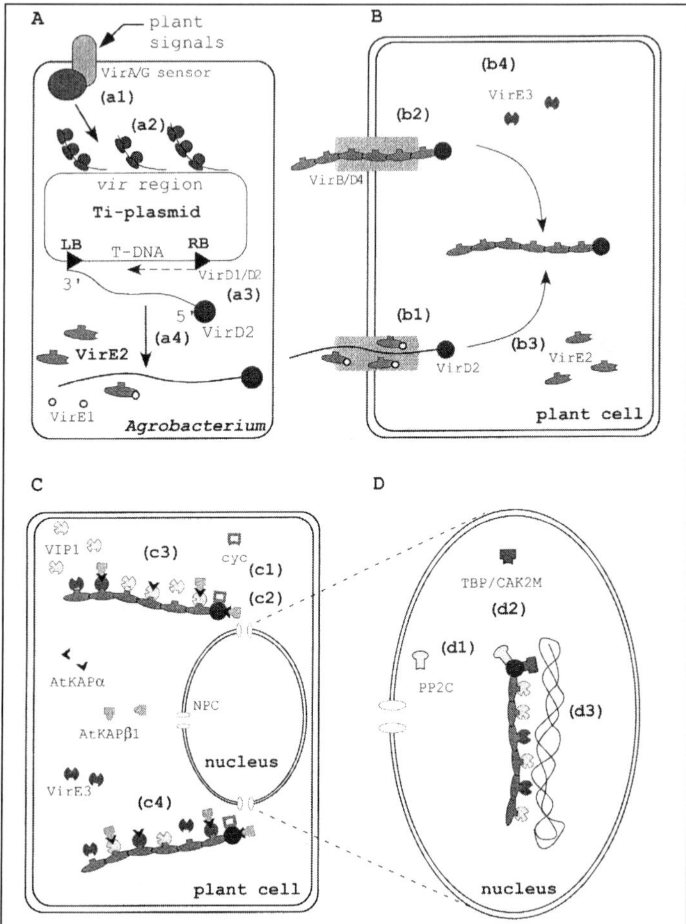

Figure 1. The process of *Agrobacterium*-mediated genetic transformation and model for the nuclear import of the *Agrobacterium* T-complex. A) Induction of the *Agrobacterium* virulence (*vir*) genes and production of the T-complex. The transformation process begins with sensing of plant signals by the *Agrobacterium* VirA/VirG two-component sensory machinery (a1) and transcriptional activation of the *vir* region (a2). The T-DNA left and right borders (LB and RB, respectively) are nicked by the VirD2/VirD1 endonuclease complex, and a single stranded T-DNA copy (T-strand) is released from the Ti plasmid (a3) to produce an immature T-complex composed of the T-strand and a single VirD2 molecule attached to its 5'-end (a4). B) Export into the host cell cytoplasm and assembly of the mature T-complex. Immature (b1) or mature (b2) T-complexes are exported into the host cell cytoplasm through the VirB/VirD4 channel. If exported individually, the immature T-complex and VirE2 molecules are later assembled into a mature T-complex in the host cell cytoplasm (b3). The VirE3 (b4) protein is also exported into the host cell cytoplasm through the same VirB2/VirD4 channel and functions later in nuclear import of the mature T-complex. C) Nuclear import of the mature T-complex. While traveling toward the nuclear pore, VirD2 interacts with cyclophilins (c1), which may maintain its proper conformation in the cytoplasm. For their nuclear import, VirD2 and VirE2 employ two different mechanisms: VirD2 is imported directly by AtKAPα (c2), while VirE2 is imported by VIP1 (c3) or VirE3 (c4) which function as molecular adaptors between VirE2 and the AtKAPα-dependent nuclear import pathway. D). Intranuclear transport to the site of integration. Once inside the nucleus, VirD2 may undergo dephosphorylation by PP2C (d1). Interaction of VirD2 with CAK2M and TBP (d2) and VirE2 with VIP1 (d3), all presumed members of the host cell transcriptional complexes, may result in the intranuclear transport of the T-complex to the point of integration in the plant chromatin.

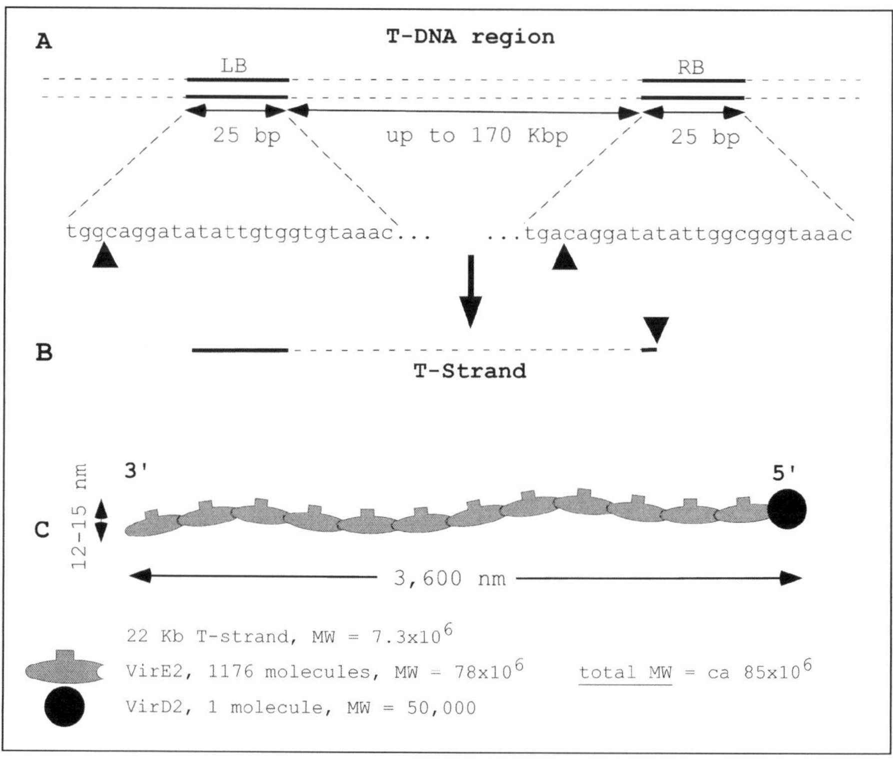

Figure 2. Molecular structure of the T-DNA region and T-complex. A) The T-strand is a ssDNA copy of the bottom (noncoding) strand of the T-DNA region. The VirD2-VirD1 endonuclease complex nicks between the third and the fourth nucleotides of both borders (filled triangles), releasing the T-strand with VirD2, which remains covalently attached to its 5'-end. B) The T-complex is a solenoid, "telephone cord"-like structure covered with numerous VirE2 molecules along its entire length, and a single VirD2 molecule at the 5'-end. C) The estimated size, molecular composition, and molecular weight of a typical 22-kb nopaline-type T-complex.

The Genetic Transformation Process

The T-DNA fraction is a specific DNA segment located on the *Agrobacterium* Ti plasmid. The T-DNA itself is not sequence-specific and is defined exclusively by its left and right borders which are two 25-bp direct repeats (Fig. 2A, and refs. 12 and 13). Since the T-DNA does not contain any specific sequences coding for its processing, translocation into the host cell cytoplasm, nuclear import or integration any sequences placed between its borders will also be transferred and integrated into the host genome. Moreover, the borders are the only *cis*-required elements needed to define and process the T-DNA,[12,13] and T-DNA molecules can therefore technically be located on other genetic elements within the *Agrobacterium* cells, rather than on the Ti plasmid itself. Indeed, the practice of using a small, low-copy binary plasmid, carrying a well-defined T-DNA region, in conjunction with large Ti plasmids which do not contain their original T-DNA (also known as "disarmed" Ti plasmids), is the molecular basis for the use of *Agrobacterium* as a vector for plant genetic engineering.[14,15]

The transformation process (Fig. 1) begins with the induction of the *Agrobacterium* VirA-VirG sensory machinery (Fig. 1A, step a1) and subsequently the virulence (Vir) proteins (a2) by host-specific small phenolic signal molecules.[16-20] The VirD1-VirD2 protein complex acts as an endonuclease which nicks both borders (Fig. 2A) on the noncoding strand of the

T-DNA region.[21-27] Through a strand-replacement process (a3), a single-stranded (ss) T-DNA molecule (the T-strand; (Fig. 2B) is released from the Ti plasmid with a single VirD2 molecule which remains attached to its 5'-end (a4 and refs. 21, 27). The T-strand-VirD2 complex (also called "immature T-complex") is then coated with numerous VirE2 molecules[28-30] to form the transported form of the T-DNA (known as a "mature T-complex", or simply, "T-complex") (Fig. 2C). The T-strand is then transported into the host cytoplasm (Fig. 1B) either as an immature (b1) or mature (b2) T-complex through a *vir*B/*vir*D4-encoded channel,[31,32] travels through the host cytoplasm (b4) and is finally imported into the host nucleus (Fig. 1C). Once inside the nucleus, the T-complex is stripped of its bacterial chaperones and is randomly integrated into the host genome (Fig. 1D) through an unknown mechanism.[33-36] Although much is known about the mechanism by which the Vir proteins mediate the production of the T-strand and the mature T-complex, little is known about the mechanism governing the T-complex transport into the host cell, its voyage through the host cytoplasm and its integration. Recent studies have begun filling this gap by producing new information about the T-complex nuclear import mechanism, especially about the role that specific host factors, interacting with the T-strand bacterial chaperones, the VirD2 and VirE2, play during the transformation process.

T-Complex Export to Plant Cells

The T-strand is a ssDNA molecule which is a copy of the coding strand of the T-DNA region.[22,27] Molecular evidences suggest that the T-strand may associate with two *vir* gene products, VirD2[26,27] and VirE2.[29,30] Following T-strand production, VirD1 is released while VirD2 remains covalently attached to the 5'-end of the border nick sites; thus, VirD2 associates with the T-strand throughout its journey to the point of integration in the host nucleus.[24-27,37,38] VirE2, a ssDNA-binding protein, is presumed to coat the T-strand along its length.[29,30] As do most ssDNA-binding proteins, VirE2 binds ssDNA cooperatively and without sequence specificity, consistent with the nonsequence-specific nature of the T-DNA itself.[29,30] Mutational analysis of VirE2 revealed that the amino-terminal part of the protein is important for its binding cooperatively, while its carboxy-terminal portion is essential for ssDNA binding.[39-41] Importantly, the carboxy-terminal part of VirE2 also contains an RPR motif, which likely functions as a signal for protein export from *Agrobacterium* into plant cells through the VirB/VirD4 channel.[31,42]

The length of the T-DNA is virtually unlimited, as it has been shown that even very large molecules can be efficiently transported and integrated into the host genome.[43,44] For example, using modified Ti plasmids, with the right border removed, the entire copy of the bacterial Ti plasmid was transported and integrated into the host genome.[45-47] Using specifically designed binary vectors engineered to carry large-size inserts, even 150-kb-long T-DNA fragments are transported and integrated into tomato plant cells.[43,44] In the latter experiment, extra copies of the *virA* and *virE2* loci were necessary to mediate the transport of such large molecules, supporting the role of the VirE2 molecule as an essential chaperone of the T-complex.[44]

It is still unclear whether the T-complex matures in the bacterial or the host cell, although there is little doubt that VirE2 is associated with the T-strand in the host cell cytoplasm.[29,48-51] Potentially, the coupling between VirE2 synthesis and T-strand generation, as well as their physical proximity in the bacterial cell, could result in an immediate interaction and the production of mature T-complex within that cell. The strong cooperative interaction between VirE2 and ssDNA,[29] and the suggestion that both are transported through the same channel[52] further support the notion that mature T-complexes are assembled early in the transformation process, within the bacterial cell. Indeed, T-strands and VirE2 are coprecipitated by anti-VirE2 antibodies from extracts of *vir*-induced *Agrobacterium* cells.[28] Nevertheless, other experimental evidence suggests that the immature T-complexes and VirE2 molecules are separately transported into the host cell cytoplasm, where they assemble to form mature T-complexes. First, cotransformation of tobacco plant cells with an *Agrobacterium* strain containing T-DNA but lacking VirE2 and a second strain lacking T-DNA but containing VirE2 restored infectivity of these individually noninfectious bacteria.[53] Second, transgenic plants over-expressing the VirE2 protein could restore the infectivity of a VirE2-deleted *Agrobacterium* mutant.[48,49] These observations suggest that immature T-complexes and VirE2 molecules are independently exported from the bacterial cells. Indeed, studies have shown that VirE2 export from *Agrobacterium* can be inhibited without affecting T-strand export.[52,54] Furthermore, VirE1, another protein product of the *virE* locus, was suggested to act as a VirE2 chaperone, preventing it from binding the ssDNA and facilitating its export into plant cells.[41,50,55,56] Recent data, however, indicate that VirE1 is neither exported into the plant cell nor it is required for VirE2 export by the VirB/VirD4 transport system, limiting the VirE1 role to stabilization of VirE2 prior to its translocation into the host cell.[42]

Mature T-complex is most likely to be the structure imported into the host cell nucleus. Several lines of experimental evidence support this assumption. First, infectivity of an avirulent VirE2 mutant *Agrobacterium* strain could be restored when inoculated onto VirE2-expressing transgenic plants.[48,49] Second, plant cells infected with the VirE2 *Agrobacterium* mutant exhibited lower accumulation of T-strand molecules than those infected with wild-type *Agrobacterium*.[57] Third, VirE2 association with ssDNA molecules in vitro was shown to protect them from exonucleolytic degradation.[29] Thus, T-strands are most probably associated with VirE2—which provides them with the necessary protection within the host cell cytoplasm—and imported into the nucleus as a mature T-complex.

Molecular Structure of the Mature T-complex

Elucidation of three-dimensional structure and computational analysis of the T-complex may assist in understanding the mechanism of its nuclear import and subsequent integration. Complexes, formed in vitro by interaction between purified VirE2 and the bacteriophage M13 ssDNA, were examined by scanning transmission electron microscopy, followed by mass analysis.[58] These analyses revealed that the VirE2-ssDNA association produces rigid and coiled filaments that are 12.6 nm-wide, with a density of 58 kDa/nm, and that each turn of the filament coil contains an average of 3.4 molecules of VirE2 and 63.6 bases of ssDNA.[58] Based on these parameters, a 22-kb T-strand of the wild-type nopaline-specific *Agrobacterium* is calculated to associate with 1,176 molecules of VirE2 (Fig. 2C and ref. 58), and a 150-Kbp T-strand derived from the longest artificial T-DNA used to transform plants[44,59] is predicted to bind 9,087 VirE2 molecules. The length of the mobilized T-strand, when coated with VirE2 molecules, is estimated to range between 40 nm and 80 microns for T-DNA regions between 20 Kb and 150 Kb.

Both the wide outer diameter (12.6 nm) and the extended length of the mature T-complex[58] suggest an active mechanism for its nuclear import. Indeed, the T-DNA outer diameter exceeds the orifice of the nuclear pore diffusion channels (9 nm, reviewed in ref 60), but it is easily compatible with the size-exclusion limit of the nuclear pore, which reaches 23-39 nm during the process of active nuclear uptake.[60-62] Thus, packaging of the T-strand into the T-complex may be essential for nuclear import because, in the absence of VirE2, free T-strand molecules would likely fold into significantly larger structures unable to penetrate the nuclear pore. For example, in vitro (i.e., in a simple solvent), consideration of long free ssDNA as a polymeric random coil, rather than an extended thread, suggests a diameter of nearly 300 nm because the typical size of a polymeric random coil is approximately the geometric mean of the extended length and the persistence length,[63,64] which, of course, is much larger than the entire nuclear pore complex.[65,66]

Nuclear import of VirE2-ssDNA complexes may also be facilitated by their apparent rigidity,[58] which would prevent their folding into random polymeric coils with large diameters. On the other hand, it remains unclear how relatively rigid T-complexes, whose length exceeds the entire diameter of a typical plant cell, move through the host cell cytoplasm and fit into its nucleus without extensive folding or supercoiling. Thus, some restructuring and bending may be required, especially for the nuclear import of long T-DNA complexes.

T-Complex Nuclear Import

T-complexes are polar molecules and their nuclear import is thought to occur in a polar fashion (reviewed in ref. 67). The VirD2, attached to the 5'-end of the T-strand, is presumed to act as the nuclear import pilot, guiding it not only to the nuclear pore[38,68-70] but also to the point of integration.[37] Various studies, using either direct immunolocalization or translational fusions with such reporters as β-glucuronidase (GUS) and GFP, have shown that VirD2 molecules specifically accumulate in the plant cell nucleus.[68,71-74] Sequence analysis of the VirD2 protein reveals that it contains two nuclear localization signals (NLSs) located at the carboxy- and amino-terminal ends.[68,72,75] The VirD2 carboxy-terminal bipartite NLS is the only sequence that appears to function during *Agrobacterium* infection,[38,73,76,77] as it was shown that T-DNA expression and tumorigenicity are reduced in plants infected with *Agrobacterium* carrying carboxy-terminal NLS-deletion mutants of VirD2.[77,78] The reduced infection efficiency, rather than complete blockage suggests that other components of the T-complex participate in the nuclear import process. Indeed, VirE2, the T-strand coating protein, was also found to accumulate in plant cell nucleus.[48,71,75] Sequence analysis of the VirE2 protein revealed a single NLS, located in the middle of the molecule.[48,71] Mutations within the central region of the VirE2 sequence decreased *Agrobacterium* tumorigenicity but did not affect the ssDNA-binding activity or stability of the protein.[39]

Several experimental approaches have been used to demonstrate the direct function of VirE2 in T-DNA nuclear import. Using microinjection of in vitro-formed T-complexes, the ability of VirE2 to direct fluorescently labeled ssDNA into the plant cell nucleus was studied.[79] First, microinjection of fluorescent single or double stranded (ds) DNA alone resulted in clear cytoplasmic fluorescence with no evident nuclear staining. This clearly demonstrated the inability of DNA alone to enter the host nucleus under these experimental conditions. Second, microinjection of in-vitro-formed VirE2-ssDNA complexes, but not of VirE2-dsDNA, resulted in efficient nuclear accumulation of the labeled DNA, thus indicating the requirement of ssDNA-VirE2 complex formation. Third, nuclear import of VirE2-ssDNA was blocked by nuclear import-specific inhibitors such as wheat germ agglutinin and nonhydrolyzable analogs of GTP,[79] indicating that VirE2-mediated nuclear import of ssDNA is an active process which requires ATP. In a different approach, a mutated *Agrobacterium* strain, lacking the entire VirE2 as well as the specific carboxy terminal NLS of VirD2,[49] was used to study the function of VirE2 during the transformation process. This mutant produced tumors on VirE2-expressing transgenic plants, but not on wild-type tobacco plants.[49] These observations demonstrated that VirE2 alone, when expressed in the host cells, can mediate the import of transferred T-strands into the nucleus, in the absence of NLS from any other known T-DNA-associated protein.[49]

A certain functional redundancy between VirE2 and VirD2 in the nuclear import of T-complex may exist. Nevertheless, it is possible that the combined action of VirE2 and VirD2 is required for the polar translocation of T-complex into the nucleus[70] in a manner which may be important for the later integration step. Various studies suggest that T-DNA integration is indeed a polar process, although it is still unclear whether it begins at the 5'- or 3'-end of the T-strand molecule.[80-82] One possibility is that the T-complex polar structure dictates the polarity of its nuclear import and integration. In such a scenario, functional variations between

VirD2, which is covalently bound to the 5'-end of the T-strand, and VirE2, which covers the entire T-strand, may determine the polarity of the transport and integration.

Several approaches have been utilized to investigate whether VirD2 and VirE2 may perform different but complementary functions during nuclear import of the T-complexes. In the first approach, using a heterologous living nonplant cell system which lacks nuclear transport machinery of the *Agrobacterium* natural host cells, differences between VirD2 and VirE2 nuclear import could be discerned. VirD2 localized in the nucleus of *Xenopus* oocytes, *Drosophila* embryos,[83] human kidney and HeLa cells,[24,75,84] and yeast cells.[75,85] In addition, VirD2 nuclear accumulation in animal cells was blocked by known inhibitors of nuclear import,[83] thus demonstrating the functionality of its evolutionarily-conserved NLS. VirE2, on the other hand, did not target to the nuclei of *Xenopus* oocytes, *Drosophila* embryos,[83] HeLa cells[75,86] or yeast cells,[75,85,86] suggesting that its nuclear import may be plant-specific.

In another approach, permeabilized human cells were used to study the differences between VirD2 and VirE2 nuclear import.[84] In this system, both VirD2 and VirE2 accumulated in the nuclei of permeabilized HeLa cells. Nevertheless, while VirD2 alone, when covalently bound to fluorescently labeled ssDNA, promoted T-strand nuclear import, VirE2 failed to do so.[84] These observations suggest that during transformation, both VirD2 and VirE2 are required for T-complex nuclear import. Indeed, a recent study showed that both VirE2 and VirD2 are required for efficient import of the T-DNA complex into plant cell nuclei.[70] Using this approach, in vitro-formed VirE2-VirD2-ssDNA complexes were tested for their import into plant nuclei in vitro. Whereas VirD2 was sufficient for the import of short ssDNA, VirE2, in combination with VirD2, was essential for the import of long ssDNA molecules.[70] Interestingly, RecA, a protein that can bind ssDNA[87] and form thin DNA-protein complexes that are only 5.2-9.3 nm-wide,[88,89] is capable of replacing VirE2 during nuclear import of long T-DNAs, but not during earlier events of T-DNA transfer to plant cells.[70] It was thus suggested that VirD2 and VirE2 perform complementary functions in T-complex nuclear import.[70] While VirD2 initially directs the T-complex into the nuclear pore, VirE2 may shape it in a transferable form and assist translocation of the entire T-complex into the host cell nucleus. Collectively, functional differences between VirD2 and VirE2 suggest that (i) in plant cells, VirE2 and VirD2 employ different cellular factors for their nuclear entry, and (ii) animal cells lack the subset of factors that recognize VirE2 and help its nuclear uptake in plant cells.

Host Cell Proteins That Interact with VirD2 and VirE2

It is well established that both VirD2 and VirE2 are required for the efficient nuclear import of the *Agrobacterium* T-strand.[70,84] During this process, VirD2 and VirE2 must function together. Their import into the host cell nucleus probably occurs *via* different cellular mechanisms, and thus VirD2 and VirE2 are expected to utilize different cellular factors for their nuclear import. In this scenario, the single VirD2 molecule, attached to the 5'-end of the T-strand, and the more abundant VirE2, which covers the remainder of the T-strand molecule, do not compete directly for the same host factors. Such physical and functional arrangements would enable VirD2 and its host interactors to initiate nuclear import of the T-complex, while VirE2 and its interactors likely drive the import process to completion. Several host factors have been identified in past years that interact with *Agrobacterium* VirD2 (see refs. 37, 90, 91) in a yeast two-hybrid protein-protein interaction assay.[92,93]

One set of VirD2-interacting host proteins are members of a large cyclophilin family of peptidyl-prolyl *cis-trans* isomerases (PPIases), which are highly conserved in plants, animals, and prokaryotes.[94-100] Specifically, VirD2 was found to interact with the *Arabidopsis* cyclophilin DIP1[91] and three more isoforms of *Arabidopsis* cyclophilins, Roc1, Roc4,[98] and CypA,[96] in both the yeast two-hybrid system and in vitro.[91] Specific inhibition of the VirD2-CypA interaction in vitro was accomplished by adding cyclosporin A,[91] which is known to bind cyclophilins and block their PPIase activity.[95,100,101] Cyclosporin A also inhibited *Agrobacterium*-mediated genetic transformation of *Arabidopsis* roots and tobacco cell suspension cultures.[91] The exact biological role of cyclophilins during *Agrobacterium* infection is still unclear. They have been proposed to maintain the proper conformation of VirD2 within the host cell cytoplasm and/or nucleus during T-DNA nuclear import and/or integration.[91] Indeed, in addition to their enzymatic activity, cyclophilins may act as molecular chaperones that aid in protein folding in animal cells.[99,102] Two additional *Arabidopsis* proteins, DIP2 and DIP3, which bind VirD2 in the two-hybrid assay, have been reported, but no information as to their identity or biological role were provided.[91] More recently, two more members of the cyclophilin family, Roc2 and Roc3, have been reported to bind VirD2. However, their biological function during *Agrobacterium* transformation has not been examined.[37]

Another host cellular factor that binds VirD2 is the tomato DIG3 protein (Y. Tao, P. Rao, and S. Gelvin, unpublished results). This protein, a type 2C serine/threonine protein phosphatase (PP2C), was found to interact specifically with the VirD2 NLS region. An *Arabidopsis abi1*[103] mutant, knocked out in a PP2C homolog,[103,104] exhibited higher sensitivity to *Agrobacterium*-mediated genetic transformation, than did wild-type plants. In addition, over-expression of DIG3 in tobacco protoplasts specifically inhibited nuclear import of a GUS-VirD2 NLS fusion protein. It was thus suggested that phosphorylation of the VirD2 NLS region may potentiate VirD2 nuclear import, while its dephosphorylation by DIG3 PP2C would negatively regulate this process (Y. Tao, P. Rao, and S. Gelvin, unpublished results).

A third type of VirD2-interacting host protein is a member of the growing karyopherin α family, known to mediate nuclear import of many NLS-containing proteins (reviewed in refs. 105, 106). Specifically, VirD2 was found to interact with AtKAPα, both in vivo and in the yeast two-hybrid system.[90] AtKAPα was found to possess the classical features typical of karyopherin a proteins: it contains eight contiguous repeats of the "arm" motif[107] and four amino-terminal clusters of basic amino acids. In animal and yeast cells, the "arm" motifs are thought to recognize the imported cargo through its NLSs while the amino terminal basic domain is thought to interact with the karyopherin β proteins.[108] Indeed, AtKAPα interaction with its cargo, the *Agrobacterium* VirD2, required the presence of the VirD2 carboxy-terminal NLS.[90] Functionally, AtKAPα showed high homology to the yeast karyopherin α, Srp1p, and could complement this gene function in a temperaturesensitive *srp1-31* yeast mutant.[90,109] In addition, the yeast Srp1p could functionally promote nuclear import of fluorescently labeled VirD2 in permeabilized yeast cells. Using this yeast-derived nuclear import assay,[110] it was also shown that VirD2 nuclear import is dependent upon its NLS. First, VirD2 lacking its NLS failed to localize in the cell nucleus and remained cytoplasmic, and second, a synthetic peptide corresponding to the VirD2 NLS blocked the nuclear import of fluorescently labeled VirD2, probably due to competition with VirD2 for interaction with Srp1p/AtKAPα.[90] In a recent report, VirD2 was found to interact with three more members of the *Arabidopsis* karyopherin α family (At1g02690, At1g09720 and At4g02150), which differed from AtKAPα,[37] and in another report an *Arabidopsis* mutant, knocked out in one of its karyopherin α genes, was resistant to *Agrobacterium* infection.[111] It was thus proposed that, during *Agrobacterium* infection, karyopherin α proteins, which naturally function as part of the nuclear import machinery, directly promote nuclear import of VirD2 and its cognate T-strand.

Recently, two additional host proteins, CAK2Ms and TBP, have been reported to interact with VirD2 in vivo and were suggested to function in the last step of the transformation process, T-DNA integration.[37] CAK2M was identified by its ability to phosphorylate transiently expressed VirD2 in alfalfa protoplasts. CAK2M, a conserved plant ortholog of cyclin-dependent kinase-activating kinases, was also found to interact and phosphorylate the C-terminal regulatory domain of RNA polymerase II largest subunit. The latter protein is known for its ability to recruit TATA box-binding proteins (TBPs),[112,113] not only for the regulation of transcription, but also for the control of transcription-coupled repeat (TCR), which ensures preferential and effective removal of DNA lesions from transcribed genes.[114] Interestingly, VirD2 was also found to tightly associate with TBP in cells that has undergone *Agrobacterium*-mediated transformation. However, it is still unknown whether CAK2M recruits VirD2 to TBP, and whether phosphorylation of VirD2 by CAK2M regulates VirD2 association with the 5'-end of the T-strand. Nevertheless, these observations show that VirD2 is recognized by widely conserved nuclear factors in eukaryotic cells and raises the possibility that CAK2M plays a function in the intra-nuclear voyage of the T-complex to the point of integration while TBP may function as a mediator between VirD2 and the host DNA repair machinery during the integration process.[37]

Since VirD2 and VirE2 utilize different cellular routes for their nuclear import, they most probably interact with different host factors. Indeed, unlike VirD2, VirE2 did not interact with AtKAPα in the yeast two-hybrid assay,[90] but was found to interact with another *Arabidopsis* protein, VIP1.[86] VIP1 contains a conserved stretch of basic amino acids (basic domain) abutting a heptad leucine repeat (leucine zipper), two structural features characteristic of the basic-zipper (b-ZIP) proteins which are known to localize to the cell nucleus.[115] Indeed, VIP1 was capable of promoting the nuclear import of GFP-tagged VirE2 in mammalian cells and, using a genetic assay for nuclear import and export,[85] VIP1 expression was shown to promote nuclear import of VirE2 in yeast cells.[86] Down-regulation of VIP1 in plant cells, using antisense transgenic plants, blocked the nuclear uptake of GUS-tagged VirE2, but not of GUS-VirD2, thus demonstrating the specific role of VIP1 in the nuclear import of VirE2 molecules in plant cells. Moreover, VIP1 antisense transgenic plants showed resistance to both transient and stable genetic transformation by *Agrobacterium*, indicating that they were blocked at the nuclear import stage of the transformation process.[86]

VIP1 is a nuclear protein which has been found to interact with *Arabidopsis* karyopherin α in the yeast two-hybrid assay.[116] Moreover, in yeast cells, VIP1 nuclear import depended on the presence of the cellular Srp1 protein, indicating that VIP1 is imported into the cell nucleus via the karyopherin α-dependent pathway.[116] It was thus suggested that VIP1 interacts with VirE2 in the cell cytoplasm and carries it into the cell nucleus by a "piggyback" mechanism, serving as a molecular adaptor between VirE2 and the host cell karyopherin α.[51,86,116]

VIP1 may represent the cellular factor involved in the plant-specific nuclear uptake of VirE2, because it induces VirE2 nuclear import in nonplant systems, and because no animal or yeast homologs of VIP1 have been found in protein databases.[86] Moreover, the low cellular levels of VIP1 found in various plant tissues[116,117] suggest that, in nature, *Agrobacterium*-mediated transformation may not occur at its maximal possible efficiency.[118,119] In fact, inoculation of various plant tissues of most *Agrobacterium*-susceptible plant species results in the transformation of an extremely low number of cells, even with very dense *Agrobacterium* inoculum. Indeed, over-expression of VIP1 in transgenic plants resulted not only in higher susceptibility to *Agrobacterium* infection, but also in faster nuclear import of the T-DNA.[86]

Because, as a b-ZIP protein, VIP1 may be involved in transcription,[115] associating with the chromosomal DNA either directly or through other components of transcription complexes, and because VirD2 was also found to interact with members of a transcriptional complex,[37] VIP1 and VirD2 interactors may function in the intranuclear transport of the T-complex. Thus, VIP1 may perform a dual function: facilitating nuclear targeting of VirE2 and playing a role in the intranuclear transport of VirE2 and its cognate T-strand to the site of integration. A similar dual role in nuclear and intranuclear transport has been suggested for the yeast Kap114p protein, of which functions are to import the TBP into the cell nucleus, and to target it to the promoters of genes to be transcribed.[120] Besides VIP1, three other VirE2-interacting proteins have been reported, but not identified or characterized.[91]

VirE3, a Bacterial Substitute for the Host Protein VIP1

Recent studies have shown that, in addition to the *Agrobacterium* immature T-complex and its VirE2 chaperone, two other bacterial proteins, VirF and VirE3, are exported from the bacterium to the host cell.[31,42,121] VirF is an F-box protein that probably participates in the targeted proteolysis of as yet unidentified substrates.[122] The function of VirE3 is just beginning to emerge. Recent studies suggest that VirE3 may act as a substitute for the cellular protein VIP1 in facilitating the nuclear import of VirE2. First, VirE3, like VIP1, targets to the nucleus of plant, mammalian and yeast cells (B. Lacroix, T. Tzfira and V. Citovsky, unpublished results). VirE3 was found to interact with AtKAPα in the yeast two-hybrid system, suggesting its nuclear import via the karyopherin α pathway. Importantly, VirE3 not only interacted with VirE2 in the yeast two-hybrid system, but also mediated its nuclear import in mammalian cells (B.L., T.T. and V.C., unpublished results), thus mimicking the VIP1 function in this heterologous nuclear import system.[86] Indeed, VirE3 complemented the VIP1 function in the nuclear import of VirE2 *in planta*, as was evidenced by its ability to direct GUS-tagged VirE2 into the cell nucleus in VIP1 antisense plants (B.L., T.T. and V.C., unpublished results). It was thus suggested that, because VIP1 is not an abundant cellular protein[116,117] and may represent one of the limiting factors for transformation,[116] *Agrobacterium* may have evolved to produce VirE3 and export it into the host cell to at least partially complement the cellular function of VIP1, necessary for the infection.

A Model for T-DNA Nuclear Import and Intranuclear Transport

Recent years have brought great progress in our understanding of the biological activities of VirD2 and VirE2, as well as in the identification of some of their cellular and bacterial interactors. A multitude of functional data has allowed developing a model for nuclear import and intranuclear transport of the *Agrobacterium* T-complex (Fig. 1B-D). The transport begins with the assembly or the entry (Fig. 1B) of the mature T-complex into the host cell cytoplasm. If the mature T-complexes form already within the bacterial cells, they are directly transported into the plant cell (b2). If, however, mature T-complexes are formed only in the host cell cytoplasm, VirE2-VirE1 complexes, as well as immature T-complexes, are transported independently into the host cell, possibly through the same VirB/VirD4 channel[52] (b1). In the latter case, before entering the cytoplasm, VirE2 must dissociate from VirE1 and attach to the VirD2-T-strand molecules to form mature T-complexes (b3). In parallel to the mature T-complex, two other *Agrobacterium* proteins are transported into the host cell cytoplasm through the same VirB/VirD4 channel, - VirE3 (b4) and VirF,[31,121] of which VirE3 most likely functions in the T-complex nuclear import (see below), and VirF probably acts in the cell nucleus to uncoat the T-complex prior to its integration (T.T. and V.C., unpublished results and ref.122).

Once inside the cytoplasm, the T-strand, shaped and protected by its chaperones VirD2 and VirE2, begins the journey to the host cell nucleus and its resident genome. During this voyage, VirE2, cooperatively coats the T-strand,[29] shapes it into a coiled filament[58] and protects it from cellular nucleases.[29] VirD2, on the other hand, acts as the T-complex pilot and guides it to the nuclear pore. Because of their very large size and rigid coiled shape,[58] T-complexes cannot move through the cytoplasm in a simple Brownian motion, let alone passively diffuse through the nuclear pore. It is unknown whether or not the T-complex undergoes structural changes as it travels through the cytoplasm, but this coiled "telephone cord"-like complex may stretch, thus reducing its outer diameter and facilitating the import process, once it arrives to the nuclear pore.

During transport, the T-complex bacterial chaperones, VirD2 and VirE2, presumably interact specifically with their respective cellular factors (Fig 1C, Y. Tao, P. Rao, and S. Gelvin, unpublished results and refs. 37, 90, 91), as well as with the *Agrobacterium* VirE3 protein

(B.L., T.T. and V.C., unpublished results). First, cellular cyclophilins may bind VirD2 (c1) to maintain its active conformation or perform some other, as yet unknown, functions as the T-complex moves in the cytoplasm, and/or during its nuclear import. Since cyclophilin binding does not involve the NLS region of VirD2,[91] it may occur concurrently with VirD2 interaction with AtKAPα (c2) (and/or other member of this protein family), which interacts directly with the carboxy-terminal NLS signal of VirD2 and mediates its nuclear import.[90] In animal and yeast cells, karyopherin α usually functions in a heterodimer with karyopherin β1. In this complex, the α subunit recognizes the NLS signal of the transported protein molecule, while the β subunit mediates docking of the entire NLS-karyopherin α/β complex at the nuclear pore and its interaction with Ran GTPase (reviewed in refs. 105, 123-126). It is still unclear whether karyopherin β1 is also involved in the nuclear import of VirD2 (c2) since, in a heterologous mammalian in vitro system, *Arabidopsis* karyopherin α has been shown to function alone, independently of karyopherin β1.[127] Nevertheless, since *Arabidopsis* karyopherin α carries the karyopherin β-binding motif, but does not contain sequences known to be required for binding to the nuclear pore or Ran,[90,128] it is highly likely that an as yet undiscovered karyopherin β1 is involved in VirD2 nuclear import. Indeed, in rice, karyopherin β1 has been isolated and shown to interact with karyopherin α,[129] and an *Arabidopsis* mutant, knocked-out in a putative karyopherin β gene, is resistant to *Agrobacterium* transformation.[111] Regardless of involvement of karyopherin β in VirD2 nuclear import, binding of AtKAPα to VirD2 at the 5'-end of the T-strand may orient the entire T-complex, initiating its directional nuclear import.

Since VirE2 does not bind directly to the host karyopherin α, but rather interacts with the host factor VIP1 which in turn binds to karyopherin α (c3), nuclear import of VirE2 and most of the T-complex molecule, may occur by a "piggyback" mechanism.[86] In this scenario, VIP1, which is a nuclear protein, interacts both with VirE2 and with the host karyopherin α.[116] Thus, nuclear import of VIP1 leads to nuclear import of its interacting VirE2 and, by implication, of the T-complex. Although the involvement of other, as yet unidentified, plant factors, such as an hypothetical VirE2-specific karyopherin α or β, in VirE2 nuclear import cannot be excluded, the ability of the *Agrobacterium* VirE3 protein to replace VIP1 (b4) further supports the crucial need for an adaptor molecule between VirE2 and the host cell nuclear import machinery during the T-complex nuclear import; because VIP1 is not an abundant cellular protein,[116,117] its function can be complemented, at least partially, by VirE3 (B.L., T.T. and V.C., unpublished results).

The combined action of AtKAPα and VIP1/VirE3 in the nuclear import of VirD2 and VirE2, respectively, further supports the notion of the T-complex polar translocation, since VirD2 and VirE2 do not directly interact with and compete for the same cellular proteins in their nuclear import. Polar translocation may be a common feature in the nuclear transport of many naturally occurring nucleic acid-protein complexes.[130] For example, nuclear export of a 75S premessenger ribonucleoprotein particle in *Chironomus tentans* initiates exclusively at the 5'-end of the RNA.[131]

Once inside the nucleus (Fig. 1D), perhaps even before the entire t-complex molecule has completely traversed the nuclear pore, the VirD2 NLS region may become dephosphorylated by PP2C (d1); this VirD2 dephosphorylation has been proposed to regulate its nuclear import (Y. Tao, P. Rao, and S. Gelvin, unpublished results). Within the cell nucleus, VirD2 may also interact with CAK2M and TBP (d2).[37] Because both CAK2M and TBP are members of the plant RNA transcription machinery,[37] their interactions with VirD2 may further guide the entire T-complex into the site of integration in the host chromosome.

Similar to VirD2, VirE2 also associates with a putative member of plant transcriptional complexes, VIP1 (d3). In addition to facilitating VirE2 nuclear import, VIP1 may also function in the intranuclear transport of the T-complex, leading it to chromosomal regions where the host

DNA is more exposed and, thus, more suitable for T-DNA integration. Here again, the combined, and noncompetitive action of VirD2 and VirE2, through their interaction with different host factors, may represent the molecular basis for the polar nature of T-DNA integration.

Future Directions

In this chapter, we describe the significant progress achieved in our understanding of nucleic acid transport during the *Agrobacterium*-host interaction. Major advances have been made in identifying host factors and their role in the nuclear import of VirD2, VirE2 and the T-complex. Further developments in this field will most likely come from the identification of additional cellular participants and regulatory components of the transport pathways. For example, plant molecular motors potentially involved in T-complex shuttling through the cell cytoplasm to the nuclear pore still need to be identified, and the mechanism by which the long T-complex molecule moves through the nuclear pore awaits better characterization. Perhaps the best way to achieve these goals is to combine biochemical, molecular, genetic, bio-physical and cell biological techniques. For example, identifying and characterizing plant mutants with altered susceptibility to *Agrobacterium* infection revealed the importance of two karyopherin proteins, as well as the possible involvement of the host cytoskeleton in the transformation process.[111] In addition, bio-physical studies, using reconstituted frog nuclei, revealed the role of dynein-like motors in the microtubule-based movement of in vitro-formed T-complexes toward the cell nucleus (M. Elbaum, personal communication). The importance of understanding the molecular mechanisms governing T-complex nuclear and intranuclear transport is difficult to overestimate. This fundamental knowledge will have a profound effect on our understanding of the general cellular mechanisms by which nucleic acids and proteins are imported into the nucleus and help us to design new strategies for the production of agronomically important plants resistant to *Agrobacterium*, and enable the development of improved genetic engineering procedures for the efficient nuclear delivery and integration of foreign genes.

Acknowledgments

The work in our laboratory is supported by grants from BRAD and HFSP to T.T. and grants from NIH, USDA, NSF, BARD, and BSF to V.C.

References

1. Gelvin SB. Agrobacterium and plant genes involved in T-DNA transfer and integration. Annu Rev Plant Physiol Plant Mol Biol 2000; 51:223-256.
2. Gelvin SB. Agrobacterium-mediated plant transformation: The biology behind the "gene-jockeying" tool. Microbiol Mol Biol Rev 2003; 67:16-37.
3. Tzfira T, Citovsky V. From host recognition to T-DNA integration: The function of bacterial and plant genes in the Agrobacterium-plant cell interaction. Mol Plant Pathol 2000; 1:201-212.
4. Tzfira T, Citovsky V. Partners-in-infection: Host proteins involved in the transformation of plant cells by Agrobacterium. Trends Cell Biol 2002; 12:121-129.
5. Zupan J, Muth TR, Draper O et al. The transfer of DNA from Agrobacterium tumefaciens into plants: A feast of fundamental insights. Plant J 2000; 23:11-28.
6. de Cleene M, de Ley J. The host range of crown gall. Bot Rev 1976; 42:389-466.
7. Piers KL, Heath JD, Liang X et al. Agrobacterium tumefaciens-mediated transformation of yeast. Proc Natl Acad Sci USA 1996; 93:1613-1618.
8. de Groot MJ, Bundock P, Hooykaas PJJ et al. Agrobacterium tumefaciens-mediated transformation of filamentous fungi [published erratum appears in Nat Biotechnol 1998; 16:1074]. Nat Biotechnol 1998; 16:839-842.
9. Gouka RJ, Gerk C, Hooykaas PJ et al. Transformation of Aspergillus awamori by Agrobacterium tumefaciens-mediated homologous recombination. Nat Biotechnol 1999; 17:598-601.

10. Kunik T, Tzfira T, Kapulnik Y et al. Genetic transformation of HeLa cells by Agrobacterium. Proc Natl Acad Sci USA 2001; 98:1871-1876.

11. Gelvin S. Improving plant genetic engineering by manipulating the host. Trends Biotechnol 2003; 21:95-98.

12. Peralta EG, Ream LW. T-DNA border sequences required for crown gall tumorigenesis. Proc Natl Acad Sci USA 1985; 82:5112-5116.

13. Wang K, Stachel SE, Timmerman B et al. Site-specific nick occurs within the 25 bp transfer promoting border sequence following induction of vir gene expression in Agrobacterium tumefaciens. Science 1987; 235:587-591.

14. Armitage P, Walden R, Draper J. Plant genetic transformation and gene expression. In: Draper J, Scott R, Armitage P et al, eds. A Laboratory Manual. London: Blackwell Scientific Publications Ltd., 1988:1-67.

15. Draper J, Scott R, Armitage P et al. Plant genetic transformation and gene expression. A Laboratory Manual. London: Blackwell Scientific Publications Ltd., 1988.

16. Lee YW, Jin S, Sim WS et al. The sensing of plant signal molecules by Agrobacterium: genetic evidence for direct recognition of phenolic inducers by the VirA protein. Gene 1996; 179:83-88.

17. McLean BG, Greene EA, Zambryski PC. Mutants of Agrobacterium VirA that activate vir gene expression in the absence of the inducer acetosyringone. J Biol Chem 1994; 269:2645-2651.

18. Stachel SE, Messens E, Van Montagu M et al. Identification of the signal molecules produced by wounded plant cell that activate T-DNA transfer in Agrobacterium tumefaciens. Nature 1985; 318:624-629.

19. Stachel SE, Nester EW, Zambryski PC. A plant cell factor induces Agrobacterium tumefaciens vir gene expression. Proc Natl Acad Sci USA 1986; 83:379-383.

20. Turk SC, van Lange RP, Regensburg-Tuink TJG et al. Localization of the VirA domain involved in acetosyringone-mediated vir gene induction in Agrobacterium tumefaciens. Plant Mol Biol 1994; 25:899-907.

21. Durrenberger F, Crameri A, Hohn B et al. Covalently bound VirD2 protein of Agrobacterium tumefaciens protects the T-DNA from exonucleolytic degradation. Proc Natl Acad Sci USA 1989; 86:9154-9158.

22. Jasper F, Koncz C, Schell J et al. Agrobacterium T-strand production in vitro: sequence-specific cleavage and 5' protection of single-stranded DNA templates by purified VirD2 protein. Proc Natl Acad Sci USA 1994; 91:694-698.

23. Pansegrau W, Schoumacher F, Hohn B et al. Site-specific cleavage and joining of single-stranded DNA by VirD2 protein of Agrobacterium tumefaciens Ti plasmids: Analogy to bacterial conjugation. Proc Natl Acad Sci USA 1993; 90:11538-11542.

24. Relic B, Andjelkovic M, Rossi L et al. Interaction of the DNA modifying proteins VirD1 and VirD2 of Agrobacterium tumefaciens: Analysis by subcellular localization in mammalian cells. Proc Natl Acad Sci USA 1998; 95:9105-9110.

25. Scheiffele P, Pansegrau W, Lanka E. Initiation of Agrobacterium tumefaciens T-DNA processing: Purified proteins VirD1 and VirD2 catalyze site- and strand-specific cleavage of superhelical T-border DNA in vitro. J Biol Chem 1995; 270:1269-1276.

26. Ward E, Barnes W. VirD2 protein of Agrobacterium tumefaciens very tightly linked to the 5' end of T-strand DNA. Science 1988; 242:927-930.

27. Young C, Nester EW. Association of the VirD2 protein with the 5' end of T-strands in Agrobacterium tumefaciens. J Bacteriol 1988; 170:3367-3374.

28. Christie PJ, Ward JE, Winans SC et al. The Agrobacterium tumefaciens virE2 gene product is a single-stranded-DNA-binding protein that associates with T-DNA. J Bacteriol 1988; 170:2659-2667.

29. Citovsky V, Wong ML, Zambryski PC. Cooperative interaction of Agrobacterium VirE2 protein with single stranded DNA: Implications for the T-DNA transfer process. Proc Natl Acad Sci USA 1989; 86:1193-1197.

30. Sen P, Pazour GJ, Anderson D et al. Cooperative binding of Agrobacterium tumefaciens VirE2 protein to single-stranded DNA. J Bacteriol 1989; 171:2573-80.

31. Vergunst AC, Schrammeijer B, den Dulk-Ras A et al. VirB/D4-dependent protein translocation from Agrobacterium into plant cells. Science 2000; 290:979-982.

32. Zupan J, Ward D, Zambryski PC. Assembly of the VirB transport complex for DNA transfer from Agrobacterium tumefaciens to plant cells. Curr Opin Microbiol 1998; 1:649-655.
33. Gheysen G, Villarroel R, Van Montagu M. Illegitimate recombination in plants: a model for T-DNA integration. Genes Dev 1991; 5:287-297.
34. Mayerhofer R, Koncz-Kalman Z, Nawrath C et al. T-DNA integration: a mode of illegitimate recombination in plants. EMBO J 1991; 10:697-704.
35. Takano M, Egawa H, Ikeda JE et al. The structures of integration sites in transgenic rice. Plant J 1997; 11:353-361.
36. Tinland B. The integration of T-DNA into plant genomes. Trends Plant Sci 1996; 1:178-184.
37. Bakó L, Umeda M, Tiburcio AF et al. The VirD2 pilot protein of Agrobacterium-transferred DNA interacts with the TATA box-binding protein and a nuclear protein kinase in plants. Proc Natl Acad Sci USA 2003; 100:10108-10113.
38. Mysore KS, Bassuner B, Deng XB et al. Role of the Agrobacterium tumefaciens VirD2 protein in T-DNA transfer and integration. Mol Plant-Microbe Interact 1998; 11:668-683.
39. Dombek P, Ream LW. Functional domains of Agrobacterium tumefaciens single-stranded DNA-binding protein VirE2. J Bacteriol 1997; 179:1165-1173.
40. Simone M, McCullen CA, Stahl LE et al. The carboxy-terminus of VirE2 from Agrobacterium tumefaciens is required for its transport to host cells by the virB-encoded type IV transport system. Mol Microbiol 2001; 41:1283-1293.
41. Zhou XR, Christie PJ. Mutagenesis of the Agrobacterium VirE2 single-stranded DNA-binding protein identifies regions required for self-association and interaction with VirE1 and a permissive site for hybrid protein construction. J Bacteriol 1999; 181:4342-4352.
42. Vergunst AC, van Lier MCM, den Dulk-Ras A et al. Recognition of the Agrobacterium VirE2 translocation signal by the VirB/D4 transport system does not require VirE1. Plant Physiol 2003; 133:978-988.
43. Frary A, Hamilton CM. Efficiency and stability of high molecular weight DNA transformation: An analysis in tomato. Transgenic Res 2001; 10:121-132.
44. Hamilton CM. A binary-BAC system for plant transformation with high-molecular-weight DNA. Gene 1997; 200:107-16.
45. Kononov ME, Bassuner B, Gelvin SB. Integration of T-DNA binary vector 'backbone' sequences into the tobacco genome: Evidence for multiple complex patterns of integration. Plant J 1997; 11:945-57.
46. Ramanathan V, Veluthambi K. Transfer of nonT-DNA portions of the Agrobacterium tumefaciens Ti plasmid pTiA6 from the left terminus of TL-DNA. Plant Mol Biol 1995; 28:1149-54.
47. Wenck A, Czako M, Kanevski I et al. Frequent collinear long transfer of DNA inclusive of the whole binary vector during Agrobacterium-mediated transformation. Plant Mol Biol 1997; 34:913-22.
48. Citovsky V, Zupan J, Warnick D et al. Nuclear localization of Agrobacterium VirE2 protein in plant cells. Science 1992; 256:1802-1805.
49. Gelvin SB. Agrobacterium VirE2 proteins can form a complex with T strands in the plant cytoplasm. J Bacteriol 1998; 180:4300-4302.
50. Sundberg C, Meek L, Carrol K et al. VirE1 protein mediates export of single-stranded DNA binding protein VirE2 from Agrobacterium tumefaciens into plant cells. J Bacteriol 1996; 178:1207-1212.
51. Ward DV, Zambryski PC. The six functions of Agrobacterium VirE2. Proc Natl Acad Sci USA 2001; 98:385-386.
52. Binns AN, Beaupre CE, Dale EM. Inhibition of VirB-mediated transfer of diverse substrates from Agrobacterium tumefaciens by the InQ plasmid RSF1010. J Bacteriol 1995; 177:4890-4899.
53. Otten L, DeGreve H, Leemans J et al. Restoration of virulence of vir region mutants of A. tumefaciens strain B6S3 by coinfection with normal and mutant Agrobacterium strains. Mol Gen Genet 1984; 195:159-163.
54. Lee LY, Gelvin SB, Kado CI. pSa causes oncogenic suppression of Agrobacterium by inhibiting VirE2 protein export. J Bacteriol 1999; 181:186-196.
55. Deng W, Chen L, Peng WT et al. VirE1 is a specific molecular chaperone for the exported single-stranded-DNA-binding protein VirE2 in Agrobacterium. Mol Microbiol 1999; 31:1795-1807.

56. Sundberg CD, Ream LW. The Agrobacterium tumefaciens chaperone-like protein, VirE1, interacts with VirE2 at domains required for single-stranded DNA binding and cooperative interaction. J Bacteriol 1999; 181:6850-6855.

57. Yusibov VM, Steck TR, Gupta V et al. Association of single-stranded transferred DNA from Agrobacterium tumefaciens with tobacco cells. Proc Natl Acad Sci USA 1994; 91:2994-2998.

58. Citovsky V, Guralnick B, Simon MN et al. The molecular structure of Agrobacterium VirE2-single stranded DNA complexes involved in nuclear import. J Mol Biol 1997; 271:718-727.

59. Hamilton CM, Frary AC, Lewis C et al. Stable transfer of intact high molecular weight DNA into plant chromosomes. Proc Natl Acad Sci USA 1996; 93:9975-9979.

60. Forbes DJ. Structure and function of the nuclear pore complex. Annu Rev Cell Biol 1992; 8:495-527.

61. Dworetzky SI, Feldherr CM. Translocation of RNA-coated gold particles through the nuclear pores of oocytes. J Cell Biol 1988; 106:575-584.

62. Pante N, Kann M. Nuclear pore complex is able to transport macromolecules with diameters of about 39 nm. Mol Biol Cell 2002; 13:425-434.

63. Briels WJ. The theory of polymer dynamics. Oxford: Clarendon Press, 1986.

64. Landau LD, Lifshitz EM. Statistical Physics. Oxford: Pergamon Press, 1980.

65. Fahrenkrog B, Aebi U. The nuclear pore complex: Nucleocytoplasmic transport and beyond. Nat Rev Mol Cell Biol 2003; 4:757-66.

66. Suntharalingam M, Wente SR. Peering through the pore. Nuclear pore complex structure, assembly, and function. Dev Cell 2003; 4:775-789.

67. Zambryski PC. Chronicles from the Agrobacterium-plant cell DNA transfer story. Annu Rev Plant Physiol Plant Mol Biol 1992; 43:465-490.

68. Howard E, Zupan J, Citovsky V et al. The VirD2 protein of A. tumefaciens contains a C-terminal bipartite nuclear localization signal: Implications for nuclear uptake of DNA in plant cells. Cell 1992; 68:109-118.

69. Tinland B, Schoumacher F, Gloeckler V et al. The Agrobacterium tumefaciens virulence D2 protein is responsible for precise integration of T-DNA into the plant genome. EMBO J 1995; 14:3585-3595.

70. Ziemienowicz A, Merkle T, Schoumacher F et al. Import of Agrobacterium T-DNA into plant nuclei: Two distinct functions of VirD2 and VirE2 proteins. Plant Cell 2001; 13:369-384.

71. Citovsky V, Warnick D, Zambryski PC. Nuclear import of Agrobacterium VirD2 and VirE2 proteins in maize and tobacco. Proc Natl Acad Sci USA 1994; 91:3210-3214.

72. Herrera-Estrella A, Van Montagu M, Wang K. A bacterial peptide acting as a plant nuclear targeting signal: The amino-terminal portion of Agrobacterium VirD2 protein directs a beta-galactosidase fusion protein into tobacco nuclei. Proc Natl Acad Sci USA 1990; 87:9534-9537.

73. Rossi L, Hohn B, Tinland B. The VirD2 protein of Agrobacterium tumefaciens carries nuclear localization signals important for transfer of T-DNA to plant. Mol Gen Genet 1993; 239:345-353.

74. Tinland B, Koukolikova-Nicola Z, Hall MN et al. The T-DNA-linked VirD2 protein contains two distinct nuclear localization signals. Proc Natl Acad Sci USA 1992; 89:7442-7446.

75. Tzfira T, Citovsky V. Comparison between nuclear import of nopaline- and octopine-specific VirE2 protein of Agrobacterium in plant and animal cells. Mol Plant Pathol 2001; 2:171-176.

76. Koukolikova-Nicola Z, Raineri D, Stephens K et al. Genetic analysis of the virD operon of Agrobacterium tumefaciens: A search for functions involved in transport of T-DNA into the plant cell nucleus and in T-DNA integration. J Bacteriol 1993; 175:723-731.

77. Shurvinton CE, Hodges L, Ream LW. A nuclear localization signal and the C-terminal omega sequence in the Agrobacterium tumefaciens VirD2 endonuclease are important for tumor formation. Proc Natl Acad Sci USA 1992; 89:11837-11841.

78. Narasimhulu SB, Deng X-B, Sarria R et al. Early transcription of Agrobacterium T-DNA genes in tobacco and maize. Plant Cell 1996; 8:873-886.

79. Zupan J, Citovsky V, Zambryski PC. Agrobacterium VirE2 protein mediates nuclear uptake of ssDNA in plant cells. Proc Natl Acad Sci USA 1996; 93:2392-2397.

80. Sheng J, Citovsky V. Agrobacterium-plant cell interaction: Have virulence proteins - will travel. Plant Cell 1996; 8:1699-1710.
81. Tinland B, Hohn B. Recombination between prokaryotic and eukaryotic DNA: integration of Agrobacterium tumefaciens T-DNA into the plant genome. Genet Eng 1995; 17:209-229.
82. Windels P, De Buck S, Van Bockstaele E et al. T-DNA integration in Arabidopsis chromosomes. Presence and origin of filler DNA sequences. Plant Physiol 2003; 133:2061-2068.
83. Guralnick B, Thomsen G, Citovsky V. Transport of DNA into the nuclei of Xenopus oocytes by a modified VirE2 protein of Agrobacterium. Plant Cell 1996; 8:363-373.
84. Ziemienowicz A, Görlich D, Lanka E et al. Import of DNA into mammalian nuclei by proteins originating from a plant pathogenic bacterium. Proc Natl Acad Sci USA 1999; 96:3729-3733.
85. Rhee Y, Gurel F, Gafni Y et al. A genetic system for detection of protein nuclear import and export. Nat Biotechnol 2000; 18:433-437.
86. Tzfira T, Vaidya M, Citovsky V. VIP1, an Arabidopsis protein that interacts with Agrobacterium VirE2, is involved in VirE2 nuclear import and Agrobacterium infectivity. EMBO J 2001; 20:3596-3607.
87. Williams RC, Spengler SJ. Fibers of RecA protein and complexes of RecA protein and single stranded *f*X174 DNA as visualized by negative-stain electron microscopy. J Mol Biol 1986; 192:110-118.
88. Flory J, Radding CM. Visualization of recA protein and its association with DNA: A priming effect of single-strand-binding protein. Cell 1982; 28:747-56.
89. Flory J, Tsang SS, Muniyappa K. Isolation and visualization of active presynaptic filaments of recA protein and single-stranded DNA. Proc Natl Acad Sci USA 1984; 81:7026-30.
90. Ballas N, Citovsky V. Nuclear localization signal binding protein from Arabidopsis mediates nuclear import of Agrobacterium VirD2 protein. Proc Natl Acad Sci USA 1997; 94:10723-10728.
91. Deng W, Chen L, Wood DW et al. Agrobacterium VirD2 protein interacts with plant host cyclophilins. Proc Natl Acad Sci USA 1998; 95:7040-7045.
92. Fields S, Song O-K. A novel genetic system to detect protein-protein interactions. Nature 1989; 340:245-246.
93. Hollenberg SM, Sternglanz R, Cheng PF et al. Identification of a new family of tissue-specific basic helix-loop-helix proteins with a two-hybrid system. Mol Cell Biol 1995; 15:3813-3822.
94. Duina AA, Chang HC, Marsh JA et al. A cyclophilin function in Hsp90-dependent signal transduction. Science 1996; 274:1713-1715.
95. Fischer G, Wittmann-Liebold B, Lang K et al. Cyclophilin and peptidyl-prolyl cis-trans isomerase are probably identical proteins. Nature 1989; 337:476-478.
96. Hayman GT, Miernyk JA. The nucleotide and deduced amino acid sequences of a peptidyl-prolyl cis-trans isomerase from Arabidopsis thaliana. Biochim Biophys Acta 1994; 1219:536-538.
97. Hunter T. Prolyl isomerases and nuclear function. Cell 1998; 92:141-143.
98. Lippuner V, Chou IT, Scott SV et al. Cloning and characterization of chloroplast and cytosolic forms of cyclophilin from Arabidopsis thaliana. J Biol Chem 1994; 269:7863-7868.
99. Marks AR. Cellular functions of immunophilins. Physiol Rev 1996; 76:631-649.
100. Takahashi N, Hayano T, Suzuki M. Peptidyl-prolyl cis-trans isomerase is the cyclosporin A-binding protein cyclophilin. Nature 1989; 337:473-475.
101. Handschumacher RE, Harding MW, Rice J et al. Cyclophilin: A specific cytosolic binding protein for cyclosporin A. Science 1984; 226:544-547.
102. Baker EK, Colley NJ, Zuker CS. The cyclophilin homolog NinaA functions as a chaperone, forming a stable complex in vivo with its protein target rhodopsin. EMBO J 1994; 13:4886-4895.
103. Meyer K, Leube MP, Grill E. A protein phosphatase 2C involved in ABA signal transduction in Arabidopsis thaliana. Science 1994; 264:1452 - 1455.
104. Leung J, Bouvier-Durand M, Morris P-C et al. Arabidopsis ABA response gene ABI1: Features of a calcium-modulated protein phosphatase. Science 1994; 264:1448-1452.
105. Nigg EA. Nucleocytoplasmic transport: Signals, mechanisms and regulation. Nature 1997; 386:779-787.

106. Powers MA, Forbes DJ. Cytosolic factors in nuclear import: what's importin? Cell 1994; 79:931-934.

107. Peifer M, Berg S, Reynolds AB. A repeating amino acid motif shared by proteins with diverse cellular roles. Cell 1994; 76:789-791.

108. Görlich D, Mattaj IW. Nucleocytoplasmic transport. Science 1996; 271:1513-1518.

109. Loeb JDJ, Schlenstedt G, Pellman D et al. The yeast nuclear import receptor is required for mitosis. Proc Natl Acad Sci USA 1995; 92:7647-7651.

110. Schlenstedt G, Hurt E, Doye V et al. Reconstitution of nuclear protein transport with semi-intact yeast cells. J Cell Biol 1993; 123:785-798.

111. Zhu Y, Nam J, Humara JM et al. Identification of Arabidopsis rat mutants. Plant Physiol 2003; 132:494-505.

112. Coin F, Frit P, Viollet B et al. TATA binding protein discriminates between different lesions on DNA, resulting in a transcription decrease. Mol Biol Cell 1998; 18:3907-3914.

113. Vichi P, Coin F, Renaud JP et al. Cisplatin- and UV-damaged DNA lure the basal transcription factor TFIID/TBP. Embo J 1997; 16:7444-56.

114. Balajee AS, Bohr VA. Genomic heterogeneity of nucleotide excision repair. Gene 2000; 250:15-30.

115. van der Krol AR, Chua N-H. The basic domain of plant B-ZIP proteins facilitates import of a reporter protein into plant nuclei. Plant Cell 1991; 3:667-675.

116. Tzfira T, Vaidya M, Citovsky V. Increasing plant susceptibility to Agrobacterium infection by overexpression of the Arabidopsis VIP1 gene. Proc Natl Acad Sci USA 2002; 99:10435-10440.

117. Avivi Y, Morad V, Ben-Meir H et al. Reorganization of specific chromosomal domains and activation of silent genes in plant cell acquiring pluripotentiality. Dev Dyn 2004; in press.

118. Newell CA. Plant transformation technology. Developments and applications. Mol Biotechnol 2000; 16:53-65.

119. Rakoczy-Trojanowska M. Alternative methods of plant transformation—a short review. Cell Mol Biol Lett 2002; 7:849-58.

120. Pemberton LF, Rosenblum JS, Blobel G. Nuclear import of the TATA-binding protein: mediation by the karyopherin Kap114p and a possible mechanism for intranuclear targeting. J Cell Biol 1999; 145:1407-1417.

121. Schrammeijer B, Dulk-Ras Ad A, Vergunst AC et al. Analysis of Vir protein translocation from Agrobacterium tumefaciens using Saccharomyces cerevisiae as a model: Evidence for transport of a novel effector protein VirE3. Nucleic Acids Res 2003; 31:860-868.

122. Schrammeijer B, Risseeuw E, Pansegrau W et al. Interaction of the virulence protein VirF of Agrobacterium tumefaciens with plant homologs of the yeast Skp1 protein. Curr Biol 2001; 11:258-262.

123. Chook YM, Blobel G. Karyopherins and nuclear import. Curr Opin Struct Biol 2001; 11:703-715.

124. Komeili A, O'Shea EK. New perspectives on nuclear transport. Annu Rev Genet 2001; 35:341-364.

125. Marte B. Passage through the nuclear pore. Nat Cell Biol 2001; 3:E135.

126. Quimby BB, Corbett AH. Nuclear transport mechanisms. Cell Mol Life Sci 2001; 58:1776-1773.

127. Hubner S, Smith HM, Hu W et al. Plant importin alpha binds nuclear localization sequences with high affinity and can mediate nuclear import independent of importin beta. J Biol Chem 1999; 274:22610-22617.

128. Hicks GR, Smith HMS, Lobreaux S et al. Nuclear import in permeabilized protoplasts from higher plants has unique features. Plant Cell 1996; 8:1337-1352.

129. Matsuki R, Iwasaki T, Shoji K et al. Isolation and characterization of two importin-beta genes from rice. Plant Cell Physiol 1998; 39:879-884.

130. Citovsky V, Zambryski PC. Transport of nucleic acids through membrane channels: Snaking through small holes. Annu Rev Microbiol 1993; 47:167-197.

131. Mehlin H, Daneholt B, Skoglund U. Translocation of a specific premessenger ribonucleoprotein particle through the nuclear pore studied with electron microscope tomography. Cell 1992; 69:605-613.

Regulation of Nuclear Import and Export of Proteins in Plants and Its Role in Light Signal Transduction

Stefan Kircher, Thomas Merkle, Eberhard Schäfer and Ferenc Nagy

The nuclear envelope separates the theatres of two major cellular processes in eukaryotes: transcription takes place in the nucleus whereas proteins are synthesized in the cytoplasm. The localization of these processes in two different compartments of the cell implies that macromolecules must be exchanged very rapidly and efficiently between the nucleus and the cytoplasm in order to ensure proper regulation of signaling and metabolism of a living cell.

All transport processes across the nuclear envelope take place at very large multi-protein complexes called nuclear pore complexes (NPCs), which provide the gates between the nucleus and the cytoplasm. It has been known for many years that nucleo-cytoplasmic transport processes of macromolecules are receptor-mediated, but only in the last few years was it revealed that most of the nuclear transport processes depend on importin β-like protein receptors.[1,2] The genes encoding these receptors constitute a small gene family, and they are named after the member that was first identified at the molecular level. These receptor proteins share limited sequence identity, they can interact with the regulatory GTPase Ran in its GTP-bound form, they are able to interact with nucleoporins, and they shuttle continuously between the nucleus and the cytoplasm.[3] For every class of the many different macromolecular transport cargos, like proteins, mRNA, tRNA, ribosomal subunits, snRNPs, there exists a specific receptor, or a combination of receptors.

Nuclear Import of Proteins

Karyophilic proteins that contain a classical basic nuclear localization signal (NLS), whether monopartite (SV40-like) or bipartite, are imported into the nucleus by the import receptor importin β (also called karyopherin β; see Fig. 1A). However, importin β does not bind directly to these import substrates. In these cases, importin α (also called karyopherin α) serves as an adapter between the cargo protein and the nuclear import receptor itself.[4] Therefore, the first step in the nuclear import of a NLS-containing protein is the specific recognition of the NLS in the cytoplasm by importin α, which constitutes the soluble NLS receptor.[5] Importin β binds co-operatively to the basic amino terminus of importin α called importin β-binding (IBB) domain.[6,7] This leads to the formation of the triple import complex consisting of importin α, importin β and the NLS-containing protein, which then docks as a single entity

---

Nuclear Import and Export in Plants and Animals, edited by Tzvi Tzfira and Vitaly Citovsky.
©2005 Eurekah.com and Kluwer Academic / Plenum Publishers.

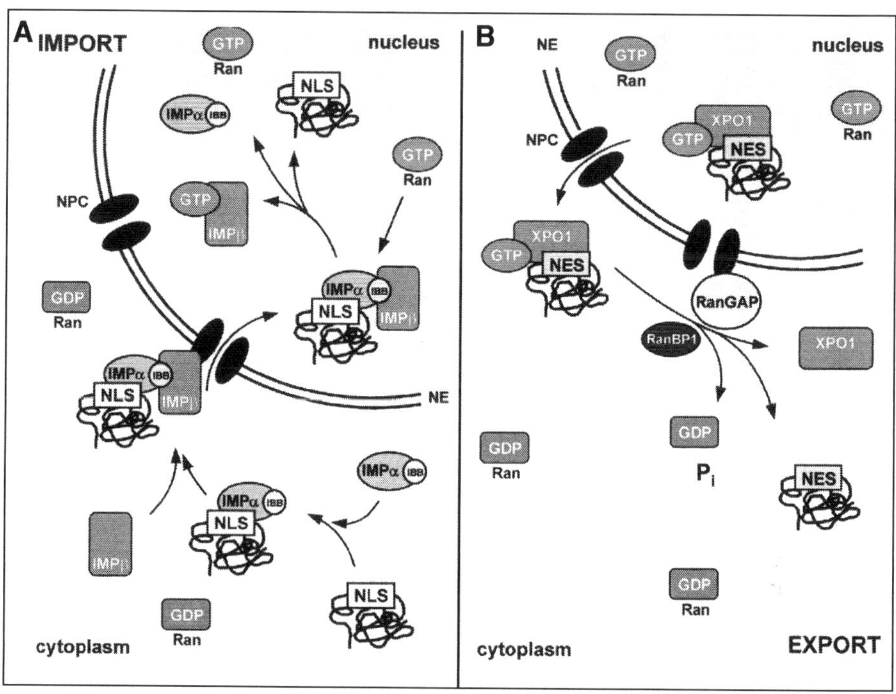

Figure 1. Comparison of nuclear import and nuclear export of proteins. A) Nuclear import of NLS-containing proteins by importin α/ importin β (IMPα/IMPβ) is initiated by the specific recognition of the NLS within a protein by the NLS receptor importin α in the cytoplasm. The nuclear import receptor importin β binds co-operatively to the importin α/cargo protein complex to form the triple import complex that docks to the nuclear pore complex (NPC) via importin β. After translocation into the nucleus, it is dissociated by the interaction of Ran-GTP with importin β, leading to the release of the import cargo protein into the nucleoplasm. B) Nuclear export of proteins that contain a leucine-rich NES is accomplished by the export receptor exportin 1 (XPO1), which directly and specifically binds to the NES and to Ran-GTP in a co-operative manner in the nucleoplasm. After translocation of this triple export complex through the NPC into the cytoplasm, the co-ordinated action of the cytosolic proteins RanBP1 and RanGAP leads to the dissociation of the export complex and to the hydrolysis of GTP on Ran and hence to the release of the export cargo into the cytoplasm.

to the NPC via importin β and is subsequently translocated into the nucleus. Within the nucleus, the concentration of Ran-GTP is high.[8] Since the binding sites on importin β for Ran-GTP and importin α overlap, Ran-GTP is competing with importin α for the binding to importin β, which leads to the dissociation of the triple import complex and hence to the release of the import cargo in the nucleus.[9] Importin β is recycled to the cytoplasm in complex with Ran-GTP, although there is also evidence that it may leave the nucleus in a Ran-independent manner.[10] After hydrolysis of GTP on Ran in the cytoplasm, Importin β is ready for a new import cycle. Importin α is exported to the cytoplasm by a specific export receptor, termed CAS, which is also a member of the importin β family.[11-14]

Although most of the proteins that are imported by importin β depend on recognition by importin α, there are also proteins that do not need this adapter. These proteins bind directly to the import receptor importin β and contain a more archaical nuclear localization signal that resembles the IBB domain of importin α. There are also many other proteins that are imported into the nucleus and that contain neither such an import signal nor a classical basic NLS (for a review see refs. 1, 2). Nuclear import of these proteins depends on other receptors, which in most cases are also members of the importin β family.

Nuclear Export of Proteins

Nuclear export of proteins was first investigated in the case of the inhibitor (PKI) of the cAMP-dependent protein kinase (PKA) in humans and in the case of the HIV protein Rev.[15,16] Like in the case of nuclear import, these proteins contain short signals with a low degree of primary sequence conservation, which are permanent and transferable. These nuclear export signals (NES) are both sufficient and necessary to label a protein for rapid nuclear export and they show a specific spacing of long-chain, hydrophobic amino acid residues, often leucines. These leucine-rich NESs are specifically recognized in the nucleus by another member of the importin β family of transport receptors, termed exportin 1 (XPO1, also called CRM1).[17-20] Here, the export receptor XPO1 binds its substrate directly and co-operatively with the GTPase Ran in its GTP-bound form (see Fig. 1B), the concentration of which is high in the nucleus. This triple export complex next interacts with nucleoporins on the nuclear side of the NPC via XPO1 and is subsequently translocated into the cytoplasm.[21,22] There, the export complex is dissociated by the co-ordinated action of two cytosolic regulatory proteins,[23,24] the Ran-binding protein 1 (RanBP1) and the GTPase-activating protein for Ran (RanGAP1). This leads to the release of the export cargo into the cytoplasm, and the hydrolysis of GTP on Ran renders this step irreversible. The export receptor XPO1 re-enters the nucleus on its own, due to its ability to interact with nucleoporins, and is ready for a new export cycle. Although this is the best-investigated scenario for the nuclear export of proteins, the leucine-rich NESs are not the only signals that confer nuclear export, as exportin 1 is not the only export receptor (for review see ref. 2).

The Regulatory GTPase Ran

The above description of the basic steps of nuclear import and export of proteins already demonstrated that the GTPase Ran plays a central role in the regulation of the directionality of nuclear transport processes that depend upon the nuclear transport receptors of the importin β family. Ran is a remarkable protein since it is the only Ras-like GTPase which is soluble and which shuttles continuously between two cellular compartments.[25] In addition, its GTPase cycle is distributed in a characteristically asymmetric fashion between the nuclear and the cytoplasmic compartments,[8] since the proteins regulating the Ran GTPase cycle show very specific localizations. The guanine nucleotide-exchange factor for Ran, called RCC1 in humans, is bound to chromatin[26] and therefore shows a strictly nuclear localization. In contrast, the GTPase-activating protein for Ran (RanGAP1) and its co-activator RanBP1 are confined to the cytoplasm.[27,28] As a consequence, the concentration of Ran-GTP is high in the nucleus and low in the cytoplasm (see Fig. 1).

Based on these characteristics of the Ran GTPase cycle, Ran is able to provide directionality to nuclear transport processes since the importin β-like transport receptors interact with Ran-GTP and their substrates in a specific manner, depending on whether they are import receptors or export receptors.[29,30] Importins, like importin β, bind to their substrates only in the absence of Ran-GTP, whereas exportins, like XPO1 and CAS, bind to their substrates only in presence of and co-operatively with Ran-GTP. This explains why import complexes form only in the cytoplasm and are dissociated in the nucleus, while export receptors bind to their cargo only in the nucleus and release it in the cytoplasm.

Shuttling of Ran between the nucleus and the cytoplasm is mainly accomplished on one hand by nuclear export of Ran-GTP complexed with an importin β-like transport receptor, as part of an export complex. On the other hand, nuclear import of Ran is ensured by a specific import receptor for Ran-GDP termed NTF2.[31-33] The latter is an example for receptor-mediated nuclear import which does not depend upon importin β-like transport receptors. Recently, the NTF2-related export protein 1 (NXT1) was identified as a nuclear transport factor that continuously shuttles between the nucleus and the cytoplasm. In contrast to NTF2, NXT1 binds Ran-GTP and regulates both XPO1-dependent and XPO1-independent nuclear export processes.[34-36] In this way, NXT1 may also contribute to nuclear export of Ran. Hydrolysis of GTP on Ran in the cytoplasm by the regulatory proteins RanBP1 and RanGAP1 is thought to ensure completeness of the Ran GTPase cycle and hence to ensure recycling all factors necessary for nuclear transport processes rather than to directly provide energy to these processes. It has been shown that GTP hydrolysis on Ran is not necessary for translocation across the NPC.[37,38] Interestingly, Ran seems to be a multi-functional protein since, in addition to its central role in regulating the directionality of nuclear transport processes, Ran-GTP also plays a major role in regulating spindle-formation during mitosis in mammalian cells.[39-42] It has been reported that the mitotic role of Ran is largely mediated by importin β, which inhibits spindle formation and sequesters protein factors required for an aster promoting activity.[43,44] In addition, Gruss et al[45] demonstrated that importin α binds and thereby inactivates a microtubule-associated protein (TPX2) that is required for spindle formation. TPX2 is displaced from importin α by Ran-GTP. Thus, Ran-GTP functions by locally releasing protein cargoes from nuclear transport factors, which serve to regulate spindle formation in mitosis.

Plant Factors and Plant-Specific Features of Nuclear Transport

At least six genes encoding importin α homologues have been described in *Arabidopsis thaliana*.[46,47] One importin α protein, AtIMP alpha, was shown to be able to bind to three different classes of nuclear import signals that are present in plants,[48] another importin α protein was shown to bind to *Agrobacterium* VirD2.[49] In addition, genes encoding importin α homologues have also been isolated from rice,[50] *Capsicum* and tomato, whereas genes encoding importin β homologues have been characterized only in rice.[51] Recently, the nuclear export receptor XPO1 that specifically binds to leucine-rich NESs has been functionally characterized in *Arabidopsis*.[52] The *Arabidopsis* protein shares 42-50% identity with its functional human, *S. cerevisiae* and *S. pombe* homologues, it interacts with *Arabidopsis* Ran and with NESs of plant and human proteins including the HIV protein Rev, and is sensitive to the cytotoxin leptomycin B. Export activity within a plant cell was demonstrated in vivo using an assay system with GFP fusion proteins, which also revealed that the Rev NES is fully functional in plants, thereby demonstrating the high conservation of this nuclear export pathway between species phyla and even kingdoms.[52]

Genes encoding the GTPase Ran have been isolated from several plant species, including 5 genes from tobacco and 3 genes from *Arabidopsis* (for a review see ref. 46). In addition, three genes encoding the Ran-specific regulator RanBP1 have been isolated from *Arabidopsis*, two of them by a yeast two-hybrid screen with a Ran mutant that is permanently blocked in its GTP-bound form.[53,54]

Although analysis of plant nuclear transport factors as well as nuclear import and export in plants has confirmed that the basic processes are highly conserved between organisms, there are some plant-specific features. The development of in vitro nuclear import systems for plants has revealed that nuclear import is not inhibited at 4°C, as in animal cells.[55,56] Also, in contrast to animal nuclear import, the lectin wheat germ agglutinin does not block this process in plants. Interestingly, At-IMPα, one of the plant importin α proteins, was reported to be able to function as nuclear import receptor without binding to importin β.[57] In addition, importin-α has been co-localized with elements of the cytoskeleton in plant cells, suggesting an implication of these structural elements in nuclear import.[58] It is not known to date, whether the latter finding is unique to plants or whether this property of importin α is also shared in other organisms.

Regulation of Nuclear Transport As a Tool to Regulate Signaling

Taken together, a cell has to invest plenty of energy to guarantee continuous and rapid exchange of macromolecules between the nucleus and the cytoplasm. All the proteins involved, from nucleoporins to nuclear transport receptors including their regulatory proteins, must be produced in great numbers. In addition, energy in the form of GTP is consumed during each and every transport cycle. However, the nucleo-cytoplasmic transport of macromolecules also provides numerous possibilities for the regulation of signal transduction processes. Not only the activity of specific factors but also their localization in a specific compartment may be subject to regulation.

Since the first step of the nuclear import of a NLS-containing protein is its binding to importin α, interference with this recognition step provides a perfect checkpoint for the regulation of the localization of such a protein. If importin α is unable to interact with the NLS of a protein, the protein in question will stay cytosolic. There are many ways to interfere with the binding of importin α to a NLS, like protein modifications such as phosphorylation, interaction of the NLS-containing protein with another protein shielding the NLS from importin α, regulated conformational changes of the NLS-containing protein which may also result in shielding of the NLS, and cytoplasmic anchoring of the NLS-containing protein by interaction with a fixed structure, over-riding nuclear import. The fact that such actions which interfere with binding by importin may be eliminated or induced (i.e., switched off and on) very quickly upon a signal makes these mechanisms perfect tools for the regulation of the localization of a protein, such as a transcription factor.

Although the regulation of nuclear import has been investigated much more extensively, the presence of a NES within a protein provides the same potential for regulatory mechanisms with respect to protein localization as does an NLS. If a protein contains both signals, combinations of different mechanisms of the regulation of protein localization become possible. One transport step may be default and the other regulated, or both may be regulated, resulting in a binary switch mechanism. In addition, not only may the localization of a protein as such be regulated, but also the half-life of its localization within the nucleus or the cytoplasm may be actively controlled. This provides an alternative to protein turnover as regulatory mechanism to control the half-life of the localization of a protein in the cytoplasm or in the nucleus.

All these possibilities for regulating protein localization are indeed operational. Together with the regulation of protein activity, they provide a network of mechanisms for the control of signaling. Signal transduction by light provides very illustrative examples for the importance of nucleocytoplasmic partitioning in plant signaling cascades. Research of the last few years has uncovered regulation of the localization of proteins at different levels of light signal transduction pathways.

Nucleocytoplasmic Partitioning in Light Signal Transduction

Plants and Light

Plants as immobile organisms have to cope with changing environmental conditions at the place where they grow. Because plants are not able to escape unfavorable conditions they depend upon reliable information about environmental factors like temperature, water supply, and light. One of the most important environmental factors for plants is light. Light not only serves as source of energy for photosynthesis, but also constitutes a morphogenic signal that is perceived by plants to sense changes in the natural environment. Light regulates a wide range of developmental processes and adaptations during the entire life cycle. The onset of seed germination, the developmental switch from skoto- to photomorphogenesis of the young seedling, the detection of neighbors competing for the incident light or the onset of the generative phase and flowering are all driven by light (for a review see refs. 59, 60). To perceive the surrounding light conditions, plants have evolved at least three different photoreceptor systems to monitor light quality and quantity ranging from UV to the infrared parts of the spectrum. These are (i) the UV-B receptors characterized by action spectroscopy,[61] (ii) the blue/UV-A receptors cry1, cry2 and phototropin[62-64] and (iii) the red/far-red reversible phytochromes (reviewed in ref. 65).

With regard to signal transduction of light, nucleocytoplasmic partitioning has been shown to play a role in the case of at least three classes of molecules. These are the phytochrome photoreceptors themselves, bZIP transcription factors, and the negative regulator of photomorphogenesis COP1.

The Phytochrome System

Phytochromes (phy), a group of plant photoreceptors involved in a number of light-dependent processes are the best characterized photoreceptors. In higher plants, phytochromes are encoded by small multigene families. In *Arabidopsis,* five members are known (*phyA* to *phyE*). Phytochromes are synthesized in darkness in their physiological inactive red light-absorbing form (P_r). The inactive P_r-form can be reversibly transformed by absorption of a photon into the physiological active, far-red light-absorbing form (P_{fr}). Each phytochrome is thought to have a different role in light signaling; PHYA and PHYB have been best characterized.[60] In *Arabidopsis* these two phytochromes are expressed in most cell types.[66,67] PHYA is most abundant in dark-grown plants, it is light-labile in its P_{fr}-form and responsible for the non-photoreversible VLFR (**v**ery **l**ow **f**luence **r**esponse) and the HIR (**h**igh **i**rradiance **r**esponse) classes of phytochrome-regulated responses. PHYB is the receptor for the classical red/far-red reversible (LFR) and continuous red light responses and is light-stable.

During the past few years insights into the mechanisms of phytochrome signaling have been substantially increased by genetic and molecular approaches. The identification of components involved in phytochrome-mediated signal transduction and recent studies about the intracellular localization of the photoreceptors implicate a tightly regulated interaction of the nuclear and cytosolic compartments.

Phytochrome-Regulated Intracellular Partitioning of Phytochromes

Until recently the dominating view has been that plant photoreceptors are localized in the cytoplasm. Physiological studies in algae, mosses, and ferns showed action dichroism for cloroplast orientation, polarotropism, and phototropism. These observations indicated that the photoreceptors regulating these responses are localized in the cytoplasm in an oriented manner, presumably in association with the plasmalemma or other membrane structures.[68] With regard to phytochromes, this hypothesis was supported by immuncytochemical studies on the P_{fr}-dependent formation of sequestered areas of phytochrome (SAPs) in the cytoplasm of monocotyledonous seedlings,[69,70] despite observations of phytochrome-dependent transcription in nuclear run-on experiments.[71] The view of an exclusively cytoplasmic localization of phytochromes was further challenged by Sakamoto and Nagatani.[72] These authors could demonstrate nuclear localization of phyB fragments fused to the GUS (β-glucoronidase) reporter in transgenic plants, pointing to functional NLS sequences in the photoreceptor. Additionally, in this study a substantial increase in the amount of phytochrome in purified nuclei of plant tissues irradiated with red light was observed. More recently, members of the same laboratory could complement a PHYB-deficient *Arabidopsis* mutant by using a protein fusion consisting of full-length PHYB and the in vivo marker protein GFP.[73] The results clearly show a light-dependent nuclear import of PHYB:GFP accompanied by the characteristic formation of speckles of PHYB-GFP inside the nuclei. A

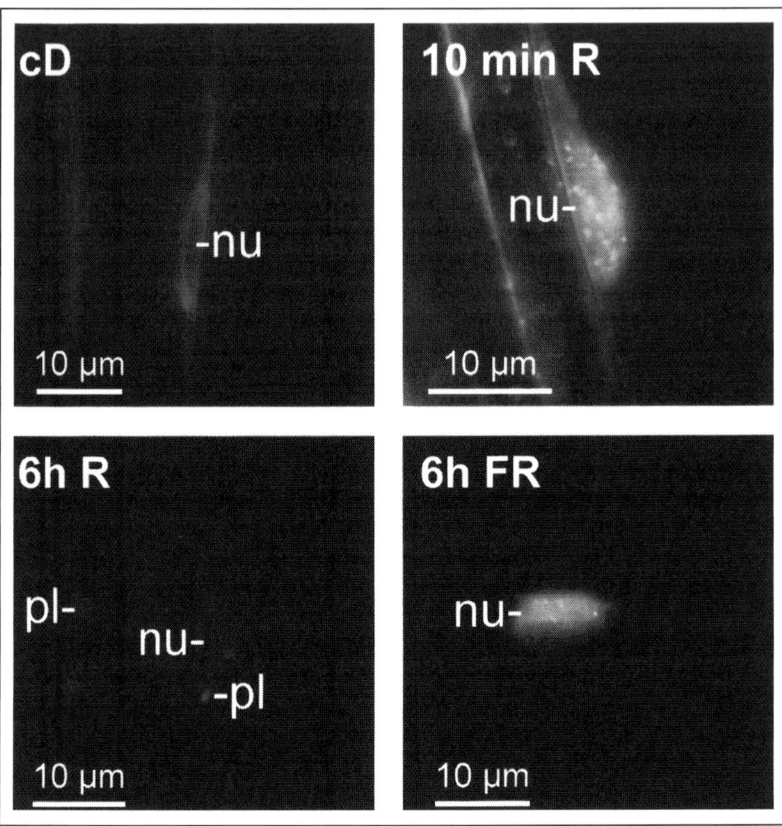

Figure 2. Intracellular localization of PHYA-GFP expressed under control of the *phyA* promoter in a *phyA* mutant of *Arabidopsis*. Seedlings were grown for 6 days in darkness and subsequently subjected to red light or far-red light prior to analysis in an epifluorescence microscope. In dark-grown seedlings, PHYA:GFP is almost exclusively localized in the cytosol (cD). The nucleus is found to be surrounded by fluorescing cytoplasm. In red light, a fast P_{fr}-dependent nuclear import occurs that is accompanied by the formation of speckles in the cytoplasm and in the nucleus (10min R). Because of the degradation of PHYA in its P_{fr} form, prolonged irradiation with red light leads to a substantial decrease of PHYA:GFP fluorescence (6h R). In contrast, 6h of far-red light treatment, which results a low ratio of P_{fr} to the light-stable P_r form, leads to a strong nuclear staining and to the formation of exclusively nuclear speckles. Nuclei (nu) and selected plastids (pl) are indicated.

more detailed study, also applying PHY-GFP fusion proteins, extended this observations to tobacco plants.[74] The authors could demonstrate nuclear uptake of phytochrome A and phytochrome B, dependent on the respective light requirements for the functionally distinct photoreceptors. Whereas nuclear import of PHYB:GFP requires red light (high amounts of P_{fr}) and is red/far-red photoreversible (LFR), nuclear uptake of PHYA:GFP can be initiated even by short far-red pulses (VLFR, low amounts of P_{fr}) and continuous far-red light (HIR; see Fig. 2). Additionally, it was shown that in continuous light the kinetics of Pfr-dependent nuclear import of PHYA is an order of magnitude faster than that of PHYB. Further studies on the light-regulated partitioning of phytochromes have refined and extended these observations. The analysis of nuclear import and speckle formation of the phytochromes of Arabidopsis with supplementary functions revealed for PHYC and PHYE similar kinetics as for PHYB. Interestingly, although PHB and PHYD are closely related genes they showed the largest difference. PHYD:GFP displayed very slow and heterogeneous nuclear import and only one or two nuclear speckles were found after 8 hours of irradiation.[74a] The capacity to complement the corresponding photoreceptor mutant in *Arabidopsis* was also demonstrated for a PHYA:GFP fusion protein.[75] The validity of the above-described light requirements of nuclear import for the endogenous protein could also be demonstrated by immunolocalization analysis on the intracellular partitioning of PHYA in pea seedlings.[76] In a reverse approach the localization of functionally impaired versions of phytochrome photoreceptors were analysed which originally were identified by genetic studies. In all cases investigated the respective amino-acid exchanges in PHYA or PHYB lead to aberrant localization patterns.[74a,76a,76b]

Taken together, these results show a strong correlation of the light requirements for physiological responses regulated by the respective phytochromes and the intracellular partitioning of these molecules and suggest a dominant role in light signal transduction for photoreceptors imported into the nucleus. This view has been strongly supported by recent findings derived from genetically characterized photo-transduction mutants in *Arabidopsis*.

Nuclear Components of the Phytochrome Signaling Pathway

The analysis of mutants defective in PHYA- and/or PHYB-mediated signal transduction has identified a number of proteins involved in the regulatory process leading to the respective physiological responses. As regards the nuclear import and function of active photoreceptors, of special interest was the identification of an *Arabidopsis* mutant that showed altered PHYB-dependent light signaling (*poc1*).[77] The corresponding gene product, also characterised as PIF3 (phytochrome interacting factor), which was isolated by a two-hybrid approach, is a basic helix-loop-helix (bHLH) transcription factor of nuclear localization. In the heterologuous yeast system PIF3 physically interacts preferentially with the P_{fr} confomers of PHYB and, to a lesser extend, of PHYA.[78,79] Recently, it was shown in planta that light signals lead in deed to a rapid and transient co-localisation of PIF3 with the P_{fr} forms of PHYA, PHYB, PHYC and PHYD in small speckles within nuclei of Arabidopsis plants expressing microscopically distinguishable CFP and YFP fusions of the respective protein pairs. Subsequently to this interaction, these early nuclear structures disappear within minutes and the negative regulator PIF3 is degraded with a half-time of about 10-20 minutes.[80a] Because the formation of the early and transient type of nuclear speckles of PHYB:GFP is lost in a *pif3* mutant it is tempting to speculate that these structures are involved in marking the bHLH factor for degradation. The identification of another bHLH transcription factor which is a positive regulator in PHYA signaling supports the hypothesis of a direct and distinct effect of phytochromes imported into the nucleus on transcriptional networks regulating photomorphogenesis (*hfr1*).[81,82] The P_{fr} forms of PHYA and PHYB can physically bind only to heterodimers of HFR1 and PIF3, but not to HFR1 homodimers.[81] Because of the low abundance of *hfr1* mRNA in plants treated with continuous red light compared to plants irradiated with far-red light it is tempting to speculate whether the abundance of this and related factors could determine the specificity of phytochrome signaling. In this light, it would be of major interest to elucidate the binding specificities of bHLH transcription factors regarding homo- and heterodimerisation on the one hand and the binding of the dimers to promoter elements of individual phytochrome-regulated genes on the other hand.

However, to elucidate the exact function of the various types of PHY-associated nuclear speckles in signal transduction, it is important to identify their components on a molecular level. In our lab, we recently purified PHYB-containing nuclear speckles from 4 week-old

Arabidopsis plants and analysed their composition by MALDI-TOF (Panigrahi, Kunkel, Klement, Medzhradszky, Nagy, Schaefer, unpublished results). About 80% of the proteins identified share high homology to proteins shown to be present in the interchromatin granual clusters (ICGs) of animal cells. The precise biological role of ICGs is not known even in animal cells but they are considered to be involved in the storage, modification and recruitment of factors necessary for transcription and splicing.[82a] We note, however, that the N-terminal fragment of PHYB fused to GUS:NLS did not show nuclear speckle formation in continuous light, yet it successfully complemented the phenotype of a *phyB* mutant.[82b] These data may indicate that nuclear import but not the formation of nuclear complexes may be essential for PHYB signal transduction. Thus we conclude that, given their multiple sizes, their transient and dynamically changing appearance, it remains a challenging task to understand the molecular functions of the light-induced nuclear protein complexes.

Other factors involved in phytochrome signal transduction have also been shown to be nuclear proteins. In the case of PHYA-signaling SPA1,[83] FAR1,[84] FIN219,[85] and EID1[86] have been characterized as nuclear factors, further underlining the importance of the nuclear compartment in light signaling.

Potential Mechanisms of Cytosolic Retention and Nuclear Import of Phytochromes

In etiolated seedlings and dark adapted plants, where phytochromes exist in their photobiologically inactive P_r forms, PHYB:GFP and PHYA:GFP are localized in the cytoplasm. A mutated version of PHYB:GFP which is not able to bind its chromophore, is confined almost exclusively to the cytoplasm of tobacco irrespective of light conditions.[87] These results indicate that the conformational change of P_r to the physiological active P_{fr} form is a necessary pre-requisite for the nuclear import of phytochromes. Localization experiments with truncated versions of PHYB:GFP fusion proteins clearly indicate the presence of a functional NLS within the C-terminus of the photoreceptor.[88,89] The light-independent exclusive nuclear localization of the C-terminal half of PHYB suggests an important role for the N-terminal part of phytochromes in cytosolic retention in darkness. In contrast to tobacco, in Arabidopsis the addition of extra NLSs to PHYB:GFP does lead to a light-independent nuclear import of this modified photoreceptor protein[82b] which physiologically results in a hypersensitive PHYB phenotype (Kirchenbauer, Kircher, Nagy and Schäfer, unpublished results). Taken together, these data suggest a cytosolic retention mechanism for phytochrome B in its P_r form. The switch making possible the interaction with the nuclear import machinery could be the release of the photoreceptor from cytosolic retention by the light-dependent conformational change to its P_{fr} form. So far no information is available on the molecular mechanism of the cytosolic retention of phytochromes, which is therefore a major target of signal transduction research at the present time.

Phytochrome-Regulated Nuclear Import of the bZIP Transcription Factor CPRF2

In addition to their suggested direct function in the nucleus, photoreceptors, especially PHYA has also been shown to mediate cytosolic events in its P_{fr} form. Besides acting via trimeric G-proteins, cGMP and calcium/calmodulin on greening and anthocyanin production as revealed by pharmacological studies,[90-92] PHYA is a phospoprotein[93] and is considered to be a protein kinase.[94] Recently, PKS1 (phytochrome kinase substrate 1), a cytosolic protein identified in a yeast two-hybrid screen, was demonstrated to be phosporylated by PHYA in vitro.[95]

In addition, P_{fr}-dependent phosporylation events in the cytoplasm could lead to nuclear import of downstream regulatory proteins, which was shown to be the case for a family of basic leucin-zipper motif (bZIP) transcription factors. CPRF proteins (common plant regulatory factors) bind to G-box promoter elements and are thought to play a role in regulating light-responsive genes in parsley, like the gene encoding chalcone synthase.[96] In an initial biochemical study analyzing cellular fractions of parsley cells, a light-regulated and phosphorylation-dependent translocation of G-box binding proteins from the cytosol into the nucleus was demonstrated. Clear evidence for light-initiated nuclear uptake of G-Box binding factors in parsley was provided by using an in vitro nuclear transport system.[97] Further characterization of several members of the CPRF gene family by immunocytochemistry and transient expression of GFP fusion proteins revealed that only CPRF2 is localized almost exclusively in the cytoplasm in the dark. The cytosolic retention of this transcription factor is released by red light treatment and is at least partially red/far-red light photoreversible. This observation points to the involvement of phytochrome in this light-regulated translocation process.[98] In good agreement with earlier observations,[97] phytochrome-dependent phosporylation within the C-terminus of CPRF2 by a cytosolic serine kinase is proposed to be a pre-requisite for nuclear import.[99] This study as well as localization experiments using truncated versions of CPRF2 point to the cytosolic retention of this transcription factor in a high molecular weight complex in darkness.[98] Transient expression of truncated CPRF2 fused to GFP in parsley protoplasts leads to the conclusion that two structural motifs in the N-terminus, distinct from the NLS-harboring bZIP domain of the factor are necessary to prevent nuclear import in darkness. Additionally, the N-terminal domain can confer cytosolic retention to another nuclear bZIP factor in domain-swap experiments. It is therefore tempting to speculate that phytochrome-dependent phosphorylation of CPRF2 leads to conformational changes within the protein that releases the factor from a cytosolic retention complex. After release, CPRF2 could interact with the nuclear import machinery, translocate into the nucleus, and bind to light-regulated target genes. Very recently, a putative retention protein of an *Arabidopsis* CPRF2 homolog was identified by a screening approach in a heterologuous system, but its function *in planta* has not been corroborated yet (Näke, Schäfer and Harter, unpublished data).

Blue Light Photoreceptors and the Intracellular Partitioning of the Transcription Factor GBF2

Phototropin and cryptochromes (CRY) belong to a class of blue light-absorbing photoreceptors controlling UV-A/blue light-dependent responses in *Arabidopsis*. Phototropin is a plasmalemma-associated flavoprotein that is thought to mainly sense the direction of the incident light and that mediates phototropism, the orientation of plants and plant organs towards light.[100] Cryptochromes mediate hypocotyl shortening, cotyledon expansion, and anthocyanin production in blue light and play an important role in the entrainment of the circadian clock. Interestingly, the latter function of setting the circadian clock is also committed by conserved cryptochrome homologues in mammals and *Drosophila*.[101]

Cryptochrome 1 (CRY1) was the first member of the long-sought blue-light-absorbing photoreceptors to be identified and cloned from the HY4 mutant of *Arabidopsis*.[102] CRY1 displays a striking similarity to photolyases, but lacks photolyase activity and has a unique C-terminal extension. FADH, the catalytically active chromophore is attached to the N-terminus of the photoreceptor; the second chromophore could be either a pterin or deazaflavin.[103] A second member of the cryptochrome class of blue light photoreceptors, CRY2, has been identified in *Arabidopsis* and is distinguished mainly by a different C-terminal extension.[104]

Microbeam irradiation indicates that fern homologues of higher-plant CRY photoreceptors are present in the cytoplasm, but they are also found to be associated with the nucleus.[105] Recent studies in the fern *Adiantum capillus-veneris* showed that at least two members of the CRY family are localized in the nucleus and that the nucleocytoplasmic distribution of these photreceptors is, at most, only slightly influenced by light.[106] With regard to higher plants, transient transformation of *Arabidopsis* CRY1 fused to GFP in plant cells indicate that the receptor is localized in the nucleus in the dark.[101] Localization of CRY1 in light-treated cells has not been reported yet; recently, however, light-dependent cytosolic enrichment of a C-terminal fragment of CRY1 fused to the GUS marker gene was shown in transgenic *Arabidopsis*.[107] It is therefore tempting to speculate about a similar light-dependent intracellular partitioning of the photoreceptor itself. Analysis of the intracellular partitioning of the second member of the cryptochrome photoreceptor family, CRY2, was performed independently by two groups, either by CRY2:GUS or CRY2:GFP fusion proteins driven by the constitutive 35S promotor.[108,109] These studies indicate a nuclear localization of the photoreceptor, but if over-expressed marker fusion proteins represent the endogenous situation properly remains to be elucidated. Blue light irradiation of cells that transiently express a fusion protein consisting of CRY2 and the marker protein RFP (red fluorescing protein) led to the formation of a speckled pattern of CRY2:RFP inside the nucleus.[110] This study also provides evidence for physical interaction between phytochrome B and CRY2 as well as co-localization of both proteins in nuclear speckles forming under irradiation conditions combining red and blue light. This finding is remarkable because phytochromes and cryptochromes often regulate similar photomorphogenic responses and interdependencies of both photoreceptor systems are well-known.[111]

Similarly to the observation of a phytochrome-dependent, red light-initiated nuclear import of CPRF2-GFP in parsley, another bZIP protein, GBF2, was demonstrated to be a candidate for nuclear import that depends upon blue light. By comparative analysis of the intracellular partitioning of GUS fusion proteins of several members of the G-box binding factor (GBF) family of *Arabidopsis* it was shown that GBF2, in contrast to the nuclear GBF1 and GBF4, is evenly distributed in the cytoplasm and the nucleus of soybean cells kept in darkness.[112] When these cells were irradiated with blue light, a substantial increase in nuclear GUS:GBF2 was detected. In contrast, irradiation with red light did not change the intracellular distribution of the bZIP protein, pointing to the involvement of blue light photoreceptors in this process. It is not yet known which of the blue light photoreceptors plays a role in the regulated intracellular partitioning of GBF2. Since binding of members of the GBF family to G-Box promoter elements is necessary for the proper expression of light-regulated genes,[113] and since the trans-acting activity of at least one member of the GBF family has been demonstrated (GBF1),[114] regulated nuclear import of GBF2 could represent an important step in blue light signaling.

The Photomorphogenic Repressor Protein Constitutive Photomorphogenesis 1 (COP1)

COP1 also provides a well-investigated example for the nucleo-cytoplasmic partitioning of a regulatory protein and its implications for signaling in plants. The COP1 protein is localized in the nucleus during seedling germination in darkness and functions as a repressor of photomorphogenesis by suppressing the expression of light-inducible transcripts in the nucleus.[115,116] Regulation of COP1 activity is negatively controlled by light and involves a light-dependent re-localization of the protein into the cytoplasm. As a result, COP1 levels are drastically reduced in nuclei of hypocotyl cells that are transferred to light as compared to those that were kept in darkness. The localization of COP1 also has a tissue-specific component, as COP1 levels are constitutively high in the nuclei of root cells, which do not undergo photomorphogenesis.[116-118] The COP1 protein contains an amino-terminal zinc binding RING finger domain, a carboxy-terminal WD-40 repeat domain, and in between a domain with the potential to form a coiled-coil structure. Detailed analysis of the localization of COP1 fragments in combination with their physiological effects revealed discrete domains that mediate light-responsive nuclear and cytoplasmic localization of the protein: a bipartite NLS within the

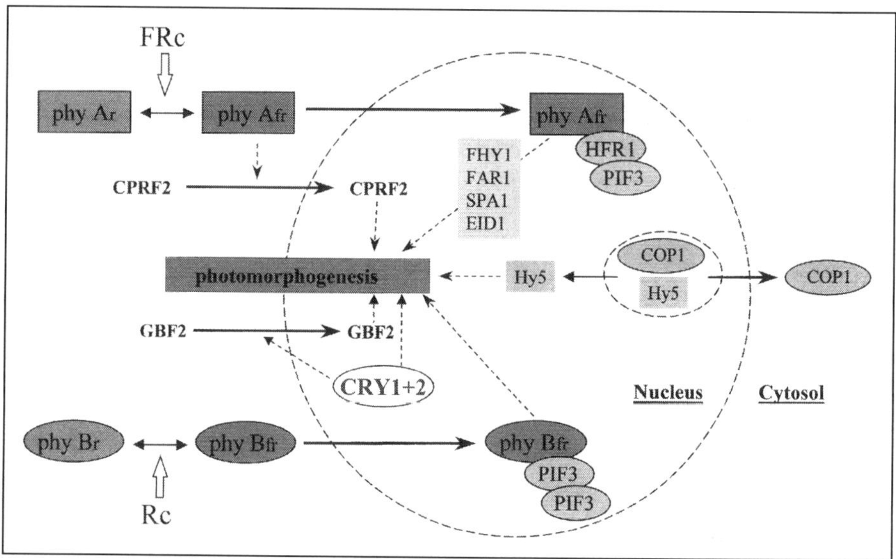

Figure 3. Simplified and speculative model of light-regulated intracellular partitioning of proteins involved in photosignaling in plants. Several classes of proteins involved in plant light signal transduction and control of photomorphogenesis show regulated intracellular partitioning. Phytochrome photoreceptors in their inactive P_r form (phy A_r; phy B_r) are cytosolic proteins in darkness. After photoconversion to the active P_{fr} forms (phy A_{fr}; phy B_{fr}) by far-red (FRc) or red light (Rc) the photoreceptors translocate to the nucleus where they physically interact with dimers of bHLH proteins (PIF3; HFR1). In concert with other nuclear factors specific for PHYA or PHYB, the development of the plant is controlled. Additionally, phy A_{fr} and/or phy B_{fr} can activate nuclear import of the bZIP protein CPRF2 which is considered as a trans-acting factor of light-regulated genes. GBF2 is another example of a bZIP transcription factor that is transported into the nucleus, possibly mediated by the nuclear blue light photoreceptors cryptochrome 1 and 2 (CRY1 + 2). COP1, a negative regulator of photomorphogenesis, is exported from the nucleus in the light. This causes release of the repression of the positively acting, nuclear bZIP transcription factor HY5. See text for further details.

core domain that mediates nuclear localization, and an amino-terminal domain including parts of the RING domain and parts of the coiled-coil domain that has cytoplasmic retention activity.[119,120] In addition, a motif that mediates targeting of the COP1 protein to subnuclear foci has been described, which overlaps the cytosolic retention signal and the putative α-helical coiled-coil domain.[121] How the switch between the cytosolic and the nuclear localization of the COP1 protein is accomplished at the molecular level is under investigation. COP1 acts as a transcriptional regulator by interaction with other regulatory proteins in the nucleus, like with the COP1-interactive protein 7 (CIP7)[122] or with the bZIP transcription factor HY5.[123,124] HY5 is constitutively nuclear, binds to G-box motifs within promoters of several light-inducible genes, and is necessary for their optimal expression.[124-126] Recently, it was shown that the nuclear interaction of COP1 and HY5 leads to the degradation of HY5, which is thereby negatively regulated by COP1 at the level of protein stability.[127]

Conclusions and Perspectives

In general, the completion of the *Arabidopsis* genome will facilitate the functional characterization of nuclear transport factors in plants. Many of the factors that are known from animals and yeast and have been identified in the *Arabidopsis* genome on the basis of sequence homology have not been investigated to date. As to the role of nucleocytoplasmic partitioning of proteins involved in the signal transduction of light, several interesting questions are waiting to be solved. To begin with the photoreceptors, the molecular mechanisms of the light-dependent re-localization of phytochromes between the nucleus and the cytoplasm are as yet unknown. Whether nuclear import of phytochromes depends upon importin α/β heterodimers or upon different import receptors is another question to be solved. The mechanism of their cytosolic retention in the dark and their release in the light may be a more challenging problem, since this is supposed to hold the key for the regulation of the nuclear import of phytochromes. In addition, apart from the nature of the speckles formed after the import of phytochromes in the nucleus and from the molecular mechanisms of the nuclear function of phytochromes, a very interesting question is whether phytochromes are degraded in the nucleus or are, at least in part, transported back to the cytoplasm. As a consequence of the latter hypothesis, at least a portion of the phytochrome pool in a cell would then show light-dependent shuttling between the nucleus and the cytoplasm. Retention mechanisms and nucleocytoplasmic shuttling are also postulated in case of the bZIP transcription factor CPRF2 and in case of the photomorphogenic repressor COP1, to give only two examples. The molecular mechanisms of these processes may differ from the corresponding processes of phytochromes. However, it would be interesting to know if some of the components that confer light-dependent regulation to nucleocytoplasmic partitioning of phytochromes, CPRF2, and COP1 are shared.

Abbreviations

IBB, importin β-binding domain; NE, nuclear envelope; NPC, nuclear pore complex; RanBP1, Ran-binding protein 1; RanGAP, Ran-specific GTPase-activating protein; Pi, inorganic phosphate.

Acknowledgements

The work of the authors was supported by the Sonderforschungsbereich 388 and 592 (E.S. and S.K.) and funded by grants from Riken, HHMI (55000325) and OTKA (T032565) to F.N. and by a grant from the Deutsche Forschungsgemeinschaft (ME1116/3-1) to T.M.

References

1. Corbett AH, Silver PA. Nucleocytoplasmic transport of macromolecules. Microbiol Mol Biol Rev 1997; 61:193-211.
2. Görlich D, Kutay U. Transport between the cell nucleus and the cytoplasm. Annu Rev Cell Biol 1999; 15:607-660.
3. Görlich D, Dabrowski M, Bischoff FR et al. A novel class of RanGTP binding proteins. J Cell Biol 1997; 138:65-80.
4. Görlich D, Kostka S, Kraft R et al. Two different subunits of importin cooperate to recognize nuclear localization signals and bind them to the nuclear envelope. Curr Biol 1995; 5:383-392.
5. Görlich D, Prehn S, Laskey RA et al. Isolation of a protein that is essential for the first step of nuclear protein import. Cell 1994; 79:767-778.
6. Weis K, Ryder U, Lamond AI. The conserved amino-terminal domain of hSRP1a is essential for nuclear protein import. EMBO J 1996; 15:1818-1825.
7. Görlich D, Henklein P, Laskey RA et al. A 41 amino acid motif in importin-α confers binding to importin-β and hence transit into the nucleus. EMBO J 1996; 15:1810-1817.
8. Izaurralde E, Kutay U, von Kobbe C et al. The asymmetric distribution of the costituents of the Ran system is essential for transport into and out of the nucleus. EMBO J 1997; 16:6535-6547.
9. Kutay U, Izaurralde E, Bischoff FR et al. Dominant-negative mutants of importin-β block multiple pathways of import and export through the nuclear pore complex. EMBO J 1997; 16:1153-1163.
10. Kose S, Imamoto N, Tachibana T et al. Beta-subunit of nuclear pore-targeting complex (importin-beta) can be exported from the nucleus in a Ran-independent manner. J Biol Chem 1999; 274:3946-3952.

11. Kutay U, Bischoff FR, Kostka S et al. Export of importin α from the nucleus is mediated by a specific nuclear transport factor. Cell 1997; 90:1061-1071.

12. Solsbacher J, Maurer P, Bischoff FR et al. Cse1p is involved in export of yeast importin α from the nucleus. Mol Cell Biol 1998; 18:6805-6815.

13. Künzler M, Hurt EC. Cse1p functions as the nuclear export receptor for importin α in yeast. FEBS Lett 1998; 433:185-190.

14. Hood JK, Silver PA. Cse1p is required for export of Srp1p/importin-α from the nucleus in Saccharomyces cerevisiae. J Biol Chem 1998; 273: 35142-35146.

15. Fischer U, Huber J, Boelens WC et al. The HIV1 Rev activation domain is a nuclear export pathway used by specific cellular RNAs. Cell 1995; 82:475-483.

16. WenW, Meinkoth JL, Tsien RY et al. Identification of a signal for rapid export of proteins from the nucleus. Cell 1995; 82:463-473.

17. Fornerod M, Ohno M, Yoshida M et al. CRM1 is an export receptor for leucine-rich nuclear export signals. Cell 1997; 90:1051-1060.

18. Fukuda M, Asano S, Nakamura T et al. CRM1 is responsible for intracellular transport mediated by the nuclear export signal. Nature 1997; 390:308-311.

19. Ossareh-Nazari B, Bachlerie F, Dargemont C. Evidence for a role of CRM1 in signal-mediated nuclear protein export. Science 1997; 278:141-144.

20. Stade K, Ford CS, Guthrie C et al. Exportin 1 (Crm1p) is an essential nuclear export factor. Cell 1997; 90: 1041-1050.

21. Fornerod M, van Deursen J, van Baal S et al. The human homologue of yeast CRM1 is in dynamic subcomplex with CAN/Nup214 and a novel nuclear pore component Nup88. EMBO J 1997; 16:807-816.

22. Neville M, Stutz F, Lee L et al. The importin-beta family member Crm1p bridges the interaction between Rev and the nuclear pore complex during nuclear export. Curr Biol 1997; 7:767-775.

23. Bischoff FR, Krebber H, Smirnova E et al. Co-activation of RabGTPase and inhibition of GTP dissociation by Ran-GTP binding protein RanBP1. EMBO J 1995; 14:705-715.

24. Bischoff FR, Görlich D. RanBP1 is crucial for the release of RanGTP from importin β-related nuclear transport factors. FEBS Lett 1997; 419:249-254.

25. Moore MS, Blobel G. The GTP-binding protein Ran/TC4 is required for protein import into the nucleus. Nature 1993; 365;661-663.

26. Bischoff FR, Ponstingl H. Catalysis of guanine nucleotide exchange on Ran by the mitotic regulator RCC1. Nature 1991; 354:80-82.

27. Mahajan R, Delphin C, Guan T et al. A small ubiquitin-related polypeptide involved in targeting RanGAP1 to nuclear pore complex protein RanBP2. Cell 1997; 88:97-107.

28. Matunis MJ, Coutavas E, Blobel G. A novel ubiquitin-like modification modulates the partitioning of the Ran-GTPase-activating protein RanGAP1 between the cytosol and the nuclear pore complex. J Cell Biol 1996; 135:1457-1470.

29. Görlich D, Pante N, Kutay U et al. Identification of different roles for RanGDP and RanGTP in nuclear protein import. EMBO J 1996; 15:5584-5594.

30. Hood JK, Silver PA. In or out? Regulating nuclear transport. Curr Opin Cell Biol 1999; 11:241-247.

31. Moore MS, Blobel G. Purification of a Ran-interacting protein that is required for protein import into the nucleus. Proc Natl Acad Sci USA 1994; 91:10212-10216.

32. Ribbeck K, Lipowsky G, Kent HM et al. NTF2 mediates nuclear import of Ran. EMBO J 1998; 17:6587-6598.

33. Smith A, Brownawell A, Macara IG. Nuclear import of Ran is mediated by the transport factor NTF2. Curr Biol 1998; 8:1403-1406.

34. Black BE, Levesque L, Holaska JM et al. Identification of an NTF2-related factor that binds Ran-GTP and regulates nuclear protein export. Mol Cell Biol 1999; 19:8616-8624.

35. Ossareh-Nazari B, Maison C, Black BE et al. RanGTP-binding protien NXT1 facilitates nuclear export of different classes of RNA in vitro. Mol Cell Biol 2000; 20:4562-4571.

36. Black BE, Holaska JM, Levesque L et al. NXT1 is necessary for the terminal step of Crm1-mediated nuclear export. J Cell Biol 2001; 152:141-155.

37. Nakielny S, Dreyfuss G. Import and export of the nuclear protein import receptor transportin by a mechanism independent of GTP hydrolysis. Curr Biol 1997; 8:89-95.

38. Ribbeck K, Kutay U, Paraskeva E et al. The translocation of transportin-cargo complexes through nuclear pores is independent of both Ran and energy. Curr Biol 1999; 9:47-50.
39. Kalab P, Pu RT, Dasso M. The Ran GTPase regulates mitotic spindle assembly. Curr Biol 1999; 9:481-484.
40. Ohba T, Nakamura M, Nishitani H et al. Self-organization of microtubule asters induced in Xenopus egg extracts by GTP-bound Ran. Science 1999; 284:1356-1358.
41. Wilde A, Zheng Y. Stimulation of microtubule aster formation and spindle assembly by the small GTPase Ran. Science 1999; 284:1359-1362.
42. Carazo-Salas RE, Guarguaglini G, Gruss OJ et al. Generation of GTP-bound Ran by RCC1 is required for chromatin-induced mitotic spindle formation. Nature 1999; 400:178-181.
43. Nachury MV, Maresca TJ, Salmon WC et al. Importin beta is a mitotic target of the small GTPase Ran in spindle assembly. Cell 2001; 104:95-106.
44. Wiese C, Wilde A, Moore MS et al. Role of importin-beta in coupling Ran to downstream targets in microtubule assembly. Science 2001; 291:653-656.
45. Gruss OJ, Carazo-Salas RE, Schatz CA et al. Ran induces spindle assembly by reversing the inhibitory effect of importin alpha on TPX2 activity. Cell 2001; 104:83-93.
46. Merkle T, Nagy F. Nuclear import of proteins: putative import factors and development of in vitro import systems in higher plants. Trends Plant Sci 1997; 2:458-464.
47. Schledz M, Leclerc D, Neuhaus G et al. Characterization of four cDNAs encoding different importin alpha homologues from Arabidopsis thaliana, designated AtIMPa1-4. Plant Physiol 1998; 116:868.
48. Smith HM, Hicks GR, Raikhel NV Importin alpha from Arabidopsis thaliana is a nuclear import receptor that Recognizes three classes of import signals. Plant Physiol 1997; 114:411-417.
49. Ballas N, Citovsky V. Nuclear localization signal binding protein from Arabidopsis mediates nuclear import of Agrobacterium VirD2 protein. Proc Natl Acad Sci USA 1997; 94:10723-10728.
50. Jiang CJ, Imamoto N, Matsuki R et al. Functional characterization of a plant importin alpha homologue. Nuclear localization signal (NLS)-selective binding and mediation of nuclear import of NLS proteins in vitro. J Biol Chem 1998; 273:24083-24087.
51. Jiang CJ, Imamoto N, Matsuki R et al. In vitro characterization of rice importin beta 1: Molecular interaction with nuclear transport factors and mediation of nuclear protein import. FEBS Lett 1998; 437:127-130.
52. Haasen D, Köhler C, Neuhaus G et al. Nuclear export of proteins in plants: AtXPO1 is the export receptor for leucine-rich nuclear export signals in Arabidopsis thaliana. Plant J 1999; 20:695-705.
53. Xia G, Ramachandran S, Hong Y et al. Identification of a plant cytosceletal, cell cycle-related and polarity-related proteins using Schizosaccharomyces pombe. Plant J 1996; 10:761-769.
54. Haizel T, Merkle T, Pay A et al. Characterization of proteins that interact with the GTP-bound form of the regulatory GTPase Ran in Arabidopsis. Plant J 1997; 11:93-103.
55. Merkle T, Leclerc D, Marshallsay C et al. A plant in vitro system for the nuclear import of proteins. Plant J 1996; 10:1177-1186.
56. Hicks GR, Smith HM, Lobreaux S et al. Nuclear import in permeabilized protoplasts from higher plants has unique features. Plant Cell 1996; 8:1337-1352.
57. Hübner S, Smith HMS, Hu W et al. Plant Importin alpha binds nuclear localization sequences with high affinity and can mediate nuclear import independent of Importin beta. J Biol Chem 1999; 274:22610-22617.
58. Smith HMS, Raikhel NV. Nuclear localization signal receptor Importin alpha associates with the cytoskeleton. Plant Cell 1998; 10:1791-1799.
59. Neff MM, Fankhauser C, Chory J. Light: an indicator of time and place. Genes Dev 2000; 14:257-271.
60. Smith H. Phytochromes and light signal perception by plants—An emerging synthesis. Nature 2000; 407:585-591.
61. Wellmann E. UV irradiation in photomorphogenesis. In: Mohr H, Shropshire W, eds. Encyclopedia of Plant Physiology, NS, 16B: Photomorphogenesis. Berlin, Heidelberg, New York: Springer Verlag, 1983:745-756.
62. Ahmad M, Cashmore AR. HY4 gene of A. thaliana encodes a protein with characteristics of a blue-light photoreceptor. Nature 1993; 366:162-166.

63. Christie JM, Reymond P, Powell GK et al. Arabidopsis NPH1: A flavoprotein with the properties of a photoreceptor for phototropism. Science 1998; 282:1698-1701.

64. Lin C, Ahmad M, Chan J et al. CRY2: A second member of the Arabidopsis cryptochrome gene family. Plant Physiol 1996; 110:1047.

65. Mathews S, Sharrock RA. Phytochrome gene diversity. Plant Cell Environ 1997; 20:666-671.

66. Somers DE, Quail PH. Temporal and spatial expression patterns of PHYA and PHYB genes in Arabidopsis. Plant J 1995; 7:413-427.

67. Adam E, Kozma-Bognar L, Kolar C et al. The tissue-specific expression of a tobacco phytochrome B gene. Plant Physiol 1996; 110:1081-1088.

68. Kraml M. Light direction and polarisation. In: Kendrick RE, Kronenberg GHM, eds. Photomorphogenesis in Plants. 2nd ed. Dordrecht: Kluwer Academic Publishers, 1994:417-443

69. Speth V, Otto V, Schäfer E. Intracellular localisation of phytochrome in oat coleoptyles by electron microscopy. Planta 1986; 168:299-304.

70. Pratt LH (1994) Distribution and localization of phytochrome within the plant. In: Kendrick RE, Kronenberg GHM, eds. Photomorphogenesis in Plants. 2nd ed. Dordrecht: Kluwer Academic Publishers, 1994:163-165.

71. Mösinger E, Batschauer A, Vierstra R et al. Comparison of the effects of exogenous native phytochrome and in vivo irradiation on in vitro transcription in isolated nuclei from barley (Hordeum vulgare). Planta 1987; 170:505-514.

72. Sakamoto K, Nagatani A. Nuclear localization activity of phytochrome B. Plant J 1996; 10:859-868.

73. Yamaguchi R, Nakamura M, Mochizuki N et al. Light-dependent translocation of a phytochrome B-GFP fusion protein to the nucleus in transgenic Arabidopsis. J Cell Biol 1999; 145:437-445.

74. Kircher S, Kozma-Bognar L, Kim L et al. Light quality-dependent nuclear import of the plant photoreceptors phytochrome-A and B. Plant Cell 1999; 11:1445-1456.

74a. Kircher S, Gil P, Kozma-Bognar L et al. Nucleo-cytoplasmic partitioning of the plant photoreceptors phytochrome A,B,C,D and E id differentially regulated by light and exhibits a diurnal rhythm. Plant Cell 2002; 14:1541-1544.

75. Kim L, Kircher S, Toth R et al. Light-induced nuclear import of phytochrome-A:GFP fusion proteins is differentially regulated in transgenic tobacco and Arabidopsis. Plant J. 2000; 22:125-134.

76. Hisada A, Hanzawa H, Weller JL et al. Light-induced nuclear translocation of endogenous pea phytochrome A visualized by immunocytochemical procedures. Plant Cell 2000; 12:1063-1078.

76a. Casal JJ, Davis SJ, Kirchenbauer D et al. The serine-rich N-terminal domain of oat phytochrome A helps regulate light responses and subnuclear localization of the photoreceptor. Plant Physiol 2002; 129:1127-1137.

76b. Yanofsky JM, Luppi PJ, Kirchbauer D et al. Missense mutation in the PAS2 domain of phytochrome A impairs subnuclear localization and a subset of responses. Plant Cell 2002; 14:1591-1603.

77. Halliday KJ, Hudson M, Ni M et al. poc1: An Arabidopsis mutant perturbed in phytochrome signaling because of a T DNA insertion in the promoter of PIF3, a gene encoding a phytochrome-interacting bHLH protein. Proc Natl Acad Sci USA 1999; 96(10):5832-5837.

78. Ni M, Tepperman JM, Quail P. PIF3, a phytochrome-interacting factor necessary for normal photoinduced signal transduction, is a novel basic helix-loop protein. Cell 1998; 95:657-667.

79. Ni M, Tepperman JM, Quail P. Binding of phytochrome B to its nuclear signalling partner PIF3 is reversibly induced by light. Nature 1999; 400:761-764.

80. Martinez-Garcia JF, Huq E, Quail P. Direct targeting of light signals to a promoter element-bound transcription factor. Science 2000; 288:859-863.

80a. Bauer D, Viczian, A, Kircher S et al. Constitutive photomorphogenesis 1 and multiple photoreceptors control degradation oh phytochrome interacting factor 3, a transcription factor required for light signalling in Arabidopsis. Plant Cell 2004; 16:1433-1445.

81. Fairchild CD, Schumaker MA, Quail PH. HFR1 encodes an atypical bHLH protein that acts in phytochrome A signal transduction. Genes Dev 2000; 14(18):2377-2391.

82. Soh MS, Kim YM, Han SJ et al. REP1, a basic helix-loop-helix protein, is required for a branch pathway of phytochrome A signaling in Arabidopsis. Plant Cell 2000; 12:2061-2073.

82a. Bubulya PS, Spector DL. Disassembly of interchromatin granule clusters alters the coordination of transcription and pre-mRNA splicing. J Cell Biol 2002; 158:425-436.

82b. Matsushita T, Mochizuki N, Nagatani A. Dimers of the N-terminal domain of phytochrome B are functional in the nucleus. Nature 2003; 424:571-574.

83. Hoecker U, Tepperman JM, Quail PH. SPA1, a WD-repeat protein specific to phytochrome A signal transduction. Science 1999; 284:496-499.

84. Hudson M, Ringli C, Boylan M et al. The FAR1 locus encodes a novel nuclear protein specific to phytochrome A signaling. Genes Dev 1999; 13(15):2017-2027.

85. Hsieh HL, Okamoto H, Wang M et al. FIN219, an auxin-regulated gene, defines a link between phytochrome A and the downstream regulator COP1 in light control of Arabidopsis development. Genes Dev 2000; 14(15):1958-1970.

86. Büche C, Poppe C, Schäfer E, Kretsch T. eid1: A new Arabidopsis mutant hypersensitive in phytochrome A–dependent high-irradiance responses. Plant Cell 2000; 12:547-558.

87. Kircher S, Kozma-Bognar L, Kim L et al. Light quality-dependent nuclear import of the plant photoreceptors phytochrome-A and B. Plant Cell 1999; 11:1445-1456.

88. Sakamoto K, Nagatani A. Nuclear localization activity of phytochrome B. Plant J 1996; 10:859-868.

89. Nagy F, Kircher S, Schäfer E. Nucleo-cytoplasmic partitioning of the plant photoreceptors phytochromes. Sem Cell Dev Biol 2000; 11:505-510.

90. Romero LC, Biswal B, Song PS. Protein phosphorylation in isolated nuclei from etiolated Avena seedlings. Effects of red/far-red light and cholera toxin. FEBS Lett 1991; 282:347-350.

91. Neuhaus G, Bowler C, Kern R et al. Calcium/calmodulin dependent- and independent-phytochrome signal transduction pathways. Cell 1993; 73:937-952.

92. Bowler C, Neuhaus G, Yamagata H et al. Cyclic GMP and calcium mediate phytochrome phototransduction. Cell 1994; 77:73-81.

93. Lapko VN, Jiang XY, Smith DL et al. Mass spectrometric characterization of oat phytochrome A: Isoforms and posttranslational modifications. Protein Science 1999; 8(5):1032-1044.

94. Yeh KC, Lagarias JC. Eukaryotic phytochromes:light-regulated serine/threonine protein kinases with histidine kinase ancestry. Proc Natl Acad Sci USA 1998; 95:13976-13981.

95. Fankhauser C, Yeh KC, Lagarias JC et al. PKS1, a substrate phosphorylated by phytochrome that modulates light signaling in Arabidopsis. Science 1999; 284:1539.

96. Weisshaar B, Armstrong GA, Block A et al. Light-inducible and constitutively expressed DNA-binding proteins recognizing a plant promotor element with functional relevance in light responsivness. EMBO J 1991; 10:1777-1786.

97. Harter K, Kircher S, Frohnmeyer H et al. Light-regulated modification and nuclear translocation of cytosolic G-box binding factors (GBFs) in parsley (Petroselinum crispum L.). Plant Cell 1994; 6:545-559.

98. Kircher S, Wellmer F, Nick P et al. Nuclear import of the parsley bZIP transcription factor CPRF2 is regulated by phytochrome photoreceptors. J Cell Biol 1999; 144:201-211.

99. Wellmer F, Kircher S, Ruegner A et al. Phosphorylation of the parsley bZIP transcription factor CPRF2 is regulated by light. J Biol Chem 1999; 274(41):29476-29482.

100. Christie JM, Reymond P, Powell GK et al. Arabidopsis NPH1: A flavoprotein with the properties of a photoreceptor for phototropism. Science 1998; 282:1698-1701.

101. Cashmore A, Jarillo JA, Yieng-Jie W et al. Cryptochromes: Blue light receptors for plants and animals. Science 1999; 284:760-765.

102. Ahmad M, Cashmore AR. HY4 gene of Arabidopsis thaliana encodes a protein with characteristics of a blue-light photoreceptor. Nature 1993; 366:162-166.

103. Malhotra K, Kim ST, Batschauer A et al. Putative blue-light photoreceptors form Arabidopsis thaliana and Sinapis alba with high degree of sequence homology to DNA photolyase contain the two photolyase cofactors but lack DNA repair activity. Biochem J 1995; 34:6892-6899.

104. Lin, CT, Yang HY, Guo HW T et al. Enhancement of blue-light sensitivity of Arabidopsis seedlings by a blue light receptor cryptochrome 2. Proc. Natl Acad Sci USA 1998; 95:2686-2690.

105. Wada M, Sugai G. Photobiology in ferns. In: Kendrick RE, Kronenberg GHM, eds. Photomorphogenesis in Plants. 2nd ed. Dordrecht: Kluwer Academic Publishers, 1994:417-443.

106. Imaizumi, T, Kanegae, T, Wada, M. Cryptochrome nucleocytoplasmic distribution and gene expression are regulated by light quality in the fern Adiantum capillus-veneris. Plant Cell 2000; 12:81-95.

107. Yang HQ, Yien-Jie W, Ru-Hang T et al. The C termini of Arabidopsis cryptochromes mediate a constititive lighr response. Cell 2000; 103:815-827.

108. Guo H, Duong H, Ma N et al. The Arabidopsis blue light receptor cryptochrome 2 is a nuclear protein regulated by a blue light-dependent post-transcriptional mechanism. Plant J 1999; 19:279-287.

109. Kleiner O, Kircher S, Harter K et al. Nuclear localization of the Arabidopsis blue light receptor cryptochrome 2. Plant J 1999; 19:289-296.

110. Más P, Devlin PF, Satchidananda P et al. Functional interaction of phytochrome B and cryptochrome 2. Nature 2000; 408:207-211.

111. Casal JJ. Phytochromes, cryptochromes, phototropin: photoreceptor interactions in plants. Photochem Photobiol 2000; 71:1-11.

112. Terzaghi WB, Bertekap RL, Cashmore AR. Intracellular localization of GBF proteins and blue light-induced import of GBF2 fusion proteins into the nucleus of cultured Arabidopsis and soybean cells. Plant J 1997; 11(5):967 982.

113. Schindler U, Menkens AE, Beckmann H et al. Heterodimerization between light-regulated and ubiquitously expressed Arabidopsis GBF bZIP proteins. EMBO J 1992; 11:1261-1273.

114. Schindler U, Terzaghi WB, Beckmann H et al. DNA binding site preferences and transcriptional activation properties of the Arabidopsis transcription factor GBF-1. EMBO J 1992; 11:1275-1289.

115. Deng XW, Caspar T, Quail PH. cop1: A regulatory locus involved in light-controlled development and gene expression in Arabidopsis. Genes Dev 1991; 5:1172-1182.

116. von Arnim AG, Deng XW. Light inactivation of Arabidopsis photomorphogenic repressor COP1 involves a cell type specific modulation of its nucleocytoplasmic partitioning. Cell 1994; 79:1035-1045.

117. von Arnim AG, Osterlund MT, Kwok SF et al. Genetic and developmental control of nuclear accumulation of COP1, a repressor of photomorphogenesis in Arabidopsis. Plant Physiol 1997; 114:779-788.

118. Osterlund MT, Deng XW. Multiple photoreceptors mediate the light-induced reduction of GUS-COP1 from Arabidopsis hypocotyl nuclei. Plant J 1998; 16:201-208.

119. Stacey MG, Hicks SN, von Arnim AG. Discrete domains mediate the light-responsive nuclear and cytoplasmic localisation of Arabidopsis COP1. Plant Cell 1999; 11:349-363.

120. Stacey MG, Kopp OR, Kim TH et al. Modular domain structure of Arabidopsis COP1. Reconstitution of activity by fragment complementation and mutational analysis of a nuclear localisation signal in planta. Plant Physiol 2000; 124:979-989.

121. Stacey MG, von Arnim AG. A novel motif mediates the targeting of the Arabidopsis COP1 protein to subnuclear foci. J Biol Chem 1999; 274:27231-27236.

122. Yamamoto YY, Matsui M, Ang LH et al. Role of a COP1 interactive protein in mediating light-regulated gene expression in Arabidopsis. Plant Cell 1998; 10:1083-1094.

123. Oyama T, Shimura Y, Okada K. The Arabidopsis HY5 gene encodes a bZIP protein that regulates stimulus-induced development of root and hypocotyl. Genes Dev 1997; 11:2983-2995.

124. Ang LH, Chattopadhyay S, Wei N et al. Molecular interaction between COP1 and HY5 defines a regulatory switch for light control of Arabidopsis development. Mol Cell 1998; 1:213-222.

125. Chattopadhyay S, Ang LH, Puente P et al. Arabidopsis bZIP protein HY5 directly interacts with light-responsive promoters in mediating light control of gene expression. Plant Cell 1998; 10:673-683.

126. Chattopadhyay S, Puente P, Deng XW et al. Combinatorial interaction of light-responsive elements plays a critical role in determining the response characteristics of light-regulated promoters in Arabidopsis. Plant J 1998; 15:69-77.

127. Osterlund MT, Hardtke CS, Wei N et al. Targeted destabilization of HY5 during light-regulated development of Arabidopsis. Nature 2000; 405:462-466.

Nuclear Export:

Shuttling across the Nuclear Pore

John A. Hanover and Dona C. Love

One of the distinguishing features of eukaryotic cells is the compartmentalization of genetic information within a membrane-enclosed nucleus. The double membrane of the nuclear envelope separates the nucleus and the cytoplasm, and all macromolecular exchange across the nuclear envelope takes place through large protein channels termed the nuclear pore complexes (NPCs). The molecules that are exchanged between these two compartments range in size from ions and other small molecules to large complexes such as the 50S ribosome and other large ribonucleoprotein complexes. In contrast to ions and small proteins that diffuse across the NPC, macromolecular movement is an active process. Active nucleocytoplasmic transport allows for the proper compartmentalization of nuclear proteins involved in transcription, replication of DNA, and remodeling of chromatin. Transport also is necessary for mRNAs, tRNAs, and rRNAs that are transcribed in the nucleus but ultimately function in the cytoplasm. This growing awareness of the role of nuclear transport in regulating gene expression has paralleled a remarkable increase in our knowledge of the nuclear transport process itself. Far from acting as static "localization signals" the sequences specifying nuclear location act in combination with other signals to alter the steady-state distribution of nuclear proteins. Thus, the concept of nuclear proteins shuttling between nucleus and cytoplasm has emerged as a dominant principle in understanding nuclear import and export. It is impossible to look at nuclear export in isolation without considering nuclear import rates. This has proven to be a barrier in understanding nuclear export as a process with distinct features and requirements. The number of import and export carriers identified has grown to include factors specific for classes of nuclear components and more general factors. In addition, it is clear that movement through the nuclear pore complex is dictated by properties of both the pore and the carrier molecules themselves. These properties are reflected in binding interactions that may facilitate movement across the nuclear pore and perhaps provide for directionality of transport. In this brief chapter, we will give a brief overview of the nuclear pore complex, methods for examining nuclear export and shuttling, the nature of the transport machinery and nuclear export carriers. We will then attempt to combine these into a coherent model for understanding nuclear import and export. We do not claim that this is a comprehensive overview. A number of excellent reviews of this type have recently appeared (see for example refs. 1, 2). What we hope to do is point out some novel aspects of nuclear export and emphasize some recent findings which suggest that factors other than traditional transport carriers may be involved and regulate the process of nuclear shuttling.

Nuclear Import and Export in Plants and Animals, edited by Tzvi Tzfira and Vitaly Citovsky. ©2005 Eurekah.com and Kluwer Academic / Plenum Publishers.

The Nuclear Pore Complex (NPC)

One key to understanding the process of nuclear import and export of proteins is the fact that the nuclear envelope is a semi-permeable membrane that excludes molecules greater than about 40 kDa if they lack specific targeting sequences. The nuclear membrane is, in fact, a double membrane that may be regarded as an extension of the endoplasmic reticulum. The nuclear envelope is interrupted at regular intervals by the nuclear pore complexes (see 3). These conserved features of eukaryotic cells can accommodate megadalton-sized particles and are themselves greater than 100 megadaltons in mass.

The number of NPCs per cell appears to depend on the need for nuclear transport and varies with cell size, growth and other cellular activity. There are approximately 200 NPCs in a yeast cell[4] and nearly 5000 nuclear pores in a rapidly growing human cell. A mature *Xenopus laevis* oocyte can have as many as 5×10^8 nuclear pores (Cordes et al 1995). An eightfold rotational symmetry and a nearly perfect two fold lateral symmetry in the plane of the nuclear membrane characterize all NPCs, regardless of origin. NPCs also exhibit cytoplasmic and nuclear extensions in the form of cytoplasmic filaments and intranuclear baskets. NPCs of higher eukaryotes have a mass of greater than 125 MDa[5] and appear to be composed of some 40 different proteins that are often collectively called nucleoporins.[3,6-8] Proteomics efforts have recently confirmed this number.[9] A large number of mammalian nucleoporins have also been molecularly identified since the major nucleoporin Nup62 was cloned in 1987.[10] These proteins share a number of characteristics. Some of the proteins and their characteristics arising from these various efforts are summarized in Figure 1. The nucleoporins often contain domains consisting of stretches of short peptide repeats containing the GFXFG and GLFG motifs. Mammalian nucleoporins also can contain repeats of the TTPST motif which appear to be sites of O-linked GlcNAc addition.[3,11] Yeast NPCs are smaller and have a mass of roughly 66 Mda.[4] Using a novel proteomics approach, approximately 30 yeast nucleoporins have been identified.[12] Like their mammalian orthologs, the majority of nucleoporins contain characteristic domains consisting of numerous short peptide repeats ending in the dipeptide FG liked the motif in mammalian nucleoporins. These repeats are now thought to play a pivotal part in the mechanism of vectorial movement across the nuclear pore complex (see below).

Methods for Analyzing Nuclear Export

Nuclear export was in essence, discovered more than 40 years ago in studies involving nuclear transplantation in amoebae.[13] As shown in Figure 2, other sensitive assays have been developed to examine nuclear export since those pioneering studies. Such approaches as antibody microinjection, heterokaryon cell fusion[14,15] r microinjection into Xenopus oocyte nuclei[16,17] have been employed. The difficulty in examining nuclear export in isolation has been the complication that nuclear import is often occurring simultaneously. Overcoming this issue has been particularly problematic.[17] The microinjection, heterokaryon fusion, and transplantation studies avoided this issue by examining redistribution of marker proteins. In more recent studies, the nuclear export of the hnRNP A1 protein or the Human Immunodeficiency Virus (HIV)-1 Rev proteins have been useful in sorting out nuclear export pathways. By sorting out the regions important to export, mutational analysis then led to the identification of the signals that mediate nuclear export. These were termed nuclear export signals (NES).[15,18] The best understood of these export signals is the leucine-rich NES present in HIV Rev and scores of other cellular or viral proteins involved in signal transduction, transcription, cell cycle (Table I). However, the description of shuttling proteins not containing leucine-rich NES points out the existence of other nuclear export signals. Our efforts have been centered upon trying to understand the export of the HIV NES using a novel GFP-based reporter construct in which nuclear import and export can be independently regulated.[19] Other in vitro systems have also

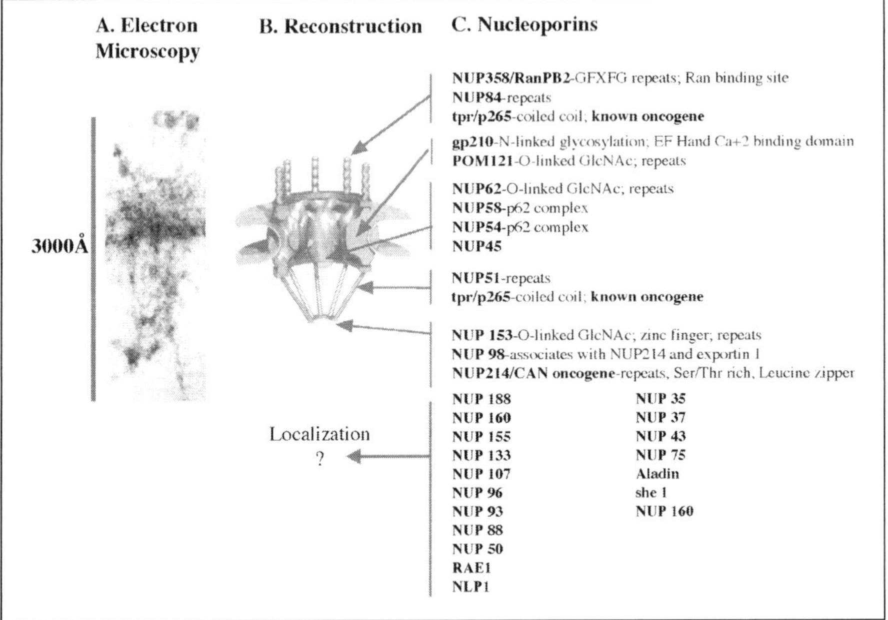

Figure 1. Structural and biochemical characterization of the mammalian nuclear pore complex. A) A thin section of the *Xenopus laevis* nuclear pore is shown with the associated nuclear and cytoplasmic fibrils evident. B) A reconstruction of the nuclear pore shows the major structural features.[58] C) A nearly complete list of the nucleoporins and their localization is given. Most of these proteins have been identified in a proteomics effort.[9]

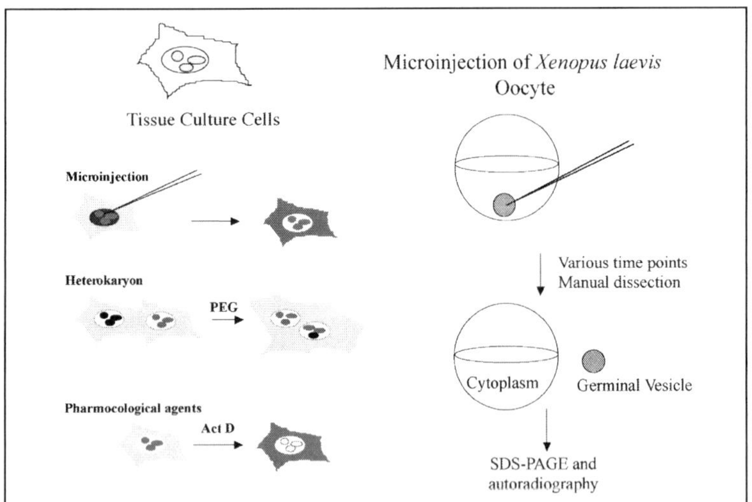

Figure 2. Methods for analyzing export of shuttling nuclear proteins. As described in the text, tissue culture cells have been microinjected, subjected to fusion to form heterokaryons and treated with pharmacological agents to allow the export process to be followed. In Xenopus, microinjection into the large nucleus (germinal vesicle) allows export of protein and RNA to be more directly examined.

Table 1. Leucine-rich nuclear export signals recognized by CRM1 (Xpo-1p). A list of proteins thought to be recognized by CRM1 and exported via a leucine-rich signal. Alignments of the signals are shown.

NES-containing proteins	Nuclear export sequence													
HIV-1 Rev				L	P	P	L	E	R	L	T	L		
PKI-α			L	A	L	K	L	A	G	L	D	I		
IκBα		R	I	Q	Q	Q	L	G	Q	L	T	L		
TFIIIA				L	P	V	L	E	N	L	T	L		
MAPKK		A	L	Q	K	K	L	E	E	L	E	L	D	E
Kap95p		L	E	G	R	I	L	A	A	L	T	L		
Gle1p					L	P	L	G	K	L	T	L		
RanBP1		K	V	A	E	K	L	E	A	L	S	V	R	
FMRP	E	V	D	Q	L	R	L	E	R	L	Q	I	D	
Zyxin		L	T	M	K	E	V	E	E	L	E	L	L	T
NF-ATc			I	V	A	A	I	N	A	L	T	T		

been developed which rely on different strategies.[20-22] These assays have led to the identification of additional nuclear export sequences and receptors for these sequences.

Rev-GR-GFP: Nuclear Export in Vitro

As stated earlier, the complication of characterizing nuclear export is that shuttling proteins such as Rev are rapidly moving between the nucleus and cytoplasm. In order to separate nuclear export from nuclear import, we designed a shuttling protein composed of full-length HIV Rev, the steroid-binding element of the glucocorticoid receptor and green fluorescent protein.[19] We termed this reporter Rev-Gr-GFP. The steady-state distribution of Rev-Gr-GFP can be regulated with the addition and removal of steroid. In the absence of steroid, the steady-state distribution is cytoplasmic. However with the addition of steroid, the reporter rapidly moves to the nucleus. Once in the nucleus, export can be induced with the removal of steroid (Fig. 3). Using this system, nuclear export can be examined in vitro simply by digitonin-permeabilization of the cells following import of the reporter. An experiment illustrating this regulated translocation in HeLa cells is shown in Figure 4. Export of Rev-Gr-GFP was dependent on the leucine-rich export signal found in Rev (Fig. 4, last panel) and could be inhibited by leptomycin B. Examining nuclear export in vitro, we found that Rev nuclear export was inhibited at low temperatures, proceeded through the NPC, required Crm1, and was dependent on ATP (Fig. 5). This assay described here makes it possible to separate nuclear export from import and allows for the identification of unique requirements in the export pathway.

The Nuclear Export Receptors (Karyopherins): Importins and Exportins

Our work has focused upon the Rev protein of HIV, whose nuclear export is leptomycin-insensitive and requires CRM1 for export. CRM1 recognizes the leucine-rich export sequence present in PKI and HIV Rev protein and is now regarded as the prototypic export receptor (Table I).[15,18] Nuclear export receptors are now often collectively called the exportins. Thus, CRM1 has been renamed Exportin 1 (Xpo1p). Table 2 briefly summarizes what is currently

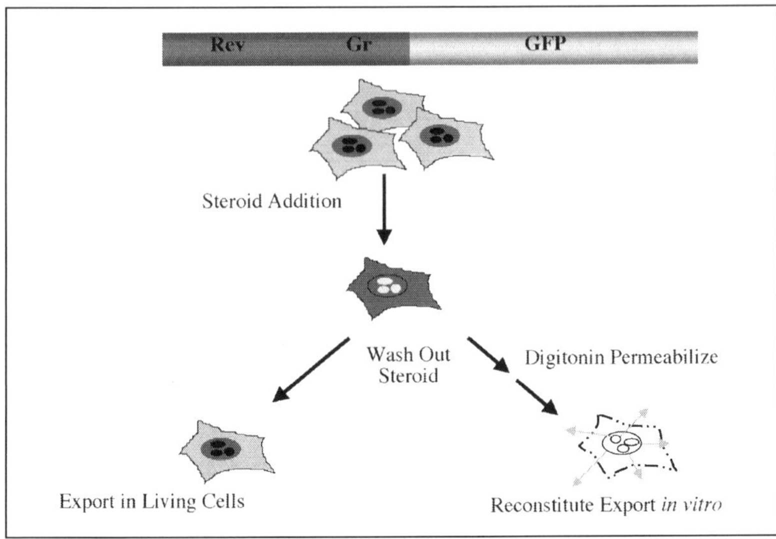

Figure 3. An assay for nuclear export in vivo and in vitro. A reporter protein composed of full-length Rev, the steroid binding element from glucocorticoid receptor and GFP (Rev-Gr-GFP) was stable expressed in HeLa cells. In the absence of steroid, the reporter is cytoplasmic and rapidly translocates to the nucleus/nucleolus with the addition of steroid. Removal of steroid induces export of the reporter to the cytoplasm. Export may be anlayzed in vitro by digitoinin-permeabilization of plasma membrane following import.

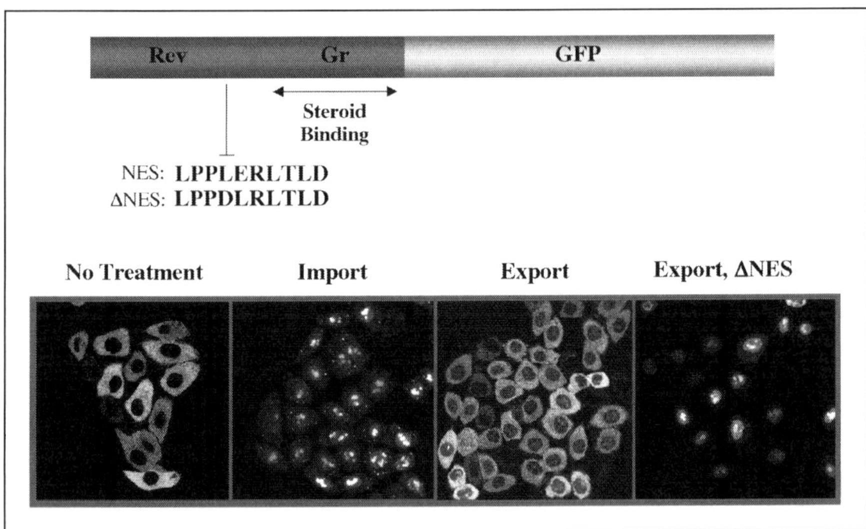

Figure 4. Nuclear translocation is regulated by steroid. A schematic of the Rev-Gr-GFP is shown along with the leucine rich nuclear export sequence (NES) and a export deficient mutation (ΔNES) of Rev. The lower panels represent nuclear transport in HeLa cells expressing Rev-Gr-GFP. Cytoplasmic localization of Rev-Gr-GFP before steroid treatment (No treatment). The reporter rapidly translocates to the nucleus/nucleolus with the addition of 1 μM dexamethasone (Import). Export is induced with the removal of steroid (Export) and is inhibited by a mutation in the leucine-rich export sequence (Export, ΔNES).

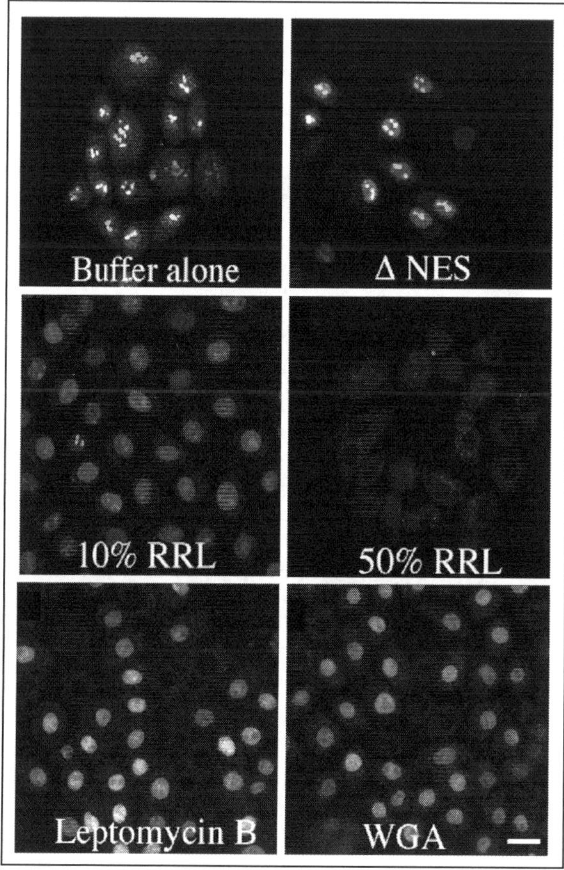

Figure 5. Reconstitution of nuclear export in vitro. Permeabilized cells were incubated in a simple isotonic buffer (buffer alone), or with transport buffer containing an energy regeneration system supplemented with either 10 % rabbit reticulocyte lysate (10% RRL), or 50% RRL. Transport in 50% RRL was inhibited by leptomycin B, wheat germ agglutinin (WGA) and mutated NES (ΔNES).

Table 2. Characterized export receptors

Vertebrate	Yeast	Cargoes
Exportins (Importin β-like)		
Cas	CSE1 (Kap lO⁹p)	Importin α
Crml (exportin 1)	Crml (Xpolp)	Leucine-rich NES containing proteins
Exportin4	Msn5 (Kapl42p)	eLF-5A, *Ste5*, Fart, Pho4
Non-Importin export receptors		
hnRNPA1	Unknown, (Balbiani ring)	
Calreticulin	Steroid hormone superfamily	

Export Carriers. The table lists most of the known export carriers, and their presumptive cargoes. For Details see accompanying text.

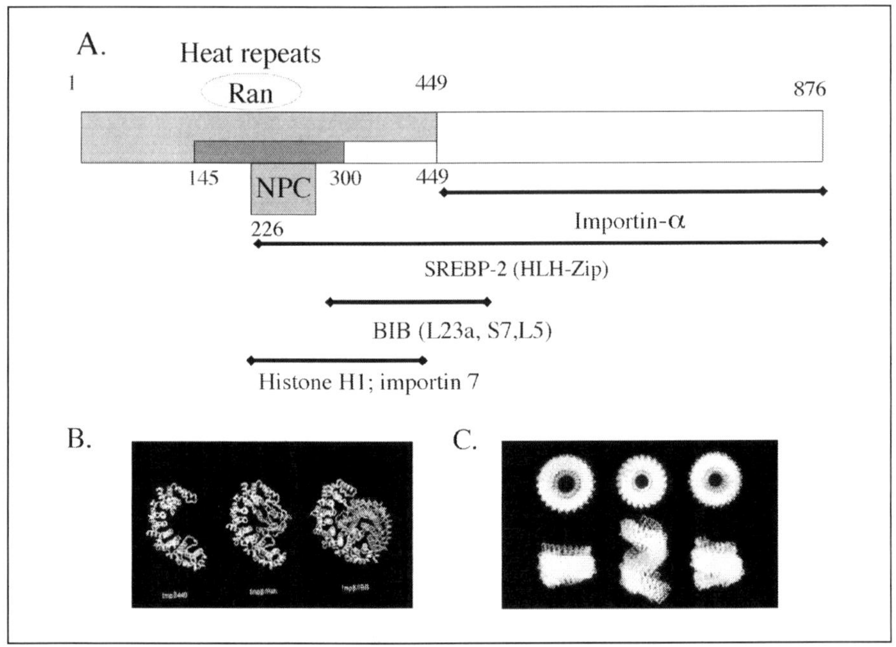

Figure 6. Interactions of importin β-family proteins. A) The interactions of importin β with known binding partners are shown. The segments of importin β required for Impotin α, transcription factor SREBP-2, BIB, Histone H1 and importin 7 are illustrated by the bars. Sites of interaction of the Heat repeats with Ran and with the NPC are also indicated. B) Crystal structures of importin β 1-449. Structural studies on the interaction of the heat domain of importin β: uncomplexed, with Ran and with IBB (left to right) are shown. C) Flexible structure of importin β. Helices of the HEAT repeat domain are superimposed. Left to right are uncomplexed, Ran-complexed and IBB complexed form. The diameter of uncomplexed importin β 1-449 is larger than the complexed forms; the pitch of importin β:Ran is roughly three times larger than uncomplexed importin β 1-449. Such dramatic conformational changes accompany Ran and cargo binding and certainly may alter the interaction of the surface of the superhelix with the NPC. Adapted from Figures 3 and 4 in reference 31.

known about nuclear export receptors. In addition to Xpo1p, is the protein CAS whose cargo includes and may be restricted to importin α. The other exportin is Exportin 4 which appears to act as both import and export carriers. The hnRNP A1 protein is also involved in export process although its precise cargoes have not been fully elucidated. As we will discuss in some detail later, calreticulin has also been demonstrated to have export activity and may act as a bone fide export receptor under some circumstances.

The exportins belong to a class of proteins with overall structural similarities to importin β, one of the first import receptors to be characterized. Several recent reviews have focused on the molecular and structural features these importin (or karyopherin) molecules[23-30] and we will simply summarize their features here. Importin β is the best characterized of this family (Fig. 6). These structural features mediate interactions with the nuclear pore complex, Ran-GTP and adapter molecules that facilitate other essential interactions (Fig. 6A). The importin β-related proteins share the homologous N-terminal amino acid sequence of importin β itself which mediates Ran-GTP interactions. The directionality of transport is largely dictated by the Ran-GTP-importin interaction; the receptors function as importins or exportins based the affinity of the interaction with Ran-GTP. Thus, affinity of the importin-Ran-GTP binding is high while the affinity of the exportin-Ran-GTP interaction is low. This means that import complexes are sufficiently stable to mediate transport themselves but export complexes require their export cargoes for stability.

Structural studies have been particularly useful in understanding the function of Exportins and Importins. We will briefly summarize these studies here (Fig. 6A,B.). Importin β has 19 copies of tandemly repeated HEAT sequence repeats. The roughly 40 residues making up the HEAT repeats form helix-turn-helix structures and the HEAT repeats themselves are linked together by short bridges to form a right-handed superhelical structure.[31] Interactions with the GFXFG repeats occur on either the convex side of the superhelix while interaction with Ran-GTP and adapters occurs at the concave surface. The known structures of the Importin β structures suggests that they undergo dramatic conformational changes and can create mutually exclusive binding interactions and cargo release upon binding of Ran-GTP to importin β. The studies have also revealed the flexible nature of importin β (Fig. 6C). The molecule has been likened to a snail; its uncomplexed, Ran-GTP- or importin-α- complexed forms show differences in both diameter and pitch of the superhelix. Thus, the molecule can bend and twist in a spring-like fashion (Fig. 6B). More detailed analysis suggests that the most prominent conformation changes occurring as a result of these interactions are in HEAT repeats 4 though 7 corresponding to the FGXFG binding domains. In fact, the crystal structure of the complex suggests that the primary binding sites for the FGXFG repeats is the hydrophobic pocket created by side chains in the HEAT repeat 5 and 6. Similar interactions almost certainly can be expected with the Exportins, which share much of this overall structure. These structural observations on Importin β may have important implications for understanding the mechanism of nuclear import and export.

A Nonclassical Export Receptor: Calreticulin

One surprising finding arose from our use of in vitro systems for detecting nuclear export receptors. In a collaborative effort with the laboratory of Bryce Paschal, we found that calreticulin could function to mediate the export of Rev-GR-GFP and several other reporter constructs.[32] Further analysis of the interaction demonstrated that the 69 amino acid DNA binding domain (DBD) of GR, which is unrelated to any known NES, is necessary and sufficient for export. In addition, the 15 amino acid sequence linking the two zinc-binding loops in the DBD of GR were sufficient to mediate nuclear export of a GFP reporter protein. These findings were intriguing since the DBD is very highly related in steroid, nonsteroid, and orphan nuclear receptors. In fact, the DBDs from several other steroid hormones were shown to function as export signals (see Fig. 7). Calreticulin is normally localized to the endoplasmic reticulum and may function as a Ca^{+2} lectin. Removal of Ca^{+2} from Calreticulin was found to limit its capacity to bind to and GR and to export GR in digitonin-permeabilized cells. Both DBD binding and nuclear export were restored by Ca^{+2} addition. Calreticulin lacks a consensus Ran interaction site and Ran GTPase is not required for Calreticulin-mediated GR export. Thus it would appear that multiple members of the nuclear receptor superfamily may utilize calreticulin as an export receptor. The findings suggest that the nuclear export pathway used by steroid hormone receptors and other DBD containing proteins is distinct from the Ran-dependent Crm1 pathway. This pathway may also complement the Ran-dependent export pathway.

Calcium-Dependent Modulation of Nuclear Transport?

The finding that calreticulin can mediate nuclear export of multiple DBD-containing cargoes has important physiological implications. Signaling events that increase Ca^{+2} could transiently alter the distribution of these molecules by inducing their nuclear export via a pathway independent of the Ran-dependent pathway (Fig. 7). There is some precedent for this proposal. In another set of studies published in 1996, we demonstrated that the ubiquitous cytosolic Ca^{+2}-binding protein calmodulin could mediate the nuclear import of a reporter

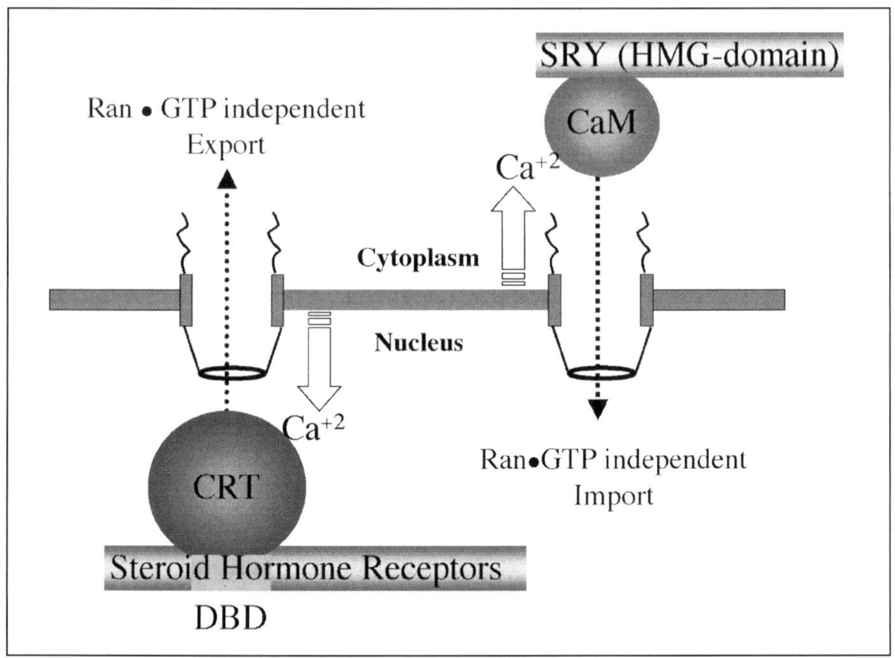

Figure 7. Calcium-dependent import and export carriers. A model summarizing our studies in which the ubiquitous calcium binding proteins calreticulin and calmodulin may function in nuclear export and import, respectively in response to elevated calcium levels during cell activation. These transport pathways appear to function without the direct involvement of Ran and may represent an alternative pathway for transport. Specific classes of import and export cargo appear to use these pathways (calreticulin for steroid hormones; calmodulin for SOX transription factor import).

protein in a Ca^{+2}–dependent, GTP-independent manner.[33] Calmodulin inhibitors blocked nuclear transport under these conditions. Recombinant calmodulin restored ATP-dependent nuclear transport. Calmodulin-dependent transport was inhibited by wheat germ agglutinin consistent with transport proceeding through nuclear pores. We proposed that the release of intracellular calcium stores upon cell activation elevated cytosolic calcium, which then acted through calmodulin to stimulate the novel GTP-independent mode of import. Although the nature of the cargo transported by the calmodulin-dependent pathway have not yet been identified, many proteins are known to interact with calmodulin and it is likely that this pathway will play a role in the transport of at least some of them. Calmodulin is known to cross the nuclear pore by a process of facilitated diffusion much like those of other transport carriers.[34] One of the strongest candidates for import mediated by calmodulin is the SOX family of HMG-box proteins (the HMG1/HMG-2 family). The best studied of these is the SRY protein.[35] A calmodulin binding site completely overlaps an N-terminal NLS in SRY and other SOX transcription factors. In addition, calmodulin appears to mediate SRY import in a manner that is independent of the Ran-dependent export pathway. Figure 7 provides a model for how such nonclassical carriers as calmodulin and calreticulin might alter nuclear shuttling in response to elevated cellular Ca^{+2} levels in response to cellular signaling. These pathways may act in parallel with the Ran-dependent pathway; the Ca^{+2}-dependent conformations of calmodulin and calreticulin appear to act as molecular switches in a manner similar to the GTP-dependent alterations that occur in the conformation of Ran.

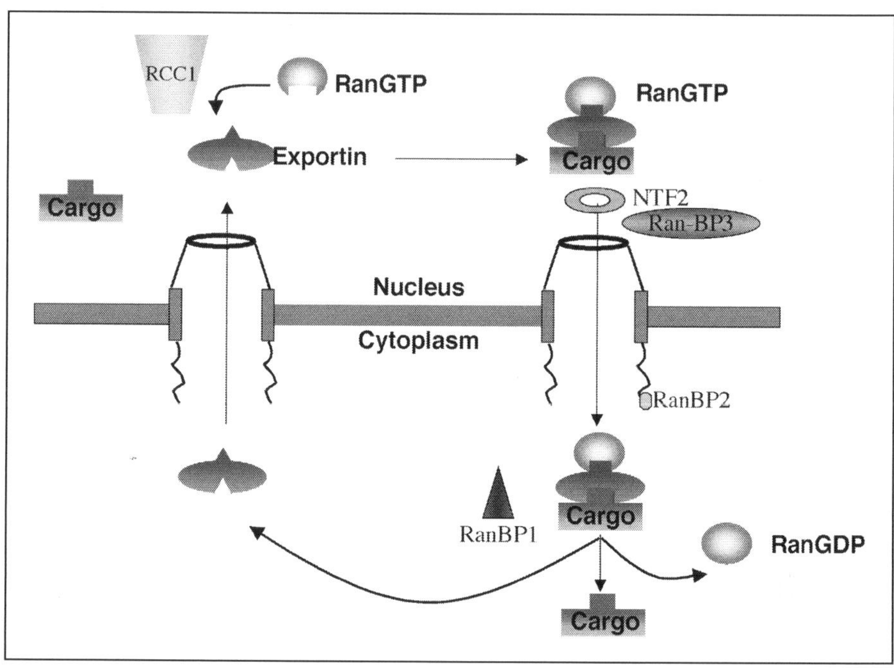

Figure 8. Export cycle regulated by Ran. The 'classical' pathway of nuclear export in which Ran-GTP alters the affinity of export receptors for their cargo. Details of the interaction are described in the accompanying text.

Mechanism of Nuclear Protein Export and Shuttling

Returning to the 'classical' mode of nuclear transport, a growing consensus among researchers in the field suggests that the small GTPase Ran orchestrates the directionality and regulation of nuclear transport. A model depicting the many interactions required for Ran-dependent export is shown in Figure 8. This pathway of nuclear export is very similar to that typically invoked for nuclear import (Fig. 9). Acting as a molecular switch, Ran is stimulated by RCC1, a RanGEF (Guanine nucleotide exchange factor) and RanGAPs (guanine nucleotide activating protein) located at the nuclear pore complex and in the cytoplasm. Other cofactors are involved which can bind Ran and modulate its activity (RanBP1, Ran BP2, and NTF2). Acting together, these regulatory proteins establish a detectable steep gradient of RanGDP/RanGTP across the nuclear pore. Ran-GTP interacts with all of the Importin β family members at a conserved N-terminal domain. However, as pointed out above, in the case of the Exportins such as CRM1, the Ran-GTP complex requires cargo to attain sufficient binding affinity. In a beautifully circular manner, the interaction of CRM1 and cargo are also dependent upon Ran-GTP. Together these properties ensure the intranuclear formation of a trimolecular export complex dependent upon cooperative binding interactions. The physical nature of these interactions must await structural analysis similar to those described above for Importin β. Another key player in the export process is the Ran-GTP Binding Protein 3 (RanBP3). RanBP3 appears to simulate export by stabilizing the export complex and preventing CRM1 from interacting with the nuclear pore complex until it has fully assembled into an export competent complex. It is very likely that this is not yet a complete list of export components; other components that remain unidentified may modulate these processes.

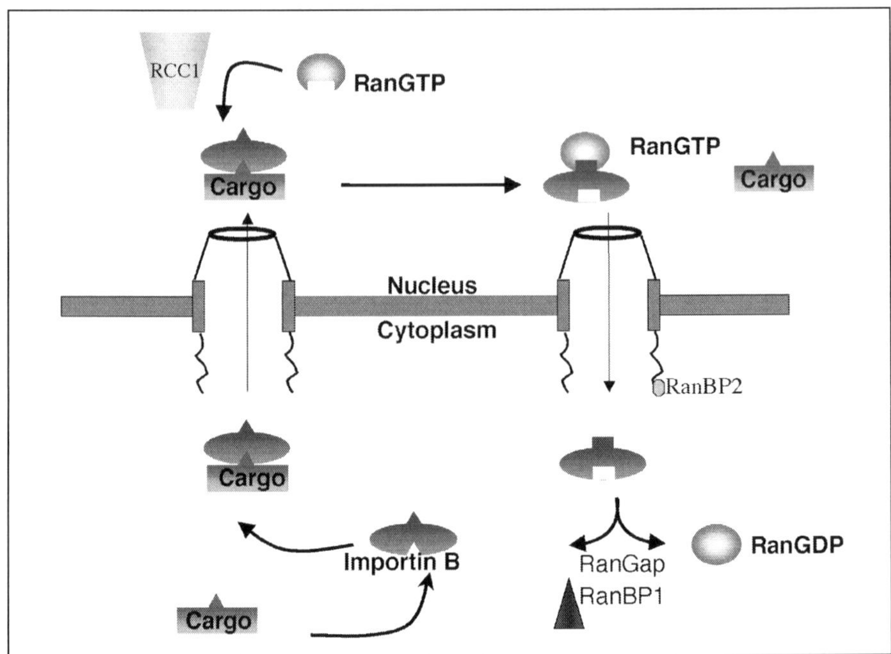

Figure 9. Import cycle regulated by Ran. The 'classical' nuclear import pathway in which Ran-GTP modulates the affinity of import receptors for their imported cargo.

Translocation of export complexes across the nuclear pore complex is now thought to occur by a process of 'facilitated diffusion' involving numerous interactions with the GFXFG motifs present in the nucleoporins (see above). As demonstrated in the structural studies mentioned above, specific interactions of the GFXFG motifs with Importin β have been directly demonstrated. The association and dissociation of these complexes with the nuclear pore is believed to facilitate movement through the pore complex. It is not clear whether a defined sequence of specific interactions with specific nucleoporins is required for such movement. At this point, too little is known about the physical properties of the internal surface of the nuclear pore complex to allow a detailed molecular explanation. Several models have been proposed (reviewed in refs. 31, 36). These models feature sieve-like exclusion of most nonnuclear molecules while allowing selective inclusion of appropriate carrier-bound nuclear cargo.

Following, or perhaps coordinately with translocation, the export complex must disassemble. This step is controlled, in large part by Ran dissociation from the complex. The dissociation step appears to require additional factors including the NXT1, an NTF2-like molecule, RanBP1 and RanBP2.[37,38] RanBP2 (a nucleoporin) may interact with the export complex by virtue of its FG repeats and zinc finger motifs. RanGAP is bound to RanBP2 as a result of its modification by the ubiquitin-like protein SUMO-1.[39] Other nucleoporins such as Nup214 that bind Exportin 1 (CRM1) may also play regulatory roles. Like the mechanism of translocation itself, the dissociation step is subject to active current investigation.

Export of RNA: Ribosomes, tRNA, snRNA and mRNA

The export of the various RNA species is more complex than protein export since transport involves RNA: protein interactions and may be coupled with other processes such as

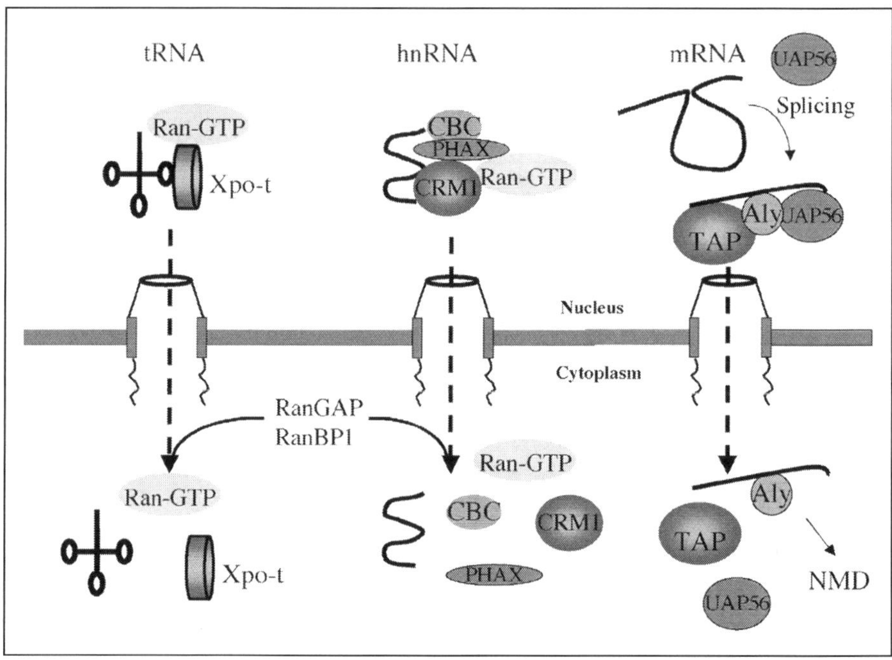

Figure 10. RNA export. The diagram indicates the interactions which must occur during the import of tRNA, hnRNA and mRNA as described in the accompanying text.

transcription and splicing and ribonucleoprotein assembly. We will briefly summarize what is known about the export of the various RNA species. Several excellent reviews focused on RNA export have recently appeared.[40-43] Our purpose here is to familiarize the reader with the general principles involved in RNA export to provide a basis for understanding the importance of nuclear transport in cell physiology. The export pathways are diagrammatically summarized in Figure 10. In a later section we will point out the interrelationships which exist between genomic organization, transcription, splicing and nuclear shuttling.

Ribosome Export

Ribosomes, the large ribonucleoprotein particles responsible for translation in the cytoplasm are made up of a catalytically important rRNA skeleton fleshed out with approximately 80 proteins to stabilize the RNA structure. These proteins are translated in the cytoplasm and enter the nucleus in a facilitated fashion by binding to specific importins. Upon entering the nucleus, the proteins associate with rRNA transcribed in the nucleolus to generate a preribosomal particle. The assembly process is enormously complex involving greater than 60 factors in yeast.[44] The processing of RNA polymerase I and RNA polymerase III transcripts to produce the rRNA species is also a complicated process. The net result is the formation of large and small ribosomal subunits that assemble in the cytoplasm as an intact ribosome on mRNA.

The export of ribosomes involves the protein Nmd3p acting as an adapter between the 60S subunit and Xpo1p.[45] The rest of the export pathway appears to rely on the standard Xpo1p (CRM1)-dependent export pathway relied upon by proteins and hnRNA (Fig. 10). Previous microinjection studies are consistent with this overall proposed pathway.[46]

tRNA Export

tRNAs, critical components of the translation machinery are transcribed in the nucleus and exported. The export process requires a specific importin β molecule which has been termed exportin-t (Xpo-t).[47] Recognition of tRNAs by exportin-t requires maturation of the tRNA (base and sugar modifications, removal of introns) and RanGTP.[47] The export of tRNA then appears to proceed much as it occurs with protein cargoes.

snRNA Export

The small nuclear RNAs (snRNAs) are essential to the formation of ribonucleoprotien complexes (snRNPs) catalyzing splicing reactions. The snRNAs are exported to the cytoplasm where they interact with the snRNP complexes that are reimported into the nucleus by Snurportin-1, and import receptor. The hnRNP proteins number approximately 20. The shuttling hnRNP proteins (particularly hnRNP A1) play key roles in mRNA export (see below).

The export of snRNAs is more complex than that of tRNA. The complex of proteins indicated in Figure 10 are needed for export (CBC, PHAX, CRM1, and Ran). Xpo1p then recognizes this assembled constellation of proteins and RNA. This organization renders the export of snRNAs exquisitely regulated since the binding interactions involved are highly cooperative.

mRNA Export

The export of mRNA is a much more complicated process than any of the other classes of RNA. Morphological analysis of the export of *Chironomous* Balbiani ring particles demonstrated that this process involves massive changes in conformation.[48] This is illustrated in Figure 11. It is also well established that maturation of mRNA must precede export. Because of the immense complexity in the mRNA export pathway, genetic studies in Saccharomyces cerevisae and Schizosaccharomyces pombe have been extremely helpful in dissecting the process. Using the technique of fluorescence in situ hybridization (FISH), the localization of mRNA could be readily examined and defects in localization were immediately identified (reviewed in

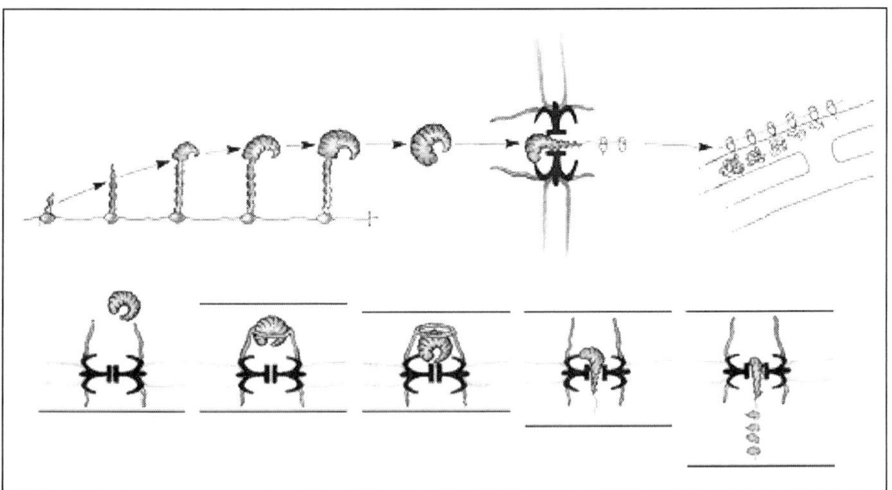

Figure 11. Morphological characterization of RNP export. This diagram demonstrates the enormous conformational alterations that accompany the export of the Balbiani ring particle in Chironomous.[48]

ref. 40). The model shown in Figure 10 is a quick summary of these findings. One molecule implicated in mRNA export is Mex67, which is well conserved from yeast to man where it is referred to as TAP.[49-53] We have been involved in studies on the S. pombe ortholog of Mex67p which interacts with Rae1p and also appears to be involved in mRNA export.[51] TAP interacts with the NTF2 homologue p15 (NXT) and with the factor Aly. TAP is a member of a family of structurally related proteins that are distinct from importins or karyopherins which are Ran-dependent.[54] However, like the importins, TAP interacts with nucleoporins (the GFXFG and GLFG repeats). Another nucleoporin termed GleI may also have a specific function in nuclear export. Thus it is currently thought that the TAP: mRNP complex mediates export of mRNA. Additional screens for export mutants have identified DEAD-box and DECD-box RNA helicases as part of the export machinery.[43] In addition, genes involved in inositol-phosphate metabolism appear to play a role in regulating nuclear export of mRNA.[55,56] Since RanGTP is not directly involved in mRNA export, the usually invoked mechanism of translocation may not be strictly applicable here. As we will discuss in the next section, the surprising finding was that components of this conserved export machinery couple splicing to export and to the process of nonsense-mediated decay or NMD.

Chromatin Organization and Transcriptional Repression

Heterochromatin is a self-assembling intranuclear structure, which can be so compacted that it may lead to the inappropriate silencing of nearby genes. However, formation of heterochromatin is necessary at telomeres and centromeres to allow for proper chromosome segregation during mitosis. There is growing awareness that some boundaries must exist to prevent the unimpeded spread of chromatin. These are termed boundary elements (or barriers). Heterochromatin is also nonrandomly positioned in the nucleus. Most heterochromatin is localized to the nuclear periphery in close proximity to the nuclear membrane. In mammals, the condensed and transcriptionally inactive Barr body (X chromosome) is also located at the nuclear envelope. Recent studies in yeast by Ishii et al[57] suggest that boundary activity may be conferred by components of the nuclear pore complex. Interestingly, various proteins involved in nuclear-cytoplasmic transport, such as exportins Cse1p, Mex67p, and Los1p, emerged from a screen designed to detect boundary activity: a 'boundary trap'. These transport proteins were envisioned to block spreading of heterochromatin by anchoring the locus to the nuclear pore complex. The authors concluded that physical tethering of genomic loci to the NPC could dramatically alter their epigenetic activity. One interpretation of these findings is depicted in Figure 12. In this model, the nuclear pore complex and associated factors involved in transport, transcription and splicing would also serve a function in anchoring this active chromatin fraction to the nuclear pore complex.

Export Machinery, Pre-mRNA Splicing, and Nonsense Mediated Decay

It has become clear that the export machinery is intimately linked to the transcription and splicing machinery (diagrammatically illustrated in Fig. 10). This has been recently reviewed and will only be summarized here.[43] Metazoan premRNAs are complex and may contain multiple exons and introns. SR proteins initially associate with exons and thereby recruit the spliceosome; the introns are packaged into hnRNP complexes. One splicing factor (UAP56) recruits the protein Aly to the spliced mRNP at a position some 20 nucleotides upstream of the newly generated exon-exon junction to form what is known as an exon junction complex (EJC). UAP56 may be release upon interacting with TAP and other factors involved in export. The spliced mRNA is then transported. Some components of the EJC remain to mediate a process known as nonsense mediated decay (NMD). NMD is a mechanism used to selectively degrade mRNAs which contain premature stop codons.[43]

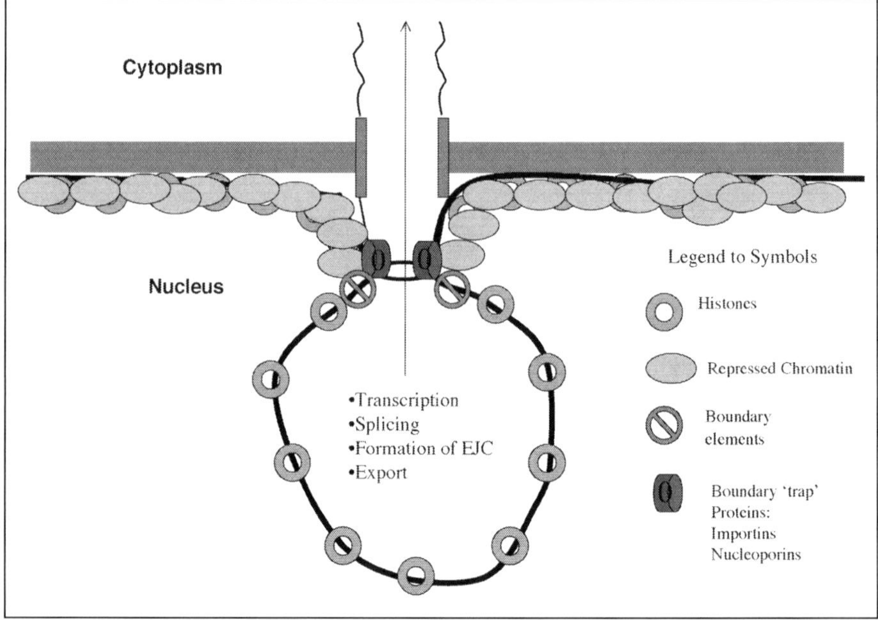

Figure 12. Nuclear pores organize active chromatin. A model for organization of active chromatin at the nuclear pore based on recent boundary trap experiments in yeast and Drosophila (see text). the nuclear pore is envisioned to organize active chromatin and coordinate the processes of transcription, splicing and formation of the Exon-Junction Complex involved in quality control and nonsense mediated decay. The mature RNA is then poised nuclear export through the pore complex.

Thus pathways involved in transcription, splicing and translation are all highly coupled and codependent upon nuclear export. When one considers that the nuclear pore and transport machinery may also organize chromatin (see above) it becomes clear that all phases of the 'central dogma' may depend upon nuclear export.

Summary and Conclusions

We hope that it is clear that this is a rapidly growing and biologically significant field. The highly conserved mechanisms involved in nuclear transport are highly regulated yet adaptable. Many pathways are poorly understood. We have pointed to transport carriers that respond to gradients across the nuclear pore complex and by interacting with molecular switches such as calmodulin and Ran (Fig. 13). The tremendous diversity in transport mechanisms is clearly evident in the export of RNA species; tRNA and snRNA export is β-importin and Ran-dependent while export of mRNA uses TAP (not an importin) and in Ran-independent pathway. The interaction of all of these carriers with the nucleoporins is clearly an important element of the regulatory paradigm, yet how those interactions are employed and how the binding is regulated is still poorly understood.

Finally, it is clear that our knowledge of nuclear export and import remains highly dependent upon inter-disciplinary approaches to analyze structure and function (Fig. 14). Biochemistry and genetic approaches have identified components of the nuclear pore complex and transport machinery. This has been combined with advanced imaging approaches involving electron microscopy, atomic force microscopy, and live cell imaging. Bioinformatics and

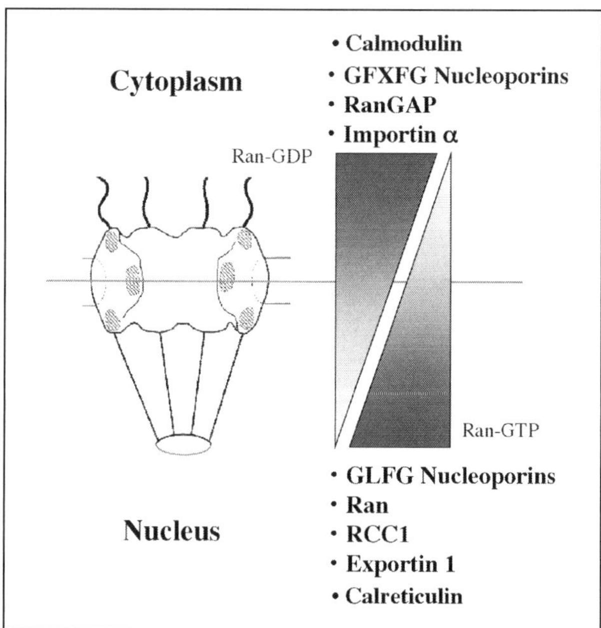

Figure 13. Summary: Gradients driving nuclear transport. The biochemical and structural gradients which exist across the nuclear pore complex are indicated as described in the accompanying text.

proteomics approaches have yielded tremendous information regarding the components of the complex supramolecular structures involved in nuclear transport. Finally biophysical approaches to understanding nuclear envelope diffusion (and electrophysiology) have begun to impact the field. Classical genetics and identification of small molecule inhibitors of transport are ongoing and likely to reveal further avenues of fruitful research. In shuttling across the nuclear pore, more surprises lie ahead.

References

1. Gorlich D, Kutay U. Transport between the cell nucleus and the cytoplasm. Annu Rev Cell Dev Biol 1999; 15:607-660.
2. Kuersten S, Ohno M, Mattaj IW. Nucleocytoplasmic transport: Ran, beta and beyond. Trends Cell Biol 2001; 11:497-503.
3. Hanover JA. The nuclear pore: At the crossroads. FASEB J 1992; 6:2288-2295.
4. Rout MP, Blobel G. Isolation of the yeast nuclear pore complex. J Cell Biol 1993; 123:771-783.
5. Holzenburg RR, Buhle A, Jarnik ELJ et al. Corelation between structure and mass distribution of the nuclear pore complex and of distinct pore complex componets. J Cell Biol 1990; 110:883-894.
6. Davis LI, Blobel G. Identification and characterization of a nuclear pore complex protein. Cell 1986; 45:699-709.
7. Davis LI, Blobel G. Nuclear pore complex contains a family of glycoproteins that includes p62: Glycosylation through a previously unidentified cellular pathway. Proc Natl Acad Sci USA 1987; 84:7552-7556.
8. Hanover JA, Cohen CK, Willingham MC et al. O-linked N-acetylglucosamine is attached to proteins of the nuclear pore. Evidence for cytoplasmic and nucleoplasmic glycoproteins. J Biol Chem 1987; 262:9887-9894.
9. Cronshaw JM, Krutchinsky AN, Zhang W et al. Proteomic analysis of the mammalian nuclear pore complex. J Cell Biol 2002; 158:915-927.

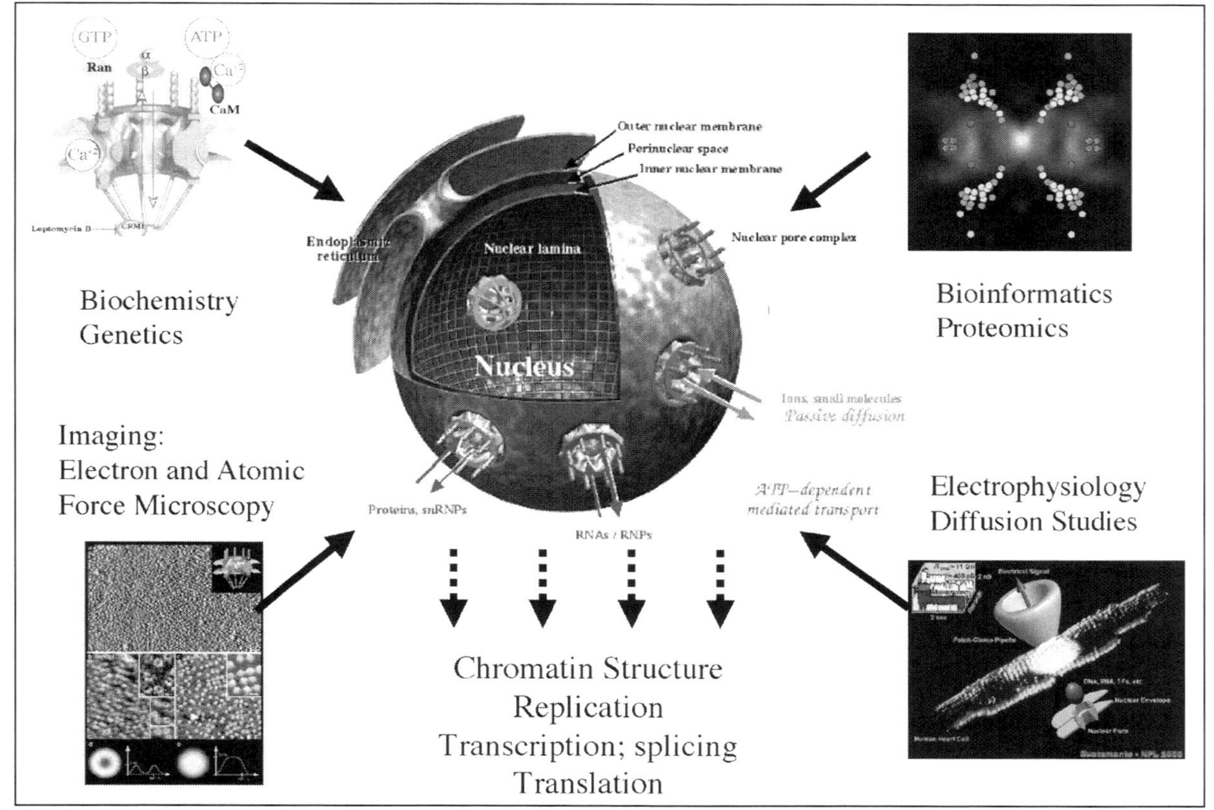

Figure 14. Nuclear export: Approaches and significance. The many approaches used to examine nuclear export are shown with a model of the nuclear envelope as described in the accompanying text.

10. D'Onofrio M, Starr CM, Park MK et al. Partial cDNA sequence encoding a nuclear pore protein modified by O-linked N-acetylglucosamine. Proc Natl Acad Sci USA 1988; 85:9595-9599.

11. Hanover JA. Glycan-dependent signaling: O-linked N-acetylglucosamine. FASEB J 2001; 15:1865-1876.

12. Rout MP, Aitchison JD, Suprapto A et al. The yeast nuclear pore complex: Composition, architecture, and transport mechanism. J Cell Biol 2000; 148:635-651.

13. Goldstein L. Localization of nucleus-specific protein as shown by transplantation experiments in Amoebae proteus. Exp Cell Res 1958; 15:635-637.

14. Borer RA, Lehner CF, Eppenberger HM et al. Major nucleolar proteins shuttle between nucleus and cytoplasm. Cell 1989; 56:379-390.

15. Fischer U, Huber J, Boelens WC et al. The HIV-1 Rev activation domain is a nuclear export signal that accesses an export pathway used by specific cellular RNAs. Cell 1995; 82:475-483.

16. Schmidt-Zachmann MS, Dargemont C, Kuhn LC et al. Nuclear export of proteins: The role of nuclear retention. Cell 1993; 74:493-504.

17. Elion EA. How to monitor nuclear shuttling. Methods Enzymol 2002; 351:607-622.

18. Wen W, Harootunian AT, Adams SR et al. Heat-stable inhibitors of cAMP-dependent protein kinase carry a nuclear export signal. J Biol Chem 1994; 269:32214-32220.

19. Love DC, Sweitzer TD, Hanover JA. Reconstitution of HIV-1 rev nuclear export: Independent requirements for nuclear import and export. Proc Natl Acad Sci USA 1998; 95:10608-10613.

20. Holaska JM, Paschal BM. A cytosolic activity distinct from crm1 mediates nuclear export of protein kinase inhibitor in permeabilized cells. Proc Natl Acad Sci USA 1998; 95:14739-14744.

21. Ossareh-Nazari B, Bachelerie F, Dargemont C. Evidence for a role of CRM1 in signal-mediated nuclear protein export. Science 1997; 278:141-144.

22. Kehlenbach RH, Dickmanns A, Gerace L. Nucleocytoplasmic shuttling factors including Ran and CRM1 mediate nuclear export of NFAT In vitro. J Cell Biol 1998; 141:863-874.

23. Chook YM, Blobel G. Karyopherins and nuclear import. Curr Opin Struct Biol 2001; 11:703-715.

24. Conti E. Structures of importins. Results Probl Cell Differ 2002; 35:93-113.

25. Conti E, Izaurralde E. Nucleocytoplasmic transport enters the atomic age. Curr Opin Cell Biol 2001; 13:310-319.

26. Fornerod M, Ohno M. Exportin-mediated nuclear export of proteins and ribonucleoproteins. Results Probl Cell Differ 2002; 35:67-91.

27. Moroianu J. Distinct nuclear import and export pathways mediated by members of the karyopherin beta family. J Cell Biochem 1998; 70:231-239.

28. Strom AC, Weis K. Importin-beta-like nuclear transport receptors. Genome Biol 2001; 2:REVIEWS3008.

29. Ullman KS, Powers MA, Forbes DJ. Nuclear export receptors: From importin to exportin. Cell 1997; 90:967-970.

30. Wozniak RW, Rout MP, Aitchison JD. Karyopherins and kissing cousins. Trends Cell Biol 1998; 8:184-188.

31. Imamoto N. Diversity in nucleocytoplasmic transport pathways. Cell Struct Funct 2000; 25:207-216.

32. Holaska JM, Black BE, Love DC et al. Calreticulin is a receptor for nuclear export. J Cell Biol 2001; 152:127-140.

33. Sweitzer TD, Hanover JA. Calmodulin activates nuclear protein import: A link between signal transduction and nuclear transport. Proc Natl Acad Sci USA 1996; 93:14574-14579.

34. Pruschy M, Ju Y, Spitz L, et al. Facilitated nuclear transport of calmodulin in tissue culture cells. J Cell Biol 1994; 127:1527-1536.

35. Forwood JK, Harley V, Jans DA. The C-terminal nuclear loclization signal of the sex-determining region Y (SRY) high mobility group domain mediates nuclear import through importin beta1. J Biol Chem 2001; 276:46575-46582.

36. Ossareh-Nazari B, Gwizdek C, Dargemont C. Protein export from the nucleus. Traffic 2001; 2:684-689.

37. Kehlenbach RH, Dickmanns A, Kehlenbach A et al. A role for RanBP1 in the release of CRM1 from the nuclear pore complex in a terminal step of nuclear export. J Cell Biol 1999; 145:645-657.

38. Black BE, Holaska JM, Levesque L et al. NXT1 is necessary for the terminal step of Crm1-mediated nuclear export. J Cell Biol 2001; 152:141-155.

39. Mahajan R, Delphin C, Guan T et al. A small ubiquitin-related polypeptide involved in targeting RanGAP1 to nuclear pore complex protein RanBP2 .Cell 1997; 88:97-107.
40. Lei EP, Silver PA. Protein and RNA export from the nucleus. Dev Cell 2002; 2:261-272.
41. Siomi MC. The molecular mechanisms of messenger RNA nuclear export. Cell Struct Funct 2000; 25:227-235.
42. Aitchison JD, Rout MP. The road to ribosomes. Filling potholes in the export pathway. J Cell Biol 2000; 151:F23-6.
43. Reed R, Hurt E. A conserved mRNA export machinery coupled to premRNA splicing. Cell 2002; 108:523-531.
44. Kressler D, Linder P, de La Cruz J. Protein tran-acting factors involved in ribosome biogenesis in Saccharomyces cedrevisiae Mol Cell Biol 1999; 19:7897-7912.
45. Ho JHN, Johnson AW. Nmd3 encodes an essential cytoplasmic protein required for stable 60S subunits in Saccharomyces cerevisiae Mol Cell Biol 2000; 19:2389-2399.
46. Bataille N, Helser T, Fried HM. Cytplsamic transport of ribosomal subunits microinjected into the Xenopus laevis ooctye nucleus: A generalized, facilitated process. J Cell Biol 1990; 1111:1571-1582.
47. Arts GJ, Kuerstein S, Romby P et al. The role of exportin-t in selective nuclear export of mature tRNAs. EMBO J 1998; 17:7430-7441.
48. Daneholt, B. A look at messenger RNP moving through the nuclear pore. Cell 1997; 88:585-588.
49. Strambio dCC, Rout MP. TAPping into transport. Nat Cell Biol 1999; 1:E31-3.
50. Thakurta AG, Whalen WA, Yoon JH et al. Crp79p, like Mex67p, is an auxiliary mRNA export factor in Schizosaccharomyces pombe. Mol Biol Cell 2002; 13:2571-2584.
51. Yoon JH, Love DC, Guhathakurta A et al. Mex67p of Schizosaccharomyces pombe interacts with Rae1p in mediating mRNA export. Mol Cell Biol 2000; 20:8767-8782.
52. Strasser K, Bassler J, Hurt E. Binding of the Mex67p/Mtr2p heterodimer to FXFG, GLFG and FG repeat nucleoporins is essential for nuclear mRNA export. J Cell Biol 2000; 150:695-706.
53. Santos-Rosa H, Morenos H, Simos G et al. Nuclear mRNA export requires complex formation between Mex67p and Mtr2p at the nuclear pores. Mol Cell Biol 1998; 18:6826-6838.
54. Herold A, Suyama M, Rodrigues JP et al. TAP (NXF1) belongs to a multigene family of putative RNA export factors with a conserved modular structure. Mol Cell Biol 2000; 20:8996-9008.
55. York JD, Odom AR, Murphy R et al. A phopholipase C-dependent inositol polyphosphate kinase pathway required for efficient messenger RNA export. Science 1999; 285:96-100.
56. Feng Y Wente SR, Majerus PW. Overexpression of the inositol phosphatase SopB in human 293 cells stimulates cellular chloride influx and inhibits nuclear mRN A export. Proc Natl Acad Sci USA 2001; 9:875-879.
57. Ishii K, Arib G, Lin C et al. Chromatin boundaries in budding yeast: The nuclear. Pore Connection Cell 2002; 109:551-562.
58. Hinshaw JE, Carragher BO, Milligan RA. Architecture and design of the nuclear pore complex. Cell 1992; 69:1133-1141.

Nuclear Protein Import:

Distinct Intracellular Receptors for DifferentTypes of Import Substrates

David A. Jans and Jade K. Forwood

Entry into the eukaryotic cell nucleus occurs through multiple pathways involving specific targeting signals, and intracellular receptor molecules of the importin/karyopherin superfamily which recognise and dock the nuclear import substrates carrying these signals at the nuclear pore. Subsequent to translocation through the pore via a series of importin-mediated docking steps at multiple sites within it, release into the nucleus is effected by the monomeric guanine nucleotide binding protein Ran. Different importins possess distinct target sequence-binding specificities, meaning that different importins mediate the nuclear import of different classes of proteins. This extends to different classes of transcription factors which are recognised by distinct importins, and whose transport to the nucleus is modulated by specific regulatory mechanisms. The first step of nuclear import is of central importance, with the affinity of the importin:targeting signal interaction being a critical parameter in determining transport efficiency. In the whole cell context, target signal recognition can be modulated through differential expression of the importins themselves, as well as through competition between different importins for the same nuclear import substrate, and between different nuclear import substrates for the same importin. In addition, there are specific mechanisms to modulate targeting sequence-importin interaction directly through phosphorylation. The fact that there are distinct nuclear import pathways for different types of nuclear import substrates enables the cell to regulate these pathways specifically, ensuring efficient nuclear import of particular proteins as and when required.

Introduction

The last few years have seen important advances in our understanding of the cellular factors that mediate signal-dependent nuclear transport. Multiple pathways have been identified, where different types of proteins are transported either into or out of the nucleus through the action of specific molecules called importins/karyopherins that recognize distinct targeting signals.[1-4] The monomeric guanine nucleotide-binding protein Ran plays a central role in release into the nucleus subsequent to translocation through the nuclear envelope-localized nuclear pore complex (NPC), with the specific Ran-transporter NTF2 (nuclear transport factor 2), and Ran-binding and -modifying proteins playing important auxiliary roles. Since the total cellular concentration of targeting signal receptors and Ran in particular is limiting,[2,5,6] the initial step of target signal recognition is critical in terms of getting proteins efficiently to their correct subcellular destination. Regulating this process by enhancing or preventing target signal recognition in response to growth and differentiation signals is a key factor detemining the nuclear entry or otherwise of particular proteins.[2,3,7] This Chapter discusses current knowledge of nuclear protein import in terms of the idea that the existence of multiple differentially regulated nuclear import pathways enables the nuclear import of particular classes of proteins to be carried out efficiently according to the dynamically changing needs of the cell, such as during development, or in response to hormonal or cytokine stimulation. Modulation of importin-target signal recognition is discussed in this context as the main mechanism of regulating nuclear import in the physiological context of the plethora of competing different importin-targeting sequence interactions in the cell.

The Transport Process

The first step of nuclear protein import involves the recognition of targeting signals by members of the importin family and translocation of the proteins carrying them to the cytoplasmic side of the NPC (Fig. 1). For certain classes of protein (see below), the importin $\alpha/\beta 1$ heterodimer is involved, whereas most other pathways require only importin $\beta 1$ or an importin β homolog; in all cases, the importin β homolog docks the importin/transport substrate complex to the NPC, and mediates interaction with Ran. The latter, dependent on the action of several key Ran-interacting and regulating proteins including NTF2, Ran binding protein 1 (RanBP1), Ran GTPase activating protein 1 (RanGAP1) and the nucleotide exchange factor RCC1,[4-6,8] mediates release into the nucleus subsequent to translocation of the import substrate through the NPC. Nucleoporins (nups), the FG (single letter amino acid code)-repeat-containing proteins present in multiple copies throughout the NPC, serve as binding sites for different transport factors including importin β homologs, Ran and NTF2, thus representing "docking bays" for transport factors and assemblies as they pass through the NPC via a succession of transient binding interactions.[9-12] Dissociation of the transport complex at the conclusion of translocation through the NPC is effected by Ran in the GTP bound form binding to importin β to trigger release of the nuclear import substrate and importin α (in the case of importin $\alpha/\beta 1$-mediated nuclear import) into the nucleoplasm. Nuclear RanGTP is maintained at sufficiently high concentration by the combined action of NTF2, which transports RanGDP from the cytoplasm to the nucleus through a series of FXFG-docking events analogous to those of importin β,[11,12] and RCC1, which converts RanGDP into RanGTP.[6,13] Cytoplasmic RanGDP is maintained through RanGAP1, which is predominantly cytoplasmic, as opposed to RCC1, which is nuclear, thus ensuring the asymmetric balance of the guanine nucleotide bound by Ran in the two subcellular compartments.[13] Although non-hydrolysable GTP analogs inhibit nuclear transport, no direct role in either import or export has been demonstrated for GTP hydrolysis,[13,14] whilst the requirement for ATP in the transport process remains controversial.[6,14,15] All transport components are recycled back to the cytoplasm subsequent to nuclear import; importin α has its own specialised nuclear export receptor (the importin β-related molecule CAS) which requires RanGTP for binding to importin α.[16,26]

α Importins

Conventional nuclear localization sequences (NLSs), the short modular peptide sequences sufficient and necessary for nuclear localization of the proteins carrying them, fall into several broad classes. Two of these are highly basic in nature:- those resembling the NLS of the SV40 large tumor antigen (T-ag: PKKKRKV[132])[17] which comprises a short stretch of basic amino acids, and bipartite NLSs which consist of two stretches of basic amino acids separated by a spacer of 10-12 amino acids.[18] Other types include NLSs resembling those of the yeast homeodomain containing protein Mat$\alpha 2$[19] where charged/polar residues are interspersed with

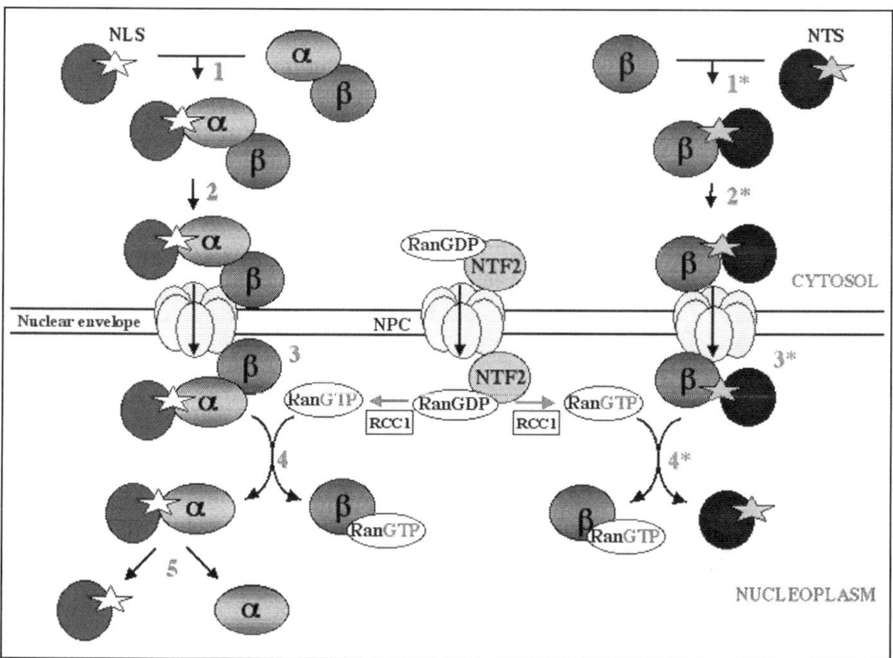

Figure 1. Importin α/β1- and importin-β1-mediated nuclear protein import pathways. 1) A protein containing a nuclear import signal (indicated by star) is recognised by Impβ1 either as a heterodimer with Impα (left) (NLS), or independently of Impα (right—indicated with asterisks) (NTS—"nuclear targeting signal", to differentiate it from Impα recognized nuclear import signals—"NLSs"—for the purposes of this figure. See also Table 1). 2) Impβ docks the transport complex at the NPC through its affinity for FXFG repeats within nucleoporin components of the NPC such as nup358. 3) The complex is translocated through the NPC through a series of transient docking interactions with FXFG repeats present in different nucleoporins throughout the NPC. 4) The transport complex is dissociated by RanGTP binding to Impβ. 5) In the case of importin α/β-mediated nuclear import, the NLS-containing protein is subsequently released from Impα into the cytoplasm by an auto-inhibitory mechanism (see Section C).[88] RanGTP is maintained in the nucleus at high levels by nuclear transport factor 2 (NTF2), which mediates nuclear import of RanGDP (middle), followed by conversion to RanGTP catalysed by the nuclear localized nucleotide exchange factor RCC1.

non-polar residues, or the protooncogene **c**-myc (PAAKRVKLD[328]) where proline and aspartic acid residues either side of the central basic cluster play a role in nuclear targeting.[20] All of these types of NLS are believed to be recognized specifically by the α/β1-importin heterodimer, as has been shown for the importins from several species.[15,21-26]

Whereas *Saccharomyces cerevisiae* contains only a single importin α isoform (SRP1/Kap60p), higher eukaryotes possess more than one, with humans possessing at least 6 distinct importin α isoforms (see Table 1). Through possession of a conserved amino-terminal "IBB" (importin β-binding) domain, all of the importin αs are able to interact specifically with importin β1 (with an apparent dissociation constant, Kd, of 2-18 nM)[24,26] to effect high affinity NLS recognition and docking at the NPC.[10,15,22-26] The fact that higher mammals have a number of distinct importin α isoforms implies specialisation in terms of their cellular role, and ability to recognise particular types of target proteins;[23,24,26,28,29,33,41,44] consistent with this is the fact that single human importin αs are not able to complement a *S. cerevisiae* SRP1 mutant.[41] In contrast, human CRM1, an importin β homolog important in nuclear protein export, can weakly complement a *Schizosaccharomyces pombe* mutant,[82] indicating that in contrast to importin αs, the

Table 1. Nuclear import receptors (importins) for which transport substrates have been defined

Nuclear Import Receptor[a]	Binding Substrates/Targeting Sequence (if known)[b,c]	Techniques Used	Import	References
Importin α1/Importin β1				
Human α1/β1 (SRP1α, Rch-1, KPNA2 / karyopherin β1, p97)	T-ag **PKKKRKV**[132]	Pull down assay (IC_{50} 100 nM) Y2H assay	Yes	26-29 27,30
	NF-κB p50 **QRKRQK**[372]	Pull down assay (IC_{50}=12.5 μM)	Yes	28
	Myc PAA**KRVK**LD[328]	Pull down assay	Yes	28
	RAG-1	Y2H assay	Yes	31
	DNA helicase Q1 **KK**- 15 a.a -**KKRK**[645]	Pull down assay	Yes	29
		Y2H assay		27
	LEF-1 **KKKKRKREK**[382]	Y2H assay		30,32
		Pull down assay		33
	EBNA1 L**KR**P**R**SPSS[386]	Y2H assay Western blot		34
	HIV-1 IN **KRK**- 22 a.a -KELQ**K**QITK[219]	Pull down assay		35
	HIV-1 MA **GKKKYKLKH**[33]	Pull down assay	Truncated forms of α1 inhibit import	36
	HIV1422 **KKK**Y**KLK**[32]	Pull down assay	Yes	28
	HIV1423 **KSKKK**[114]	Pull down assay	Yes	28
	RCP 4.1R EED- 350 a.a -**KKKRER**LD[413]	Resonant mirror detection (Kd 30 nM)	Yes	37
	NP **KR** - 9 a.a. - **KKKKL**[171]		Yes*	26
	hnRNP K **KR** - 11 a.a. - **KRSR**[37]		Yes*	26
	P/CAF		Yes*	26
	RCC1		Yes*	26
	Heat Shock TF3 **KRKR**[243]	Pull down assay		38

continued on next page

Table 1. Continued

Nuclear Import Receptor[a]	Binding Substrates/Targeting Sequence (if known)[b,c]	Techniques Used	Import	References
Mouse α1/β1 (PTAC58, Pendulin, Impα-P1, kpna2/PTAC97)	T-ag + CK2 site SSDDEATADSQHSTP**PKKKRKV**[132]	ELISA K_d 2.7 nM	Yes	23
	T-ag **PKKKRKV**[132]	ELISA K_d c. 50 nM	Yes	25
		Pull down assay		29
	N1N2 **RKKRK**- 9 a.a -**KAKKSK**[554]	ELISA K_d 5.4 nM	Yes	23
	RB **KR**- 11 a.a -**KKLR**[877]	ELISA K_d 45 nM	Yes	15
	Dorsal + PK-A site RRPS- 22 a.a -**RRKRQK**[340]	ELISA K_d 26 nM	Yes	22
	CBP80 **RRR**- 11 a.a -**KRRK**[20]	Pull down assay	Yes	29
	DNA helicase Q1 **KK**- 15 a.a -**KKRK**[645]	Pull down assay	Yes	29
	LEF-1 **KKKKRKREK**[382]	Y2H assay	Yes	32
		Pull down assay		33
	HMG14	ELISA K_d 15.5 nM		39
	HMG17	ELISA K_d 3.5 nM		39
Importin α3/Importin β1				
Human α3/β1 (QIP1/KPNA4)	T-ag **PKKKRKV**[132]	Pull down assay	Yes	26,27
		Y2H assay		27
	DNA helicase Q1	Pull down assay	Yes	27,29
	CYFQKKAANMLQQSGSKNTGA**KKRK**[645 d]	Y2H assay		27
	tTG DIL**RR**- 323 a.a. -PKQ**KRK**[602]	Y2H assay		40
		Immunoprecipitation		
	NP **KR** - 9 a.a. - **KKKKL**[171]		Yes*	26
	hnRNP K **KR** - 11 a.a. - **KRSR**[37]		Yes*	26
	P/CAF		Yes*	26
	RCC1		Yes*	26

continued on next page

Table 1. Continued

Nuclear Import Receptor[a]	Binding Substrates/Targeting Sequence (if known)[b,c]	Techniques Used	Import	References
Human α4/β1 (hSRP1g, KPNA3)	T-ag **PKKKRKV**[132]		Yes	26,41
	NP **KR** - 9 a.a. - **KKKKL**[171]		Yes*	26
	hnRNP K **KR** - 11 a.a. - **KRSR**[37]		Yes*	26
	P/CAF		Yes*	26
	RCC1		Yes*	26
Importin α5/**Importin** β1				
Human α5/β1 (hSRP1, NPI-1, KPNA1)	T-ag **PKKKRKV**[132]	Pull down assay	Yes	26-29
		Y2H assay		27,42
	RAG-1	Y2H assay		43
	STAT1	Pull down assay	Yes	44
	CBP80 **RRR**- 11 a.a.-**KRRK**[20]	Pull down assay	Yes	29
	DNA helicase Q1 **KK**- 15 a.a. -**KKRK**[645]	Pull down assay	Yes	29
		Y2H assay		27
	BRCA1 **KRKRRP**[508], **PKKNRLRRK**[615]	Y2H assay	Yes	45
		Pull down assay		42
	Mitosin **KRQR**- 20 a.a. -**KKSKK**[2958]	Pull down assay		42
	Myc PAA**KRVK**LD[328]	Pull down assay	Yes	28
	NF-*k*B p50 **QRKRQK**[372]	Pull down assay	Yes	28
	NF-*k*B p65 HRIEE**KRKR**TYETFKSI[345]	Pull down assay	Yes	28
	HIV1422 **KKK**YKL**K**[32]	Pull down assay	Yes	28
	HIV1423 **K**SKKKAQ[116]	Pull down assay	Yes	28
	NP **KR** - 9 a.a. - **KKKKL**[171]		Yes*	26
	hnRNP K **KR** - 11 a.a. - **KRSR**[37]		Yes*	26
	P/CAF		Yes*	26
	RCC1		Yes*	26

continued on next page

Table 1. Continued

Nuclear Import Receptor[a]	Binding Substrates/Targeting Sequence (if known)[b,c]	Techniques Used	Import	References
Mouse α5/β1 (SRP1, Impα-S1; kpna1)	LEF-1 **KKKKRKREK**[382]	Y2H assay Pull down assay		32 33
Yeast α/β1 (SRP1/Kap60p/Kap95p, RSL1p)	T-ag **PKKKRKV**[132] T-ag + CK2 site <u>SSDDE</u>ATADSQHSTP**PKKKRKV**[132] N1N2 **RKKRK**- 9 a.a -**KAKKSK**[554] HIV-1 IN **KRK**- 22 a.a -KELQ**KQITK**[219]	Pull down assay ELISA K_d 2.1 nM ELISA K_d 5.7 nM Pull down assay	Yes Yes Yes	10 23 23 35
Human α7/β1 (KPNA6)	T-ag **PKKKRKV**[132] NP **KR** - 9 a.a. – **KKKKL**[171] hnRNP K **KR** - 11 a.a. spacer- **KRSR**[37] P/CAF RCC1		Yes* Yes* Yes* Yes* Yes*	26 26 26 26 26
Snurportin/Importin β1	SnRNPs, U1 snRNP, U5 snRNP: m3-G cap	UV cross-linking	Yes	46
XRIPα/Importin β1	Replicon Protein A p70	Y2H Pull down assay	Yes	47
Importin β1				
human β1	TCPTP **RKRIR**ED**RK**ATT AQKVQQMKQRLNENE**RKRKR**[381] Cyclin B1 Ribosomal proteins: rpS7/rpL5/rpL23a BIB domain: VHS**HKKKK**IRTSPTFTTPK**T**L**RL**RRQPKYP**RK**SAP**RRNK**LDHY[74] HIV-1 Rev **RQARRNRRR**WR[46] HTLV-1 Rex MP**KTRRR**PRRSQ**RKR**PPT[18] SMAD3 KKLKK[44]	Immunoprecipitation Pull down assay Pull down assay Pull down assay Pull down assay Pull down assay Pull down assay	 Yes Yes Yes Yes	48 49 50 51 52,53 54

continued on next page

Table 1. Continued

Nuclear Import Receptor[a]	Binding Substrates/Targeting Sequence (if known)[b,c]	Techniques Used	Import	References
	SREBP2 RSSINDKIIELKDLVMGTDAKMHKSGV LRKAIDYIK YLQQVNHKLRQENMVLKLANQKNKL[403] (dimer)	Pull down assay	Yes	55,56
	PTHrP TNKVETYKEQPL**KTPGKKKKGK**[93]	Pull down assay	Yes	57
mouse β1 (PTAC97)	PTHrP TNKVETYKEQPL**KTPGKKKKGK**P[94]	ELISA Kd 1.6 nM	Yes	57,58
	GAL4 RLKKLKCSKEKPKCAKCLKNNWECRYSPKTKR[47]	ELISA Kd 21 nM		59
	bZIP TFs Fos, Jun, AP-1, CREB	ELISA Kd 6 nM (CREB); FP; native PAGE	Yes	60
	HMG-box containing TFs			
	SOX9	ELISA Kd 1.7 nM		61
	SRY	ELISA Kd 3.1 nM; FP; native PAGE	Yes	62
	Histones			
	H1	ELISA Kd 2.7 nM		39
	H2A/H2B	ELISA Kd 5.5 nM		39
	H2Az/H2B	ELISA Kd 4 nM		39
	H3/H4	ELISA Kd 8.2 nM		39
	H3/H4 (acetylated)	ELISA Kd 29 nM		39
	Telomere binding protein TRF1	ELISA Kd 23 nM		63
	IGFBP-3 NLS KKGFYKKKQCRPSKGRKR[232]		Yes	64
yeast β1 (Kap95p, RSL1p)	PTHrP YLTQETNKVETYKEQPL**KTPGKKKKGK**P[94]	ELISA Kd 2 nM	Yes	58
	GAL4 RLKKLKCSKEKPKCAKCLKNNWECRYSPKTKR[47]	ELISA Kd 15 nM		59
	bZIP TFs AP-1, Jun, Fos, CREB	ELISA Kd 38 nM (CREB); FP; native PAGE	Yes	60

continued on next page

Table 1. Continued

Nuclear Import Receptor[a]	Binding Substrates/Targeting Sequence (if known)[b,c]	Techniques Used	Import	References
Importin β2				
human β2 (transportin, karyopherin β2)	mRNA binding proteins: hnRNP A1 M9 sequence: NQSSNFGPMKGGNFGGRSSGPYGGGGQ YFAKPRNQGGY[305 e]	Pull down assay	Yes	65,66
	hnRNP F	Pull down assay	Yes	67
	hnRNP A2, hnRNP M, FBRNP, TAF$_{II}$[68]	Phage display screening		68
	hnRNP DO	Western blotting		69
	Ribosomal proteins: rpS7, rpL5, rpL23a BIB domain: VHSHKKKKIRTSPTFTTPKTLRLRRQPKYPRKSAPR RNKLDHY[74]	Pull down assay	Yes	50
yeast β2 (Kap104p)	mRNA binding proteins: Nab4p/Nab2p, NAB35: APVDNSQRFTQRGGGAVGKNRRGGRGGNRG GRNNNSTRFNPLAKALGMAGESNMN[252]	Pull down assay Y2H	Yes	68
Importin β3				
human β3 (Kapβ3, RanBP5)	Ribosomal proteins: RL13, rpS7, rpL5, rpL23a BIB domain: VHSHKKKKIRTSPTFTTPKTLRLRRQPK YPRKSAPRRNKLDHY[74]	Overlay assay Pull down assay	Yes	50
yeast β3 (Kap121p, Pse1p)	Ribosomal proteins L25: MAPSAKATAAKKAVVKGTNGKKALKVRTSATFRLP KTLKLAR[41]	Pull down assay	Yes	70
	Pho4: SANKVTKNKSNSSPYLNKRKGKPGPDS[166]	Pull down assay	Mislocalization in mutant	71,72
	TATA binding protein	Pull down assay		73
	Pdr1 TYKSWTDMNKILLDFDNDYSVYRSFAH	Pull down assay	Mislocalization in mutant	74

continued on next page

Table 1. Continued

Nuclear Import Receptor[a]	Binding Substrates/Targeting Sequence (if known)[b,c]	Techniques Used	Import	References
Importin β4				
yeast β4 (Kap123p, YRB4p)	Ribosomal proteins: RP10Ap, RP28A, RPL15A, RL32p, RPL4A, RPL16A, RPL41A, L25: MAPSAKATAAKKAVVKGTNGKKALKVRTSATFRLPKTLKLAR[41]	Pull down assay	Mutation/ over-expression causes mislocalization	70
	TATA binding protein	Pull down assay	Yes	73
Importin β7				
Xenopus β7 (RanBP7)	Ribosomal proteins: rpL5, rpS7, rpL23a BIB domain: VHSHKKKKIRTSPTFTTPKTLRLRRQPKYPRKSAPRRNKLDHY[74]	Pull down assay	Yes	50
yeast β7 (Nmd5, Kap119p, yjr132w)	TFIIs	Pull down assay	Mislocalization in mutant	75
	HOG1 (MAPK)		Mislocalization in mutant	76
yeast Kap122 (Pdr6p)	TFIIA Toa1p and Toa2p subunits	Pull down assay	Mislocalization in mutant	77
yeast SXM1 (Kap108p, Ydr395)	Lhp1p; ribosomal proteins RPL16p, Rpl25p, Rpl34p	Immunoprecipitation	Mislocalization in mutant	78
yeast Mtr10 (Kap111p, Yor160w)	Np13p: RGG box + C-terminal tail	Pull down assay	Mislocalization in mutant	79
yeast Kap114 (Hrc1004/ygl241w)	TATA-binding protein	Pull down assay	Mislocalization in mutant	73

continued on next page

Table 1. Continued

Nuclear Import Receptor[a]	Binding Substrates/Targeting Sequence (if known)[b,c]	Techniques Used	Import	References
human TRN-SR (transportin SR)	S/R-domain-containing proteins: ASF/SF2, SC35	Pull down assay	Yes	80
Importin β1/Importin β7	Histone H1[0], H1[t]	Pull down assay	Yes	81

[a]Importin α numbering as per Koehler et al[26]

[b]Abbreviations: aa, amino acid: RAG-1, recombination activating protein 1; RCP, red cell protein; RB, Retinoblastoma protein; P/CAF p300/CBP-Associated Transcriptional Co-Factor; STAT, signal transducer and activator of transcription; CBP80, Cap-binding protein; LEF, Lymphocyte enhancer factor; EBNA, Epstein-Barr virus nuclear antigen; IN, HIV-1; tTG, tissue transglutaminase; snRNP, small nuclear ribonucleoprotein particale; m3G, trimethylguanosine; ICP: Infected cell protein; FBRNP, fetal brain RNA-binding protein; TBP-associated factor 68.

[c]The single letter amino acid code is used, with critical NLS residues in bold type; phosphorylation sites implicated in enhancing importin binding are underlined.

[d]The upstream sequences from the Q1 NLS are required for efficient recognition by importin α3.[22]

[e]Bogerd et al[66] have determined the minimal importin β2-recognized sequence to be YNNQSSNFGPMK[227], with only the underlined sequences being absolutely essential, with residue Y[266] able to be any aromatic amino acid, F[273] any hydrophobic amino acid, and P[272] able to be substituted by K, M[276] by L or V and K[277] by R, without any loss of function.

*Transport only in the presence of exogenous cytosol.

specific function of different importin β homologs is likely to be conserved across eukaryotes. Conservation in the nuclear transport system across eukaryotes is further indicated by the fact that the nuclear targeting signals from diverse organisms such as SV40 (the T-ag NLS) are functional in distinct cell types from a range of different species,[23,24,26,83] and that components of the nuclear transport machinery from different species can function efficiently in concert to mediate nuclear import.[23,24,26,81] Significantly, the regulation of nuclear import through the modulation by phosphorylation of NLS recognition by importins is also conserved across eukaryotes.[22,84]

Direct evidence indicates that different importin α isoforms have distinct NLS binding properties (see Table 1).[23,26-30,33,41,44] Importin α5 (NPI-1) but not α1 (Rch1) recognizes the transcription factor (TF) STAT1, for example,[44] whilst, in contrast to α1 and α5, importin α3 (Qip1) requires specific additional flanking sequences either side of conventional T-ag-like basic amino acid cluster to mediate binding and nuclear import in the presence of importin β1/Ran/NTF2.[29] The high sequence specificity of NLS binding is further indicated by the fact that importin α1 and α5 recognize the lymphoid enhancer factor-1 (LEF-1) NLS (KKKKRKREK[382]) but not the highly similar T-cell factor-1 NLS (KKKRRSREK[237]), as shown using yeast two hybrid (Y2H) and protein binding assays.[33]

A qualitative comparison of the nuclear import properties of 5 of 6 human importin α isoforms[26] indicated that α5/α3/α7/α2 were markedly more efficient in mediating nuclear import of a T-ag NLS-containing import substrate than α4, which in turn was markedly better than α1 (see also ref. 41). Analogously, nucleoplasmin was imported into the nucleus more efficiently by importin α5 than α2-α4, and even more so than α1/α7, and hnRNP K was imported by α5/α1-α3 markedly better than by α4/α7. Interestingly, RCC1 was found to be transported to the nucleus by importin α3 more efficiently than α4, but not by any of the other isoforms. These apparent preferences were altered quite strikingly in several cases when the importin αs were confronted with two substrates simultaneously (see also below).[26] Studies comparing importin αs from several species[23,24,26] support the idea that the different α isoforms have distinct NLS-binding affinities, and accompanying nuclear import properties. Table 1 lists observations for the different importin αs from several species in the context of their ability to bind to and mediate nuclear transport conferred by different NLSs, with quantitative results indicated where possible.

X-ray crystallographic data for yeast importin α (closest in homology to the human/mouse α5 isoform) lacking the IBB domain[85] indicates a structure of 10 tandem Armadillo (ARM) repeats (each comprising 3 α-helices) in a right hand superhelix of helices which effectively constitute a binding face, with two specific sites (repeats 2-4 and 7-8; residues 121-247 and 331-417) for the T-ag NLS. T-ag NLS binding was found to be through a complex series of electrostatic and hydrogen-bonding interactions, as well as hydrophobic interactions involving Asn and Trp residues;[85] that there are two NLS binding sites is the basis of bipartite NLS recognition.[86,87] Kobe[88] showed that residues KRRNV[53] of the IBB domain of full length mouse importin α1 bind to the first NLS-binding site (amino acids 146-235), constituting an autoinhibitory mechanism. The fact that importin β1 binding to the IBB domain releases these residues from the NLS-binding site, explains the switch between the cytoplasmic high affinity NLS-binding form of importin α bound to importin β, and the low affinity form that results upon release from importin β1 in the nucleus effected by RanGTP binding to the latter.[88] This implies strongly that NLS binding by importin α is much less likely in the absence than in the presence of importin β1 (see, however, ref. 24).

Two other nuclear import proteins appear to bind import substrates and mediate nuclear import in concert with importin β1 in analogous fashion to importin α. Through an IBB domain highly homologous to that of the importin α isoforms, the m3G-cap receptor snurportin,[46] interacts with importin β1 and mediates nuclear import of snRNPs, whilst

XRIPα,[47] completely unrelated in sequence to importin α, heterodimerises with importin β1 to mediate nuclear import of the single stranded DNA binding protein replication protein A (RPA). Neither snurportin nor XRIPα contain ARM repeats, and hence are not members of the importin superfamily.

Since the different importin α isoforms possess distinct NLS-recognition properties, it is clear that the importin α repertoire of a particular cell/tissue type will determine what types of proteins are targeted to the nucleus. The efficiency of nuclear import of particular nuclear import substrates can thus be regulated by differential modulation of the level of expression of different importin isoforms. Nadler et al[28] found that importin α5 and α1 showed marked differences in protein expression in a range of human leukemia lines, the trend being that more differentiated lines contained higher levels of both importins; less differentiated lines contained only one isoform, and lower amounts thereof. Lipopolysaccharide, concanavalin A, or phorbol ester/ionomycin treatment was able to increase importin α expression in normal blood lymphocytes, indicating that importin levels can be regulated in response to cellular signals. Importantly, the differences observed in the importin repertoire correlated with differential nuclear import efficiency for proteins containing different NLSs.[28] Other studies indicate quite marked differences in the level of mRNA expression for different importin α isoforms across different tissues; whereas importin α5 shows low to medium level expression in most tissues, mouse and human importin α4 and α5/α6 are highly expressed in testis and to a lesser extent in spleen.[32,90] Importin α7 appears to be the most widely and highly expressed human isoform, although it is absent from thymus.[26]

Assessment of protein levels in different tissues supports the idea of differences in expression between the various importin αs.[41,89] Human importin α4, for examples, makes up >1% of protein in skeletal muscle (100-fold higher levels than importin α5), but appears to be absent from heart/spleen/kidney.[41] Importin α1 protein levels are high in heart, testis skeletal muscle and ovary, whilst importin α3 is most highly expressed in ovary.[89] Koehler et al[26] compared the human importin α isoforms using Western analysis in a range of tissues, concluding that importin α6 is almost certainly a testis-specific isoform. Importin α7 was the most widely and highly expressed isoform, with only spleen and liver exhibiting low levels. The other isoforms were all expressed at reasonable levels in most tissues, with large amounts of all of them in ovary, lung and small intestine, lower amounts in testis and heart, and small amounts in other tissues; importin α4 and α7 appeared to the only major forms in brain, with importin α5 the only reasonably well expressed isoform in liver. Spleen showed low levels of expression of importin α1, α3 and to a lesser extent α5, with importin α5 and α3, in addition to α7, being the predominant forms in kidney.

That importin levels are regulated in lower eukaryotes is indicated by a study by Torok et al,[90] who found that oho31 from *Drosophila* (importin α1/pendulin) is supplied maternally, and rapidly degraded during the first 13 nuclear divisions, and then expressed at reduced levels in proliferating tissues; reduction of expression through P-element insertion or excision led to malignant development of haematopoietic organs and the genital disc. The fact that the levels of different importins are able to change according to tissue-type, differentiation/developmental state and the stage of the cell cycle represents a mechanism by which the cell, according to need, can regulate nuclear transport of certain proteins or types of proteins. More recent studies in *Drosophila*[91] and *Caenorrhabditis elegans*[92] suggest that importin α3 is expressed ubiquitously throughout development, but may also play a specialized role in oogenesis. The expression of importin α3 increases during *Drosophila* embryogenesis, in contrast to α1 which is expressed in the early embryo, and shows reduced expression later in embryogenesis and in pupae and adults (see also above), giving the idea that importin α3 is the main "house-keeping" importin α, whereas α1 has a more specialised role during embryogenesis.[91,92] A role for importin α3 in *Drosophila* oogenesis is implicated by its increased expression in the requisite tissues during oogenesis, as well as the fact that a mutant in importin α3 is poorly viable and female sterile. Analogously, importin α3 shows high levels of expression during development in *C. elegans* larvae and adults,[92] with RNAi inhibition of expression leading to a sterile germline due to a failure in oogenic meiosis. Thus, although importin α3 would appear to be the importin α isoform ubiquitously expressed in a range of eukaryotes, it has an essential role in oogenesis, presumably during which it is required specifically to effect nuclear import of a critical mediator(s) of the oogenic process.[91,92]

Importin β1 and Homologs

Signal-mediated nuclear protein import is not mediated exclusively by the α/β1-importin heterodimer, but also by an array of importin β homologs, all of which appear able to function in the absence of importin α to bind targeting sequences present in transport substrates, dock them at the NPC, and interact with Ran to mediate translocation into the nucleus. Table 1 lists the various importin β homologs for which nuclear import substrates have been identified. Importin β1 itself directly recognises targeting sequences present in transport substrates such as the T-cell protein tyrosine phosphatase (TCPTP),[48] human immunodeficiency virus (HIV-1) Rev protein,[51,52] and parathyroid hormone-related protein (PTHrP).[57,58] Although the generally arginine-rich NLSs[51-53] of HIV-1 Rev and HTLV-A Rex may bind importin β through simulating binding by the IBB domain of importin α, it seems clear that binding of the other substrates is through an alternative mechanism.[56-58,60] In the case of PTHrP nuclear import, the lack of a need for importin α has been formally demonstrated by reconstituting nuclear import in vitro using only importin β and Ran;[57,58] significantly, importin α inhibits nuclear import markedly. The pathway by which importin β1 mediates nuclear import is shown schematically in Figure 1 (right side); the other importin β-homolog-mediated pathways (Table 1) are believed to be analogous in all respects.

There are at least 12 (yeast) different importin β homologs in eukaryotic cells additional to importin β1,[1,2,4] many of which are involved in mediating nuclear export (see Chapter in this volume on Nuclear Export). The different importin β homologs appear to have quite specific transport roles with respect to particular classes of proteins; importin β4 (Kap123p/Yrb4p), for example, mediates the import of ribosomal proteins into the nucleus,[70] as can β3 (Kap121p/Pse1p/ Imp5/RanBP5),[50,70] Sxm1p (Kap108p)[78] and Imp7 (RanBP7/Nmd5p/Kap119p).[50] Interestingly, a region from ribosomal protein rpL23a, the β-like importin receptor binding (BIB) domain (see Table 1), has been shown to be recognized by each of importin β1, β2, β3 and Imp7.[50] rpL5 and rpS7 can similarly be transported to the nucleus by all of these importin β homologs, as well as by importin α/β1; import of rpL5 mediated by importin β1-mediated import is reported to be "variable", and by α/β1- and β2 "weak", implying different transport properties on the part of different importin β homologs.[50] In analogous fashion to the BIB domain, amino acids 1-41 of rpL25 can be recognized by either importin β3 or β4.[70] Importin β2 (Kap104p/transportin)[65,67-69] and Mtr10p (Kap111p)[79] mediate the nuclear import (as well as export in the former case) of mRNA-binding proteins. The hydrophobic 38 amino acid M9 nuclear targeting (shuttle) sequence, first identified in the large heterogeneous human mRNA-binding protein hnRNP A1,[65] has been identified in a number of other transport substrates of transportin,[66,67] with a consensus sequence (YNNQSSNFGPMK277) defined by Bogerd et al.[66]

The crystal structure of importin β1 bound to the IBB domain[93] reveals 19 HEAT repeats which are structurally related to ARM repeats, but consist of two α-helices, one of which is kinked by a proline residue, rather than three. Importin β1 possesses a snail-like winding structure with repeats 7-11 and 12-19 wrapping around the IBB domain. The structure implies that significant conformational changes occur when importin β1 binds to or releases the IBB domain (see above), suggesting how dissociation of the importin α/β1 heterodimer may be achieved subsequent to nuclear entry and binding of RanGTP. The structure of importin β2 bound to RanGTP[94] shows 18 HEAT repeats arranged in 2 continuous orthogonal arches, with Ran clamped in the amino terminal arch and substrate binding activity in the C-terminal arch (repeats 7-17 bind hnRNP A1); the two arches are spanned by HEAT repeat 7 which is implicated in RanGTP-mediated dissociation of the import substrate through binding of Ran to the importin β2 C-terminus. By analogy with the ARM repeats of importin α, the array of HEAT repeats in importin β homologs provides a large surface for interaction with a variety of different proteins; e.g., importin β1 amino acids 160-340 bind to the NPC,[95] 1-462 to cyclin B1,[49] 148-401 to Imp7 (see below),[81] 200-876 to histone H1,[81] 280-450 to the BIB domain,[50] 1-364 to RanGTP, and so on. Bayliss et al[11] showed that a binding face of importin β1 within HEAT repeats 5 and 6, not involved in binding Ran or the IBB domain, is involved in binding to FXFG repeats through a series of hydrophobic interactions; mutations specifically reducing the efficiency of importin β1 binding to FXFG nucleoporins lead to impaired nuclear import. In terms of the release end of nuclear transport, RanGTP binding is believed to induce conformational change in importin β1[96] (see also ref. 94) such that it twists to release the IBB domain; RanGTP binding similarly modulates binding to FXFG sequences.[11]

As alluded to above, importin β1 has been reported to heterodimerize with other importin β homologs such as RanBP7 (Imp7) and RanBP8 (Imp8).[50,81,97] The heterodimer of β1/Imp7 has been reported to bind to linker histone H1 variants, with H1 nuclear import being able to be reconstituted in vitro using human β1 and *Xenopus* Imp7.[81]

At this stage in our understanding, it seems that differential regulation of importin expression as a mechanism to regulate nuclear import may apply more to importin α isoforms rather than to importin β homologs, although only limited information is available with respect to the latter. Western analysis of importin β1 and CAS in various human tissues[26] indicates expression in all tissues, with the exception of spleen, and low amounts of the former in ovary; highest expression levels of both proteins were observed in testis and brain. Interestingly, lacZ reporter analysis indicates that importin β1 (Ketel) expression in Drosophila is ubiquitous during embryogenesis, but apart from in the central nervous system, appears to be largely restricted to mitotically active cells in larval and adult tissues, being particularly high in ovary and testis.[98,99] Dominant negative Ketel[D] mutations induce sterility by preventing cleavage nuclei formation, implying a key role in embryogenesis.[98]

Competition between Target Sequences/Receptors

Since there are nuclear targeting signals that are able to confer interaction with more than one importin such as the T-ag NLS which is recognised by all the different importin αs (see Table 1), and the BIB domain which can confer interaction with four different importin β homologs,[50] it seems very likely that, in the context of the whole cell, there is competition between different importins for nuclear import substrates, and that the highest affinity binding interaction should be dominant;[2] this should also hold true for competition for the same importin between different targeting signal-containing proteins (see ref. 26). The cellular concentration of total importin β family members (c. 20 μM)[2] in *Xenopus* oocytes is somewhat higher than that for Ran (c. 10 μM), with even lower concentrations in HeLa cells,[5] meaning that the intrinsic affinity of transport substrate binding by importins must play an important role in subsequent competition for Ran. Estimated to be at a cellular concentration of 3 μM similar to other importin β homologs (the cellular importin αs are estimated at c. 6 μM), importin β1 would also appear likely to be in intense demand as a binding partner, not only by proteins recognized by heterodimers in which importin β1 is involved such as conventional NLS-containing proteins, snRNPs, and histone H1 variants, but also by directly bound import substrates such as PTHrP or TCPTP (see Table 1; see also Fig. 2). Of relevance is also the fact that the amount of import substrates such as snRNPs, histones and ribosomal subunits required in high, more or less constitutive demand, is almost

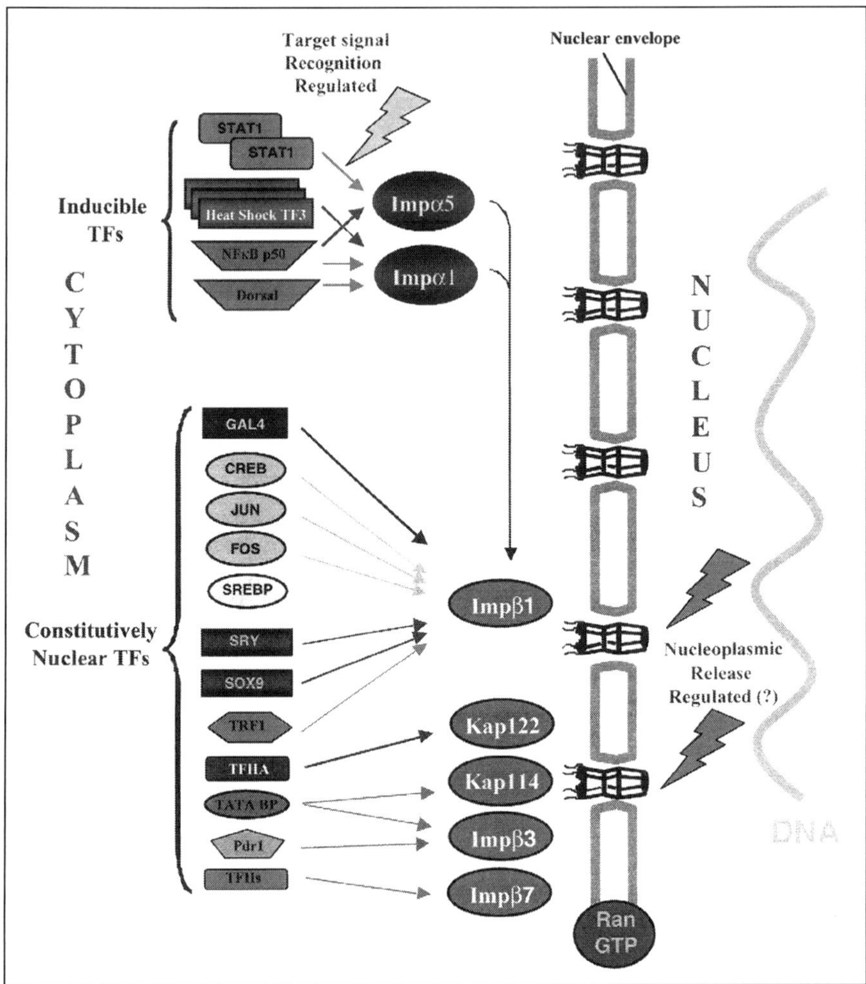

Figure 2. Distinct nuclear import pathways for inducible and constitutively nuclear TFs. Inducible TFs are imported to the nucleus through interaction with Impα/β1 with the recognition step regulated directly or indirectly in response to cellular signals (see Table 2). Many constitutively nuclear TFs are imported into the nucleus through direct interaction with Impβ1 which is not regulated at the NLS recognition step, but may be regulated at the nuclear release end of the transport process, through modulation by binding to DNA; general TFs are transported to the nucleus analogously by importin β homologs.

certainly in excess of that of substrates such as inducible TFs which are only required in the nucleus in small amounts in response to specific cellular signals and for short periods. This imbalance in terms of the sheer amount of transport cargoes entering the nucleus through pathways mediated by the importin β homologs, and those mediated by importin α/β1, means that mechanisms to effect rapid increases in signal binding affinity (e.g., through phosphorylation—see below)[2,22,25,100,101] are essential for importin α/β1-recognized substrates to compete successfully for importin β1 and Ran, and achieve subsequent nuclear entry. Görlich and coworkers[50,81] have reported that the binding of importin α/β1 to nuclear import substrates such as ribosomal proteins rpL23a and histone HI is "non-productive" in terms of not leading to nuclear import, adding an additional dimension to the above scenario of competition between substrates for transport receptors; clearly, the highest affinity combination of substrate and receptor will prove the most efficient in terms of transport, but if the former is non-productive for transport, nuclear import will in fact be inhibited. Analogous effects are implied by the observations that importin α inhibits importin β1-specific binding to transport substrates such as PTHrP and GAL4.[58,59]

Regulation of targeting sequence accessibility, or of the affinity of recognition by importins through phosphorylation or other mechanisms, ultimately determines the precise level of nuclear accumulation or otherwise.[2] Specific mechanisms by which target sequence recognition can be modulated to effect regulation of signal-dependent nuclear protein transport include the enhancement (10-100-fold) of both NLS recognition by importin α/β1 and nuclear import in the case of T-ag and the *Drosophila* morphogen and inducible TF Dorsal,[22,25,100,101] and the modulation of nuclear import targeting sequence accessibility through masking either within the same molecule or by a heterologous protein such as in the case of NF-κB subunits (see ref. 2),[102,103] or by phosphorylation, as in the case of the yeast TF Pho4 where phosphorylation at Ser^{152} within the Pho4 nuclear targeting signal (see Table 1) prevents Pse1p (Impβ3) binding.[71,72]

Distinct Nuclear Import Receptor for Different Types of TFs; Differential Regulation?

Analysis thus far indicates that NLS-dependent nuclear import of inducible TFs (TFs that are cytoplasmic in the basal, uninduced state, but are induced to translocate to the nucleus upon the activation of specific signalling pathways) such as the NF-κB family members Dorsal[22] (see above), NF-κB p50 and p65 (RelA),[28] and the STAT (signal transducer and activator of transcription) TF family member STAT1[44] are mediated by the importin α/β1 heterodimer (see Table 2; see also Fig. 2). As already alluded to, transport is regulated, in the case of Dorsal[22] (and presumably NF-κB p50 and p65) by specific phosphorylation which modulates the interaction with importin α/β1, as well as NLS masking by inhibitor molecules such as IκB[102,103] to prevent importin:NLS interaction. In the case of STAT1, dimerization, dependent on JAK kinase phosphorylation, is a prerequisite for nuclear import.[104]

In stark contrast, several different types of constitutively nuclear TFs, including the cAMP-response element binding protein (CREB) and related AP-1 and jun and fos constituents, SREBP2, GAL4, and SOX9/SRY, are all recognized specifically by importin β1 rather than the importin α/β1 heterodimer (see Table 2; Fig. 2). This implies that constitutively nuclear TFs may utilise a nuclear import pathway mediated by importin β1, which is distinct from the importin α/β1-mediated pathway used by inducible TFs. Other constitutively nuclear TFs including those of the general transcription machinery, such as the TATA BP (Kap114 and Impβ4),[73] TFIIS (Impβ7)[75] and TFIIA (Kap122),[77] are imported by other Impβ homologs (see Table 2), again independently of importin α. The clear implication is that constitutively nuclear TFs are transported to the nucleus by nuclear import receptors distinct from those used by inducible TFs which appear largely to be recognised by importin α/β1, and with the corollary of direct or indirect regulation of the recognition interaction (Table 2, Fig. 2). Pho4 is an exception in that although it is an inducible TF whose interaction with its nuclear import receptor importin β3 (Pse1) is modulated directly by phosphoryation, its nuclear import receptor is not importin α/β1. There are also importin β1-recognized TFs that appear to show conditional nuclear localisation in response to cellular signals; in the case of SREBP-2, proteolytic processing as well as dimerisation is critical to nuclear transport so that it is not clear that direct modulation of nuclear targeting sequence recognition is involved. SMAD-3 is similar in showing regulation of its subcellular localisation, but it is unclear whether modulation of the importin β interaction with SMAD-3 is the key event in triggering its nuclear translocation in response to growth factor signalling. Despite these exceptions, inducible TFs are generally localised in the nucleus through the action of importin α/β1 where there is direct or indirect

Table 2. Import receptors and regulatory mechanisms for the nuclear import pathways of inducible and constitutively nuclear TFs

TF	Nuclear Import Receptor	Regulation	Binding Interaction[1]	References
Inducible TFs				
Dorsal/NF-κB	Impα/β1	Phosphorylation regulates NLS recognition by importin α/β1	nM	22
Heat Shock TF3	Impα/β1	Trimer formation determines NLS recognition by importin α/β1		38
NF-κB	Impα/β1	NLS masking by IkBα/phosphorylation	nM	28,102,103
STAT-1	Impα/β1	Phosphorylation/dimerisation is a prerequisite for importin α/β1 binding		44,104
BRCA1	Impα/β1	Cytoplasmic retention controls NLS accessibility		42,45
Pho4	Impβ3	Phosphorylation regulates NLS recognition by importin β3		71,72
Constitutively nuclear TFs				
AP-1	Impβ1		nM	60
CREB	Impβ1		nM	60
Fos	Impβ1		nM	60
Jun	Impβ1		nM	60
GAL4	Impβ1	Specific DNA-binding – assists nucleoplasmic release ?	nM	59
SOX9	Impβ1	Specific DNA-binding**2**– assists nucleoplasmic release ?	nM	61
SRY	Impβ1	Specific DNA-binding – assists nucleoplasmic release ?	nM	62
SREBP-2	Impβ1			55,56
TRF1	Impβ1		nM	63
Pdr1	Impβ3			74
TATA BP	Kap114, Impβ3	Specific DNA-binding – assists nucleoplasmic release ?		73
TFIIS	Impβ7			75
TFIIA	Kap122			77

[1] See also Table 1.

[2] Based on close homology to SRY, particularly in the Impβ1 binding sequence, it is predicted that DNA binding will inhibit Impβ1 binding to SOX9.

regulation of NLS-importin $\alpha/\beta 1$ interaction, whilst constitutively nuclear TFs are transported into the nucleus largely by importin $\beta 1$, or other importin β homologs in the case of general TFs, and mostly show no regulation of the initial importin-targeting signal interaction.

Interestingly, regulation of importin $\beta 1/\beta$ homolog-import substrate interactions may, however, occur at the nucleoplasmic release end of the transport process. This relates to observations that DNA binding on the part of GAL4[59] and SRY[62] competes with importin $\beta 1$ binding. TATA site-containing DNA is similarly able to enhance dissociation of the TATA-binding protein from its nuclear import receptor Kap114p.[73] Since the DNA and RNA binding regions of many nuclear proteins appear to overlap with or be in close vicinity to NLSs,[105] it is not inconceivable that comparable regulatory mechanisms may exist with respect to the nuclear import of a number of other DNA- and RNA-binding proteins. It is known that the promoters of active genes are generally localised very closely to NPCs[106] so that it does not seem unreasonable to speculate that competition by specific promoter sequences for Imp$\beta 1$ and/or Kap114p binding sites on the requisite TFs may assist in transport substrate release on the nucleoplasmic side of the NPC, possibly as a "fail-safe" mechanism (see also Fig. 2).[59,62,73] In the context of competition between NLS-carrying substrates for NLS-receptors (see above), this may be of importance, especially in terms of the competition for transport factors such as Ran and Imp$\beta 1$ utilized/required by multiple nuclear import substrates/pathways;[2] e.g., in a situation where nuclear RanGTP may be limiting, DNA binding-effected release from the NPC might be critical in overcoming a potential bottleneck in TF nuclear import. Thus, whilst the regulation of nuclear import of inducible TFs largely occurs at the initial step of importin recognition, the nuclear import of constitutive TFs may be regulated at the release end, representing a fundamental difference in the nuclear import mechanism (Fig. 2). An analogous intranuclear release mechanism appears to operate for the mRNA-binding protein Npl3p, which is only able to be dissociated by RanGTP from its import receptor Mtr10p in the presence of RNA; i.e., intranuclear release of Npl3p from its specific import receptor requires the cooperative action of RanGTP and newly synthesized mRNA.[79]

Importantly in the context of the plethora of competing nuclear import receptors and pathways (see above), it is obvious that, where analysed (see Table 2), the importin-targeting sequence interaction is of high affinity; clearly, nM binding affinity is a prerequisite to targeting to the NPC, and subsequent translocation through it and release within the nucleoplasm.[2] Through their different regulatory mechanisms, specialised nuclear import pathways for different classes of proteins enable the cell to regulate nuclear entry of particular types of proteins specifically and efficiently.

Unanswered Questions

As expounded here, nuclear import revolves around a series of related but distinct nuclear import pathways mediated by specific targeting signals and receptor molecules of the importin superfamily. Modulation of target sequence recognition, either through enhancement of binding affinity or inhibition of accessibility, is central to regulating signal-dependent nuclear transport.[2] This, together with the differential expression of importin subunits with distinct targeting signal recognition capabilities, and competition between different targeting signal-containing proteins and importins, ensures that proteins are targeted to the nucleus as and when required. It is critical to understand how the multitude of nuclear import pathways are coordinated such that each functions efficiently in the cellular context. The affinity of target sequence recognition appears to determine not only whether or not proteins are targeted to the nucleus, but also the rate at which they are imported, and hence is clearly central to the cellular "selection process" determining which proteins are imported into the nucleus, and in what amounts, to carry out particular nuclear functions.

The binding affinities and kinetic parameters for many of the nuclear import pathways mediated by the importins listed in Table 1 are not available, and until this basic information is obtained and put in the context of the nuclear transport efficiencies at the single cell level in vivo, it is almost impossible to understand how they may be regulated coordinately. Some answers may be obtained using current high resolution imaging techniques at the level of the single NPC[107,108] in conjunction with photobleaching approaches,[109-111] which enable transport kinetics to be determined at the level of a single living cell. This sort of resolution, of course, ultimately needs to be applied to transport processes in higher order structures such as tissue in a developmental context (perhaps in conjunction with microarray analysis to monitor changes in gene expression of components of the nuclear import machinery), or the whole animal. Only with this detailed information will we be able to appreciate fully how the various nuclear import pathways, targeting sequences, import receptors and regulatory mechanisms, integrate and function in ordered and efficient fashion in the context of the whole cell.

Abbreviations

NPC, nuclear pore complex; NTF2, nuclear transport factor 2; RanBP1, Ran binding protein 1; RanGAP1, Ran GTPase activating protein 1; NLS, nuclear localization sequence; nup, nucleoporin; T-ag, SV40 large tumour antigen; IBB, importin-β-binding; TF, transcription factor; LEF-1, lymphoid enhancer factor-1; ARM, Armadillo; TCPTP, T-cell protein tyrosine phosphatase; PTHrP, parathyroid hormone related protein; BIB, β-importin-like binding; MAPK, mitogen-activated protein kinase; dsDNA-PK, double stranded DNA-dependent protein kinase; PK-A, cAMP-dependent protein kinase; Y2H, yeast two hybrid

Acknowledgments

The support of the National Health and Medical Research Council Australia (grants ID# 128821, 143710 and 143790) is gratefully acknowledged.

References

1. Wozniak RW, Rout WP, Aitchison JD. Karyopherins and kissing cousins. Trends Cell Biol 1998; 8:184-188.
2. Jans DA, Xiao C-Y, Lam MHC. Nuclear targeting signal recognition: A key control point in nuclear transport? BioEssays 2000; 22:532-544.
3. Hood JK, Silver PA. In or out? Regulating nuclear transport. Curr Opin Cell Biol 1999; 11:241-247.
4. Gorlich D, Kutay U. Transport between the cell nucleus and the cytoplasm. Annu Rev Cell Dev Biol 1999; 15:607-660.
5. Ribbeck K, Lipowsky G, Kent HM et al. NTF2 mediates nuclear import of Ran. EMBO J 1998; 17:6587-6598.
6. Bischoff FR, Ponstingl H. Catalysis of guanine nucleotide exchange on Ran by the mitotic regulator RCC1. Nature 1991; 354:80-82.
7. Jans DA, Chan CK, Hübner S. Signals mediating nuclear targeting and their regulation: Application in drug delivery. Med Res Rev 1998; 18:189-223.
8. Melchior F, Gerace L. Two-way trafficking with Ran. Trends Cell Biol 1998; 8:175-179.
9. Doye V, Hurt E. From nucleoporins to nuclear pore complexes. Current Opin Cell Biol 1997; 9:401-411.
10. Rexach M, Blobel G. Protein import into nuclei: Association and dissociation reactions involving transport substrate, transport factors, and nucleoporins. Cell 1995; 83:683-692.
11. Bayliss R, Littlewood T, Stewart, M. Structural basis for the interaction between FxFG nucleoporin repeats and importin-beta in nuclear trafficking. Cell 2000; 102:99-108.
12. Stewart M. Insights into the molecular mechanism of nuclear trafficking using nuclear transport factor 2 (NTF2). Cell Struct Funct 2000; 25:217-225.
13. Izaurralde E, Kutay U, von Kobbe C et al. The asymmetric distribution of the constituents of the Ran system is essential for transport into and out of the nucleus. EMBO J 1997; 16:6535-6547.

14. Engelmeier L, Olivio J-C, Mattaj IW. Receptor-mediated substrate translocation through the nuclear pore complex without nucleotide triphosphate hydrolysis. Current Biol 1999; 9:30-41.
15. Efthymiadis A, Shao H, Hübner S et al. Kinetic characterization of the human retinoblastoma protein bipartite nuclear localization sequence in vivo and in vitro: A comparison with the SV40 large T-antigen NLS. J Biol Chem 1997; 272:22134-22139.
16. Kutay U, Bischoff FR, Kostka S et al. Export of importin α from the nucleus is mediated by a specific nuclear transport factor. Cell 1997; 90:1061-1071.
17. Kalderon D, Roberts BL, Richardson WD et al. A short amino acid sequence able to specify nuclear location. Cell 1984; 39:499-509.
18. Robbins J, Dilworth SM, Laskey RA et al. Two interdependent basic domains in nucleoplasmin nuclear targeting sequence: Identification of a class of bipartite nuclear targeting sequence. Cell 1991; 64:615-623.
19. Hall NM, Craik C, Hiraoka Y. Homeodomain of yeast repressor alpha 2 contains a nuclear localization signal. Proc Natl Acad Sci USA 1990; 87:6954-6958.
20. Makkerh JS, Dingwall C, Laskey RA. Comparative mutagenesis of nuclear localization signals reveals the importance of neutral and acidic amino acids. Current Biol 1996; 6:1025-1027.
21. Smith HM, Hicks GR, Raikhel NV. Importin alpha from Arabidopsis thaliana is a nuclear import receptor that recognizes three classes of import signals. Plant Physiol 1997; 114:411-417.
22. Briggs LJ, Stein D, Goltz J et al. The cAMP-dependent protein kinase site (Ser^{312}) enhances dorsal nuclear import through facilitating nuclear localization sequence/importin interaction. J Biol Chem 1998; 273:22745-22752.
23. Hu W, Jans DA. Efficiency of importin α/βmediated NLS recognition and nuclear import: Differential role of NTF2. J Biol Chem 1999; 274:15820-15827.
24. Hübner S, Smith HMS, Hu W et al. Plant importin α binds nuclear localization sequences with high affinity and mediates nuclear import independent of importin β. J Biol Chem 1999; 274:22610-22617.
25. Hübner S, Xiao C-Y, Jans DA. The protein kinase CK2 site (Ser$^{111/112}$) enhances recognition of the SV40 large T-antigen nuclear localization sequence by importin. J Biol Chem 1997; 272:17191-17195.
26. Kohler M, Speck C, Christiansen M et al. Evidence for distinct substrate specificities of importin alpha family members in nuclear protein import. Mol Cell Biol 1999; 19:7782-91.
27. Seki T, Tada S, Katada T et al. Cloning of a cDNA encoding a novel importin-α homologue, Qip1: discrimination of Qip1 and Rch1 from hSrp1 by their ability to interact with DNA helicase Q1/RecQL. Biochem Biophys Res Commun 1997; 234:48-53.
28. Nadler SG, Tritschler D, Haffar OK et al. Differential expression and sequence-specific interaction of karyopherin α with nuclear localization sequences. J Biol Chem 1997; 272:4310-4315.
29. Miyamoto Y, Imamoto N, Sekimoto T et al. Differential modes of nuclear localization signal (NLS) recognition by three different classes of NLS receptors. J Biol Chem 1997; 272:26375-26381.
30. Herold A, Truant R, Wiegand H et al. Determination of the functional domain organization of the importin α nuclear import factor. J Cell Biol 1998; 143:309-318.
31. Cuomo CA, Kirch SA, Gyuris J et al. Rch1, a protein that specifically interacts with the RAG-1 recombination-activating protein. Proc Natl Acad Sci USA 1994; 91:6156-6160.
32. Prieve MG, Guttridge KL, Munguia J et al. The nuclear localization signal of lymphoid enhancer factor-1 is reconized by two differentially expressed Srp1-nuclear localization sequence receptor proteins. J Biol Chem 1996; 271:7654-7658
33. Prieve MG, Guttridge KL, Munguia J et al. Differential importin-α recognition and nuclear transport by nuclear localization signals within the high-mobility-group DNA binding domains of lymphoid enhancer Factor 1 and T-cell factor 1. Mol Cell Biol 1998; 18:4819-4832.
34. Fischer N, Kremmer E, Lautscham G et al. Epstein-Barr virus nuclear antigen 1 forms a complex with the nuclear transporter karyopherin α2. J Biol Chem 1997; 272:3888-4005.
35. Gallay P, Hope T, Chin D et al. HIV-1 infection of nondividing cells through the recognition of integrase by the importin/karyopherin pathway. Proc Natl Acad Sci USA 1997; 94:9825-9830.
36. Gallay P, Stitt V, Mundy C et al. Role of the karyopherin pathway in human immunodeficiency virus type 1 nuclear import. J Virol 1996; 70:1027-1032.
37. Gascard P, Nunomura W, Lee G et al. Deciphering the nuclear import pathway for the cytoskeletal red cell protein 4.1R. Mol Biol Cell 1999; 10:1783-1798.

38. Nakai A, Ishikawa T. Nuclear localisation signal is essential for stress-induced dimer-to trimer transition of heat shock transcription factor 3. J Biol Chem 2000; 275:34665-34671.

39. Johnson-Saliba M, Siddon NA, Clarkson MJ et al. Distinct importin recognition properties of histones and chromatin assembly factors. FEBS Lett 2000; 467:169-174

40. Peng X, Zhang Y, Zhang H et al. Interaction of tissue transglutaminase with nuclear transport protein importin-α3. FEBS Lett 1999; 446:35-39.

41. Nachury MV, Ryder UW, Lamond AI et al. Cloning and characterization of hSRP1 gamma, a tissue-specific nuclear transport factor. Proc Natl Acad Sci USA 1998; 95:582-587.

42. Li S, Ku C-Y, Farmer AA et al. Identification of a novel cytoplasmic protein that specifically binds to nuclear localization signal motifs. J Biol Chem 1998; 273:6183-6189.

43. Cortes P, Ye ZS et al. RAG-1 interacts with the repeated amino acid motif of the human homologue of the yeast protein SRP1. Proc Natl Acad Sci USA 1994; 91:7633-7637.

44. Sekimoto T, Imamoto N, Nakajima K et al. Extracellular signal-dependent nuclear import of Stat1 is mediated by nuclear pore-targeting complex formation with NPI-1, but not Rch1. EMBO J 1997; 16:7067-7077.

45. Chen C-F, Li S, Chen Y et al. The nuclear localization sequences of the BRCA1 protein interact with the importin-α subunit of the nuclear transport signal receptor. J Biol Chem 1996; 271:32863-32868.

46. Huber J, Cronshagen U, Kadokura M et al. Snurportin 1, an m_3G-cap-specific nuclear import receptor with a novel domain structure. EMBO J 1998; 17:4114-4126.

47. Jullien D, Görlich D, Laemmli UK et al. Nuclear import of RPA in Xenopus egg extracts requires a novel protein XRIPα but not importin α. EMBO J 1999; 18:4348-4358.

48. Tiganis T, Flint AJ, Adam AS et al. Association of the T-cell protein tyrosine phosphatase with nuclear import factor p97. J Biol Chem 1997; 272:21548-21557.

49. Moore JD, Yang J, Truant R et al. Nuclear import of cdk/cyclin complexes: identification of distinct mechanisms for import of cdk2/cyclin E and cdc2/cyclin B1. J Cell Biol 1999; 144:213-224.

50. Jaekel S, Görlich D. Importin β, RanBP5 and RanBP7 mediate nuclear import of ribosomal proteins in mammalian cells. EMBO J 1998; 17:4491-4502.

51. Truant R, Cullen BR. The arginine-rich domains present in human immunodeficiency virus type 1 Tat and Rev function as direct importin β-dependent nuclear localization signals. Mol Cell Biol 1999; 19:1210-1217.

52. Henderson, BR, Percipalle G. Interactions between HIV Rev and nuclear import and export factors: the Rev nuclear localisation signal mediates specific binding to human importin-beta. J Mol Biol 1997; 274:693-707.

53. Palmeri D, Malim MH. Importin β can mediate the nuclear import of an arginine-rich nuclear localization signal in the absence of importin α. Mol Cell Biol 1999; 19:1218-1225.

54. Xiao Z, Liu X, Lodish HF. Importin beta mediates nuclear translocation of Smad 3. J Biol Chem 2000; 275:23425-23428.

55. Nagoshi E, Imamoto N, Sato R et al. Nuclear import of sterol regulatory element-binding protein-2, a basic helix-loop-helix-leucine zipper (bHLH-Zip)-containing transcription factor, occurs through the direct interaction of importin beta with HLH-Zip. Mol Biol Cell 1999; 10:2221-2233.

56. Nagoshi E, Yoneda Y. Dimerization of sterol regulatory element-binding protein 2 via the helix-loop-helix-leucine zipper domain is a prerequisite for its nuclear localization mediated by importin beta. Mol Cell Biol 2001; 21:2779-2789.

57. Lam MHC, Hu W, Xiao CY et al. Molecular dissection of the Importin β1-recognized nuclear targeting signal of parathyroid hormone-related protein. Biochem Biophys Res Commun 2001; 282:629-634.

58. Lam MHC, Briggs LJ, Hu W et al. Importin β recognizes parathyroid hormone-related protein (PTHrP) with high affinity and mediates its nuclear import in the absence of importin α. J Biol Chem 1999; 274:7391-7398.

59. Chan CK, Hübner S, Hu W et al. Mutual exclusivity of DNA binding and nuclear localization signal recognition by the yeast transcription factor GAL4: Implications for nonviral DNA delivery. Gene Therapy 1998; 5:1204-1212.

60. Forwood JK, Lam MHC, Jans DA. Nuclear import of the CREB and AP-1 transcription factors is dependent on importin β1 and Ran, and independent of importin α. Biochemistry 2001; 40: 5208-5217.

61. Preiss S, Argentaro A, Clayton A et al. Compound effects of point mutations causing campomelic dysplasia/autosomal sex reversal upon SOX9 structure, nuclear transport, DNA binding and transcriptional activation. J Biol Chem 2001; 276:27864-27872.

62. Forwood JK, Harley V, Jans DA. The C-terminal nuclear localization signal of the SRY HMG-domain mediates nuclear import through importin β1. J Biol Chem 2001; 276:46575-46582.

63. Forwood JK, unpublished observations.

64. Schedlich LA, Le Page SL, Firth SM et al. Nuclear import of insulin-like growth factor binding protein-3 (IGFBP-3) and IGFBP-5 is mediated by the importin β subunit. J Biol Chem 2001; 275:23462-23470.

65. Pollard VW, Michael WM, Nakielny S et al. A novel receptor-mediated nuclear protein import pathway. Cell 1996; 86:985-994.

66. Bogerd HP, Benson RE, Truant R et al. Definition of a consensus transportin-specific nucleocytoplasmic transport signal. J Biol Chem 1999; 274:9771-9777.

67. Siomi H, Eder PS, Kataoka N et al. Transportin-mediated nuclear import of heterogeneous nuclear RNP proteins. J Cell Biol 1997; 138:1181-1192.

68. Nakielny S, Shaikh S, Burke B et al. Nup153 is an M9-containing mobile nucleoporin with a novel Ran-binding domain. EMBO J 1999; 18:1982-1995.

69. Siomi MC, Fromont M, Rain J-C et al. Functional conservation of the transportin nuclear import pathway in divergent organisms. Mol Cell Biol 1998; 18:4141-4148.

70. Schlenstedt G, Smirnova E, Deane R et al. Yrb4p, a yeast Ran-GTP-binding protein involved in import of ribosomal protein L25 into the nucleus. EMBO J 1997; 16:6237-6249.

71. Kaffman A, Rank NM, O'Shea EK. Phosphorylation regulates association of the transcription factor Pho4 with its import receptor Pse1/Kap121. Genes Dev 1998; 12:2673-2683.

72. Komeili A, O'Shea EK. Roles of phosphorylation sites in regulating activity of the transcription factor Pho4. Science 1999; 284:977-980.

73. Pemberton LF, Rosenblum JS, Blobel G. Nuclear import of the TATA-binding protein: mediation by the karyopherin Kap114p and a possible mechanism for intranuclear targeting. J Cell Biol 1999; 145:1407-1417.

74. Delahodde A, Pandjaitan R, Corral-Debrinski M et al. Pse/Kap121-dependent nuclear localisation of the major yeast mulitdrug (MDR) transcription factor Pdr1. Mol Microbiol 2001; 39:304-312.

75. Albertini M, Pemberton LF, Rosenblum JS et al. A novel nuclear import pathway for the transcription factor TFIIS. J Cell Biol 1998; 143:1447-1455.

76. Ferrigno P, Posas F, Koepp D et al. Regulated nucleo/cytoplasmic exchange of HOG1 MAPK requires the importin β homologs NMD5 and XPO1. EMBO J 1998; 17:5606-5614.

77. Titov AA, Blobel G. The karyopherin Kap122p/Pdr6p imports both subunits of the transcription factor IIA into the nucleus. J Cell Biol 1999; 147:235-245.

78. Rosenblum JS, Pemberton LF, Blobel G. A nuclear import pathway for a protein involved in tRNA maturation. J Cell Biol 1997; 139:1655-1661.

79. Senger B, Simos G, Bischoff FR et al. Mtr10p functions as a nuclear import receptor for the mRNA-binding protein Npl3p. EMBO J 1998; 17:2196-2207.

80. Kataoka N, Bachorik JL, Dreyfuss D. Transportin-SR, a nuclear import receptor for SR proteins. J Cell Biol 1999; 145:1145-1152.

81. Jaekel S, Albig W, Kutay U et al. The importin β/importin 7 heterodimer is a functional nuclear import receptor for histone H1. EMBO J 1999; 18:2411-2423.

82. Kudo N, Khochbin S, Nishi K et al. Molecular cloning and cell cycle-dependent expression of mammalian CRM1, a protein involved in nuclear export of proteins. J Biol Chem 1997; 272:29742-29751.

83. Shiozaki K, Yanagida M. Functional dissection of the phosphorylated termini of fission yeast DNA topoisomerase II. J Cell Biol 1992; 119:1023-1036.

84. Jans DA, Moll T, Nasmyth K et al. Cyclin-dependent kinase site-regulated signal-dependent nuclear localization of the SW15 yeast transcription factor in mammalian cells. J Biol Chem 1995; 270:17064-17067.

85. Conti E, Uy M, Leighton L et al. Crystallographic analysis of the recognition of a nuclear localization signal by the nuclear import factor karyopherin α. Cell 1998; 94:193-204.

86. Conti E, Kuriyan J. Crystallographic analysis of the specific yet versatile recognition of distinct nuclear localization signals by karyopherin alpha. Structure Fold Des 2000; 8:329-338.

87. Fontes MR, Teh T, Kobe B. Structural basis of recognition of monopartite and bipartite nuclear localization sequences by mammalian importin-alpha. J Mol Biol 2000; 297:1183-1194.

88. Kobe, B. Autoinhibition by an internal nuclear localization signal revealed by the crystal structure of mammalian importin α. Nature Struct Biol 1999; 6:388-397.

89. Tsuji L, Takumi T, Imamoto N et al. Identification of novel homologues of mouse importin α, the α subunit of the nuclear pore-targeting complex, and their tissue-specific expression. FEBS Lett 1997; 416;30-34.

90. Torok I, Strand D, Schmitt R et al. The overgrown hematopoietic organs-31 tumor suppressor gene of Drosophila encodes an Importin-like protein accumulating in the nucleus at the onset of mitosis. J Cell Biol 1995; 129:1473-1489.

91. Mathe E, Bates H, Huikeshoven H et al. Importin-alpha3 is required at multiple stages of Drosophila development and has a role in the completion of oogenesis. Dev Biol 2000; 223:307-322.

92. Geles KG, Adam SA. Germline and developmental roles of the nuclear transport factor importin α3 in C. elegans. Development 2001;128:1817-1830.

93. Cingolani G, Petosa C, Weis K et al. Structure of importin-β bound to the IBB domain of importin α. Nature 1999; 399:221-229.

94. Chook YM, Blobel G. Structure of the nuclear transport complex karyopherin-β-2-Ran.GppNHp. Nature 1999; 399:230-237.

95. Chi NC, Adam EJH, Adam SA. Different binding domains for Ran-GTP and Ran-GDP/RanBP1 on nuclear import factor p97. J Biol Chem 1997; 272:6818-6822.

96. Lee SJ, Imamoto N, Sakai H et al. The adoption of a twisted structure of importin-beta is essential for the protein-protein interaction required for nuclear transport. J Mol Biol 2000; 302:251-264.

97. Görlich D, Dabrowski M, Bischoff FR et al. A novel class of RanGTP binding proteins. J Cell Biol 1997; 138:65-80.

98. Tirian L, Puro J, Erdelyi M et al. The Ketel^D dominant-negative mutations identify maternal function of the Drosophila importin-β gene required for cleavage nuclear formation. Gentics 2000; 156:1901-1912.

99. Lippai M, Tirian L, Boros I et al. The Ketel gene encodes a Drosophila homologue of importin-β. Genetics 2000; 156:1889-1900.

100. Xiao C-Y, Hübner S, Jans DA. SV40 large tumor-antigen nuclear import is regulated by the double-stranded DNA-dependent protein kinase site (serine 120) flanking to the nuclear localization sequence. J Biol Chem 1997; 272:22191-22198.

101. Xiao C-Y, Jans P, Jans DA. Negative charge at the protein kinase CK2 site enhances recognition of the SV40 large tumor-antigen nuclear localization signal by importin; Effect of proline near the CK2 site. FEBS Lett 1998; 440:297-301.

102. Huxford T, Huang D-B, Malek S et al. The crystal structure of the IkBα/NF-κB complex reveals mechanisms of NF-κB inactivation. Cell 1999; 95:759-770.

103. Henkel T, Zabel U, Van Zee K et al. Intramolecular masking of the nuclear localization signal and dimerization domain in the precursor for the p50 NF-κB subunit. Cell 1992; 68:1121-1133.

104. Strehlow I, Schindler C. Amino-terminal signal transducer and activator of transcription (STAT) domains regulate nuclear translocation and STAT deactivation. J Biol Chem 1998; 273:28049-28056.

105. Lacasse EC, Lefebvre YA. Nuclear localization signals overlap DNA- or RNA-binding domains in nucleic acid-binding proteins. Nucleic Acids Res 1995; 23:1647-1656.

106. Grasser KD, Feix G. Isolation and characterization of maize cDNAs encoding a high mobility group protein displaying a HMG-box. Nucleic Acids Res 1991; 19:2573-2577.

107. Kubitscheck U, Wedekind P, Zeidler O et al. Single nuclear pores visualized by confocal microscopy and image processing. Biophys J 1996; 70:2067-2077.

108. Keminer O, Siebrasse JP, Zerf K et al. Optical recording of signal-mediated protein transport through single nuclear pore complexes. Proc Natl Acad Sci USA 1999; 96:11842-11847.

109. Kruhlak MJ, Lever MA, Fischle W et al. Reduced mobility of the alternate splicing factor (ASF) through the nucleoplasm and steady state speckle compartments. J Cell Biol 2000; 150:41-51.

110. McNally JG, Muller WG, Walker D et al. The glucocorticoid receptor: rapid exchange with regulatory sites in living cells. Science 2000; 287:1262-1265.

111. Lam MH, Thomas RJ, Loveland KL et al. Nuclear transport of parathyroid hormone (PTH)-related protein is dependent on microtubules. Mol Endocrinol 2002; 16(2):390-401.

The Molecular Mechanisms of mRNA Export

Tetsuya Taura, Mikiko C. Siomi and Haruhiko Siomi

A general paradigm for nuclear transport was established primarily through studies of protein import and export. Until recently, this paradigm was generally presumed also to apply to the process of RNA export from the nucleus. In particular, it was assumed that general mRNA export was mediated by one or more transport receptors of the importin-β family and that the RanGTP/GDP gradient was required to impart directionality to the process. The highly abundant class of nuclear RNA-binding proteins—the hnRNP proteins—were regarded as primary candidates for mRNA export adapter proteins that could link mRNAs to importin-β family export factors. Within the past few years, however, an explosion of data has largely disproven prior assumptions about the mechanisms of mRNA export, permanently changing the face of the field. The dust is still settling, but what we now see, albeit incompletely, is the outline of a probable major route of mRNA export that is independent of the importin-β family and the Ran GTPase system.

Introduction

The distinguishing feature of eukaryotic cells is the segregation of RNA biogenesis and DNA replication in the nucleus, separate from the cytoplasmic machinery for protein synthesis. Communication between the nucleus and the cytoplasm occurs through aqueous channels in the nuclear envelope called nuclear pore complexes (NPCs).[22,92] Small molecules can pass through NPCs by diffusion, but there is a permeability barrier for larger molecules—those with a relative molecular mass of >40 kDa—which permits transport only of selected cargo with the help of transport receptors. The NPC is a gigantic proteinous complex ranging in size from approximately 50 MDa in the yeast *Saccharomyces cerevisiae* to 125 MDa in higher eukaryotes, and possessing an eight-fold symmetric structure. All nucleocytoplasmic transport occurs via the central aqueous channel found in NPCs. As the maximum diameter of this channel may be only ~25 nm, achieving nuclear transport of large complexes such as ribosomes and viral genomes presents a potentially formidable challenge and must involve a considerable change in the three-dimensional conformation of the transport cargo or of the pore itself.

Active transport through the NPC is a signal-mediated process involving recognition of cargo molecules by a large class of soluble transport factors.[30,66] Transport is bidirectional, energy dependent, and highly regulated. Research into the molecular mechanisms of nucleocytoplasmic transport has been initiated by studies of nuclear protein import. Protein export from the nucleus has been shown to utilize similar mechanisms as protein import, and now researchers are focusing on the transport of another class of macromolecules—RNAs. RNA export, especially export of mRNA, has been less understood because the process is more complex than that of protein export, involving the coordination of several post-transcriptional

Nuclear Import and Export in Plants and Animals, edited by Tzvi Tzfira and Vitaly Citovsky.
©2005 Eurekah.com and Kluwer Academic / Plenum Publishers.

processing events with the formation of RNA-protein complexes (RNPs) that are the actual export cargoes. In the past few years, however, researchers have accumulated a vast amount of information that reveals the existence of a distinct transport system dedicated to mRNA export.[17,79,80] In this chapter, we review recent studies of the molecular mechanisms of nucleocytoplasmic transport, from 'classical' protein transport to the modern view of the mRNA export system.

Ran Dependent Nucleocytoplasmic Transport

Considerable progress has recently been made in understanding the mechanisms underlying the sequence-specific transport of proteins between the nucleus and the cytoplasm and the critical role played by the NPC in this process.[30,66] Nucleocytoplasmic protein transport is promoted by signal-receptor recognition process. Generally, cargo proteins contain peptide motifs that function as nuclear localization signals (NLSs) and/or nuclear export signals (NESs). These include the so called 'classical NLSs', such as SV40 large T antigen NLS, which consists of a short cluster of basic amino acids, the nonclassical M9 signal, which is a 38- amino acid domain of hnRNP A1 used both for nuclear import and export, and HIV Rev-type NESs consisting of an approximately 10-amino acid stretch rich in leucine residues that are found in many nuclear export cargos. Evolutionarily conserved transport signal receptors specific for each transport signal have been identified. These members form a protein family known as the 'importin-β family' or as 'karyopherins' that includes more than 20 members in metazoans and 14 members in yeast belong (from now on, these will be referred to as β family receptors). The family members share a partial similarity in sequence and structure as well as a biochemical property of interaction with Ran, a Ras-related small GTP binding protein. Like other GTP-binding proteins, Ran has both GTP- and GDP-bound forms, and the switch between these two forms plays a crucial role in regulating transport by promoting the association and dissociation of transport receptors and their cargoes, as well regulating their interactions with the NPC. Ran requires two regulatory factors, the GTPase activating RanGAP and the guanine nucleotide exchange factor RanGEF, to switch between its two nucleotide bound states. At steady state, these regulators localize to the cytoplasm and the nucleus, respectively; this asymmetric distribution generates a RanGTP/GDP gradient across the nuclear envelope, which is essential for most nuclear transport pathways. Cooperation with the Ran GTPase system allows transport receptors to bind and subsequently release their substrates on opposite sides of the nuclear envelope, which in turn ensures directed nucleocytoplasmic transport.

The most well-characterized β family receptors are importin-β itself and CRM1/ exportin-1.[30,66] Importin-β was the first nuclear transport receptor to be identified. It forms a heterodimeric complex with the adapter protein impotin-α, which mediates recognition of classical NLSs in nuclear import cargoes. The importin-α/-β-cargo complex is formed in the cytoplasm, then travels through the NPC to the nuclear interior, where the cargo is released from the receptor upon binding of RanGTP to importin-β. In contrast, RanGTP binding to the export receptor CRM1/exportin-1 is required for its association with proteins that contain leucine-rich NESs. After the ternary complex of CRM1-RanGTP-cargo is translocated to the cytoplasm, the cytoplasmic Ran activators RanGAP and Ran-binding protein 1/2 (RanBP1/2) stimulate GTP hydrolysis, resulting in the conversion of RanGTP to RanGDP. The switch to RanGDP results in the dissociation of the export complex and release of the export cargo. The export receptor then returns to the nucleus for another round of export. Thus, import and export are essentially reverse processes, with their directionality maintained by the presence of RanGTP in the nucleus and RanGDP in the cytoplasm. The interactions between soluble transport factors described in these examples suggested that it was not unlikely that a given receptor could function in the import of some substrates and in the export of others. In fact, recent studies have demonstrated that some of the β family receptors do function both in import and in export.[67,101]

It is interactions between β family receptors and components of the NPC that mediate translocation through the pore, while adapters and substrates seem to behave as inert cargo. The β family receptors preferentially bind to FG (phenylalanine-glycine) dipeptide repeats contained in nucleoporins that line the aqueous channel of the NPC. Current models[9,81,83] propose that the consecutive array of FG nucleoporins provides transient docking sites for moving receptor-cargo complexes through the pore. It is also suggested that these FG repeats form a sieve-like structure that restricts the flow of large molecules unless they can compete for direct binding to the motifs. The interactions of these motifs with nuclear transport receptors would allow infiltration into the barrier and facilitate fast transport. This 'selective phase model' could explain how NPCs function as a permeability barrier for inert molecules and yet become selectively permeable for nuclear transport receptor-cargo complexes.[81] The two nucleotide-bound forms of Ran modulate the affinity of receptors for binding to FG motifs during the translocation process. For example, CRM1/exportin-1 efficiently interacts with the FG repeat-containing nucleoporins, Nup214/CAN and p62 in a RanGTP dependent manner.[3,50] In contrast, Pse1p/Kap121p, a yeast import receptor, shows increased affinity for binding to FG nucleoporins when the cellular RanGDP level is elevated.[88] In both cases, enhanced receptor-nucleporin interactions facilitate transport by these receptors. Thus, Ran acts as a molecular switch to regulate two key events of the transport process: formation of transport complexes and their interaction with the NPC.

RanGTPase Dependent RNA Exports

Nuclear export of some RNA species also requires the RanGTPase system. In addition to exporting NES substrates, CRM1/exportin-1 also exports U small nuclear RNA (U snRNA). In this case, CRM1/exportin-1 does not recognize U snRNA directly, but through an adapter protein called PHAX, which carries a leucine-rich NES. PHAX interacts with the cap-binding complex (CBC) bound to the cap structure of U snRNA, and phosphorylation of PHAX promotes the formation of the export complex.[75] In general, signals for nuclear export of RNAs are thought not to reside in the RNA molecules themselves, but rather in proteins that decorate the RNA. A notable exception is that of tRNAs. Exportin-t,[1,54] another member of the β family receptors, directly binds to the TΨC and acceptor arm structures of tRNAs,[2] which act as NES, to form an exportin-t-RanGTP-tRNA ternary export complex similar to that of CRM1-RanGTP-NES. Involvement of the RanGTPase system in export of 5S rRNA in higher eukaryotes,[20] and of both the 40S and 60S ribosomal subunits in yeast has been suggested,[27,37,40,69] however, further studies will be required to understand the precise mechanisms of these export processes.

Export of mRNA

The transport field had initially assumed that one or more members of the well-studied importin-β receptor family would be responsible for the bulk of mRNA export. In fact, one such family member, CRM1/exportin-1, was the first mRNA export receptor to be identified. Human immunodeficiency virus (HIV) Rev protein binds to a structured RNA element called the Rev-responsive element (RRE) within an HIV intron, and CRM1/exportin-1 forms an export complex via a leucine-rich NES contained in Rev, thereby facilitating export of HIV pre-mRNA.[4,25] It was proposed that this mechanism might also apply to endogenous cellular mRNA export, but experiments with leptomycin B, a specific antibiotic inhibitor of CRM1/exportin-1, suggested that this was unlikely, since leptomycin B did not affect the export of bulk mRNA.[73] There are individual mRNAs, however, whose export requires the RanGTPase

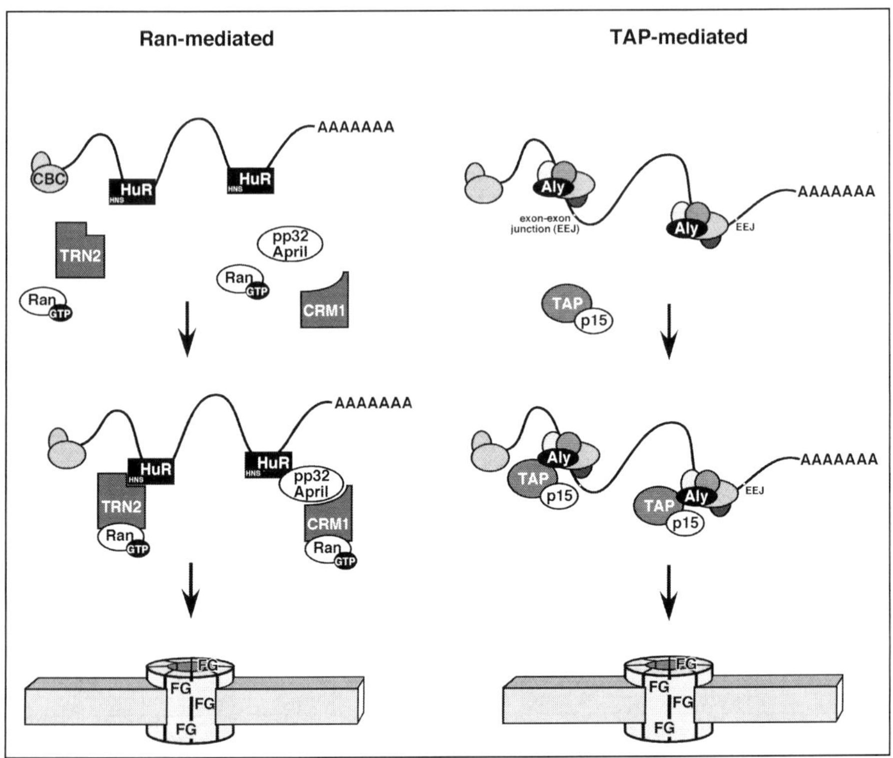

Figure 1. Diverse mechanisms of cellular mRNA export. Cellular mRNAs are exported by at least two independent mechanisms after maturation in nucleus. In the case of the Ran-mediated pathway of c-fos mRNA, export receptors CRM1/exportin-1 or TRN2/transportin-2 and GTP-bound form of Ran may organize a ternary export complex with mRNPs being exported. An adaptor molecule HuR has a critical function to connect the mRNA to the Ran-receptor complex by associating with specific sequence in mRNA and with receptors either directly or indirectly through ligand proteins pp32 and APRIL (Ran-mediated). 'Spliced' mRNAs are exported by TAP-mediated mechanism (TAP-mediated). During splicing, RNA-protein complexes are formed at 20-22 nucleotides upstream of exon-exon junction (EEJ). A RNA binding protein Aly is recruited into the the exon-exon junction complex (EJC) by function of U2AF associating factor UAP56, and Aly deposits export receptor TAP onto the EJC. Both in Ran- and TAP-mediated pathways, mRNA-export machinery complexes recognize the phenylalanine-glycine (FG) dipeptide repeat in nucleoporins. TAP forms a heterodimer with p15 in order to recognize the FG repeats, whereas CRM1/exportin-1 and TRN2/transportin-2 can directly bind to these repeats.

system and importin-β family members. For example, the protein HuR, which is involved in stabilization of RNAs containing AU-rich elements (AREs) containing RNA, acts as a transport adaptor for the mRNA transcript of the early response gene *c-fos* (Fig. 1). The *c-fos* mRNA/ HuR complex is either directly recognized by TRN2/transportin-2,[91] another β family receptor, via an export signal in HuR called HNS,[23] or indirectly by CRM1/exportin-1 via leucine-rich NESs in the HuR ligands pp32 and APRIL.[12,28]

Until recently, the primary candidates for cellular mRNA export factors were the highly abundant heterogeneous nuclear ribonucleoproteins (hnRNP) proteins. Nascent pre-mRNAs are bound by a subset of hnRNP proteins. The approximately 20 classes of hnRNP proteins (hnRNP A to U) are divided into two types: those that shuttle between the nucleus and the cytoplasm and those that are restricted to the nucleus.[53,72] hnRNP A1 was thought to be the major player in mRNA export because this protein belongs to the shuttling class of hnRNP proteins and contains an export signal M9. In addition, hnRNP A1-like proteins have been shown to bind to the giant Balbiani ring mRNA in *Chironomus tentans* and to accompany this mRNA to the cytoplasm.[98] Furthermore, excess hnRNP A1 or M9 peptides inhibit export of DHFR mRNA.[41] To date, however, no export receptor has been identified for hnRNP A1, although its M9-dependent import by a β family receptor TRN1/transportin-1, has been well-characterized.[77] Thus, no direct connection between hnRNP A1 and mRNA export still has yet been established.

From Gene to Nuclear Pore to Cytoplasm

Studies over the past several years indicate that there is extensive coupling between the different steps in gene expression.[21,35] Before mRNA can leave the nucleus, proper processing events, such as capping, polyadenylation, and splicing must occur. These processing events are thought to occur cotranscriptionally as the mRNA is synthesized. As will be discussed in the following sections, a splicing-dependent mRNP complex specifically targets mature mRNA for export.[43] There is also evidence linking proper 3'-end formation to mRNA export. Thus, it is likely that very early mRNA maturation events determine export competence of the transcript.[59,65] In addition, various proteins become associated with transcripts concomitant with transcription and behave differently during RNA transport according to the particular function of each protein. These functions are not limited to early nuclear events, but may even affect the cytoplasmic fate of exported mRNA molecules, impinging on such aspects as transport of mRNA within the cytoplasm, translational efficiency, and mRNA turnover. Thus, proteins loaded cotranscriptionally on pre-mRNA determine to a large extent the fate of mRNA in both the nucleus and the cytoplasm.

TAP-Mediated mRNA Export: Ran Independent Nucleocytoplasmic Transport

While it is clear that the RanGTPase system is required for some types of mRNA export, new studies have shifted the focus of attention to several highly conserved proteins that function in general mRNA export independent of Ran. The major export receptor is TAP (also called as NXF1), which connects messenger ribonuleoprotein particles (mRNPs) to nuclear pores. TAP was originally identified as a cellular factor that stimulated the nuclear export of RNAs containing constitutive transport elements (CTE) from type D retroviruses.[11,31,45] These studies revealed that TAP binds directly to the CTE, and that titration of TAP by excess CTE causes nuclear retention of bulk cellular mRNA, suggesting that TAP is likely to be the primary export receptor for cellular mRNA. TAP is a nucleocytoplasmic shuttling protein that associates with cellular poly(A)+ RNA and interacts with the NPC.[5,8,45,47,87] Genetic experiments involving the *S. cerevisiae* TAP homolog Mex67p and TAP homologs in *Caenorhabditis elegans* and *Drosophila melanogaster* (called NXF-1) have provided evidence of a crucial role for TAP in bulk mRNA export.[34,89,97,100] TAP is now recognized as a major cellular mRNA export receptor, and these TAP homologs comprise a family called the nuclear export factor (NXF) family.

The importance of TAP for bulk mRNA export was unexpected because the protein is structurally distinct from the Ran-dependent β family receptors, and the TAP export pathway is in fact independent of the RanGTPase system.[16] TAP forms a heterodimer with a small protein designated p15 (also called as NXT1).[47] Recent work has shown that the TAP-p15 heterodimer directly stimulates the export of cellular mRNAs.[10,32,47] p15 has sequence similarity to NTF2, a protein that mediates nuclear import of Ran and interacts with nucleoporin FG repeats[15,76]

TAP alone can bind weakly to nucleoporins carrying FG repeats, but Tap-p15 heterodimer formations significantly stimulates this association.[60,99] Structural studies have revealed that p15 and a NTF2-like domain found in TAP form a single structural domain that binds to an FG motif.[26] Thus, p15 can modulate the affinity of the TAP-mRNA export complex for binding to the NPC. The Ran GTPase system is not required for nuclear shuttling of TAP, but the TAP-p15 interaction is essential for the process.[16,46] These findings suggest that transport mediated by TAP-p15 may also involve sequential nucleoporin docking and release steps as proposed for translocation of β family receptors. However, given that Ran is not required for general mRNA export, directionality of mRNA transport must be determined by another mechanism. In principle, the TAP-p15 heterodimer can translocate across the nuclear pores in both directions, and it remains to be determined how directionality of mRNA export is established.

As in higher eukaryotes, the yeast TAP ortholog Mex67p also interacts with FG repeat-containing nucleoporins as a heterodimer, but interestingly, its partner protein, Mtr2p,[44,85] does not show any sequence similarity to p15. However, the *C. elegans* TAP-p15 heterodimer can replace the function of Mex67p-Mtr2p in yeast cells that lack these endogenous export factors, indicating that the function of these heterodimers is evolutionarily maintained.[47]

An Adaptor Protein and Other Conserved mRNA Export Factors

TAP itself does not bind strongly to RNA, but interacts predominantly with other components of the mRNP, which will be referred to as adapters. The most conspicuous adapter is a shuttling protein called Aly/REF (Yra1p in yeast). Aly was originally discovered as a hnRNP-type coactivator of the transcription factors LEF-1 and AML-1 in human cells and is evolutionarily conserved.[14,96] In *Xenopus* oocytes, microinjected recombinant Aly enhances cellular mRNA export and, conversely, inhibition of Aly function by microinjected Aly antibody specifically blocks mRNA export. These data strongly suggest that Aly is directly involved in the export of cellular mRNA. Aly simultaneously interact with both TAP and mRNA through distinct domains, further supporting the conclusion that Aly serves as an adapter molecule to link mRNA to its export receptor, TAP.

Another important aspect of Aly function is to link splicing events to the downstream process of mRNA export. Aly was originally found to colocalize with structures called nuclear speckles, which are sites of accumulation of factors required for pre-mRNA maturation.[82,103] Aly and several other proteins are recruited into mRNP particles coincident with pre-mRNA splicing, forming a complex that is situated about 20-24 nucleotides upstream of the exon-exon junction (EEJ) and is termed the exon-exon junction complex (EJC).[57] The EJC is thought to serve as a marker that designates mRNP particles as export-competent. Subsequent to EJC deposition, Aly is presumed to recruit the export receptor TAP, thus directly coupling the processes of pre-mRNA splicing and export of mature mRNAs.[94,103]

A component of the spliceosome, the evolutionarily conserved DEAD-box RNA helicase UAP56, binds directly to Aly and plays a critical role in its recruitment to the spliced mRNA. UAP56 associates with the splicing factor U2AF65, a large subunit of U2AF, and is required for the U2 snRNP interaction with pre-mRNA.[24] Although Aly has some intrinsic RNA binding ability, an Aly mutant lacking the UAP56 binding domain failed to be recruited into spliced mRNA.[62] A mutation in the yeast *S. cerevisiae* UAP56 ortholog *SUB2* is defective for export of poly(A)+ RNA, confirming a role for UAP56/Sub2p in mRNA export through the coupling of the splicing and export machineries.[42,95] In addition to the Aly-UAP56 interaction, TAP itself appears to interact directly with the spliceosome via the smaller U2AF subunit U2AF35.[104] Thus, U2AF is a key player that recruits export factors to spliced mRNP complexes. Interestingly, Mex67p and Sub2p compete for binding to a domain of the Yra1p protein, a yeast counterpart of Aly,[95] suggesting that Mex67p may promote a release of the mRNP from the spliceosome and target it to the NPC for export.

It was surprising that Sub2p is essential for general mRNA export in *S. cerevisiae* since only a limited number of intron-containing genes exist in the genome of the organism (only 250 out of the ~6,000 genes in *S. cerevisiae* contain introns).[61] This indicates that mechanisms of recruiting export factors onto intronless mRNA must exist in yeast and probably also in metazoans. An interesting observation is that Yra1p associates with coding regions of the yeast genes *PMA1* and *GAL10* in a transcription-dependent manner, suggesting that Yra1p might also be recruited cotranscriptionally.[58] A 22-nucleotide element of the intronless histone H2a gene recruits the shuttling splicing factors SRp20 and 9G8 to promote export of its transcribed mRNA in higher eukaryotic cells.[39] In this case, these splicing factors may directly mediate export of the intronless mRNA or may act as recruiters of export machinery, similar to the role of U2AF and Sub2p in export of spliced mRNAs.

Interactions between mRNA Export Machineries and Nucleoporins

In contrast to the study of soluble transport factors, little progress has been made toward elucidating the events that occurs in the NPC during mRNA export. Although the importance of interactions between of export receptors and nucleoporin FG repeats is clear, the pathway of transit through the NPC and the events that occur in the pore channel are not yet well-understood. Genetic screens in yeast have implicated several nucleoporins as important participants in mRNA export (reviewed in ref. 22). Most of members of two major NPC subcomplexes, Nup84-85-120-145C-Seh1-Sec13[90] and Nup159-82-116-Nsp1,[6,36] have been shown to be essential for the export of poly(A)+ RNA. The former subcomplex is found at the central core of NPC and the latter on the cytoplasmic fibrils, suggesting that they may constitute a mRNP docking site and a terminal release site, respectively. Some mammalian homologs of these nucleoporins, including CAN/Nup214, Nup98 and p62, physically interact with TAP,[5,47,87] suggesting that the route of mRNA export in the nuclear pore is conserved among various organisms. Interestingly, TAP seems to share NPC binding sites with β family receptors, but TAP does not depend on RanGTP/GDP for these interactions.[5]

The yeast genetic screens have also identified several NPC associating soluble factors implicated in mRNA export and the studies have been extended onto their mammalian homologs. An ATP-dependent DEAD-box helicase, Dbp5p, is mainly located in the cytoplasm where it interacts with cytoplasmic fibrils of the NPC, probably by associating with the FG nucleoporin Nup159p.[38,86] Gle1p also localizes to the cytoplasmic side of the NPC, where it interacts with the FG-repeat nucleoporin Rip1p/Nup42p as well as with Dbp5p.[71,93] By these multiple interactions, Gle1p may provide a platform for mRNP disassembly and mRNA release at the cytoplasmic side of the NPC. Gle2p/Rae1 (also known as mrnp 41 in mammals) is a shuttling poly(A)+RNA associating factor.[13,52,70] It associates with the NPC by recognizing specific peptide sequence called GLEBS, which are found in nucleoporins such as Nup116p in yeast and Nup98 in mammals.[7,78] These properties of Gle2p/Rae1 meet all the prerequisites for an mRNA export receptor, suggesting that Gle2p/Rae1 may act in a second pathway of mRNA export that is redundant with the TAP-p15 pathway.

Links between mRNA Quality Control and Nuclear Export

As previously mentioned, the loading of the exon-exon junction complex (EJC) onto spliced mRNAs serves as a marker for export-competent mRNPs, thus playing a role in mRNA quality control. The release of various EJC components from RNPs takes place at different times during the export process. Whereas two splicing factors, SRm160 and DEK, leave the mRNA inside nucleus,[64] other members of the EJC, including the export adapter Aly, remain to bound to the mRNA during its export to the cytoplasm.[49,56,64] Aly as well as TAP then dissociate from the complex when the mRNA reaches the cytoplasm and they shuttle back to the nucleus for the next round of export. Although no direct confirmatory evidence has reported, an attractive hypothesis is that the helicase activity of Dbp5, perhaps in concert with activities of Dbp5-associating factors at the cytoplasmic fibrils of NPC, mediates the release of the export machinery from the exported RNA.

The remaining members of the EJC—RNPS1, Y14, hUpf3 and Magoh—seem not to be inert piggy-back factors, but rather they appear to have active functions as determiners of the cytoplasmic fate of an mRNA. In particular, nuclear deposition of these factors appears to direct the specific interaction of an mRNP with cytoplasmic machineries involved in such processes as translation and mRNA degradation (see Fig. 2). One example of this is the involvement of EJC components in nonsense-mediated mRNA decay (NMD), a process by which mRNAs containing a premature termination codon are targeted for degradation. The molecular mechanisms of NMD were originally studied in the yeast *S. cerevisiae*,[19] but a conserved mechanism has been shown to exist in humans. An shuttling human EJC component, hUpf3, is a homolog of the yeast Upf3p protein, which is an essential factor of the NMD pathway. As part of the EJC, hUpf3 interacts with two other components: RNPS1,[105] which was originally identified as a general splicing activator, and a novel protein called Y14.[49] These interactions are important for the recruitment of hUpf3 into the complex.[51,64] The role of hUpf3 may be to recruit two cytoplasmic NMD components, hUpf1 and hUpf2, onto the RNA after it has been exported to cytoplasm.[51,63,64] It is hypothesized that an actively translating ribosome would displace this reorganized EJC; if the ribosome dissociates from the template at a nonsense mutation, however, the remaining EJC, along with the translation release factors eRF1 and eRF3, may serve as a tag for sorting the mRNA to a degradation pathway. In the yeast *S. cerevisiae*, a specific sequence element called DSE (downstream sequence element,[84]) functions like the EEJ, since most transcripts in this organisms are not spliced. In this case, termination codons located upstream of the DSE would be regarded as nonsense mutations. The DSE appears to be recognized by a shuttling hnRNP protein called Hrp1p.[29] The Hrp1p-DSE association is thought to trigger the deposition of other yeast NMD factors such as several *UPF* and *MOF* gene products[18,19] onto the RNA, forming a complex which has similar function to the EJC.

Some mRNAs require further transport to specific sites in the cytoplasm before being translated. A well-known example occurs in *Drosophila* embryogenesis, where transcripts of genes essential for oogenesis, such as *oskar* and *bicoid*, are synthesized in nurse cells and transported to the posterior pole to allow exclusive expression of their gene products at this site. The *Drosophila* EJC component Y14 colocalizes with *oskar* mRNA at the posterior pole; functional Y14, as well as the Mago Nashi protein, a *Drosophila* counterpart of the human Magoh protein, are essential for this asymmetric localization of *oskar* and *bicoid* Mrna.[33,74] Magoh was isolated as a Y14 associating factor and it binds to Y14 and TAP as part of the EJC.[48,55,102] The mechanism of selective translocation of specific mRNAs and the anchoring of these RNAs at their destinations is still unclear, but the Y14-Magoh complex, probably along with other additional associating factors such as the *Drosophila* protein Tsunagi,[68] may act as a 'pass' for further transport in the cytoplasm.

Conclusion

The discovery of the TAP-mediated mRNA export system has been a major leap forward in our understanding of mRNA export. It is also a perfect example of how physiological cellular functions are connected and organized: the export reaction is integrated into a system of total mRNA metabolism that also includes mRNA biogenesis and degradation. The splicing event increases the diversity of gene information dramatically. At the same time, however, it has the risk of producing aberrant, deleterious peptides that may cause fatal cellular defects. By incorporating mechanisms of quality control, such as NMD, the system seems to have been

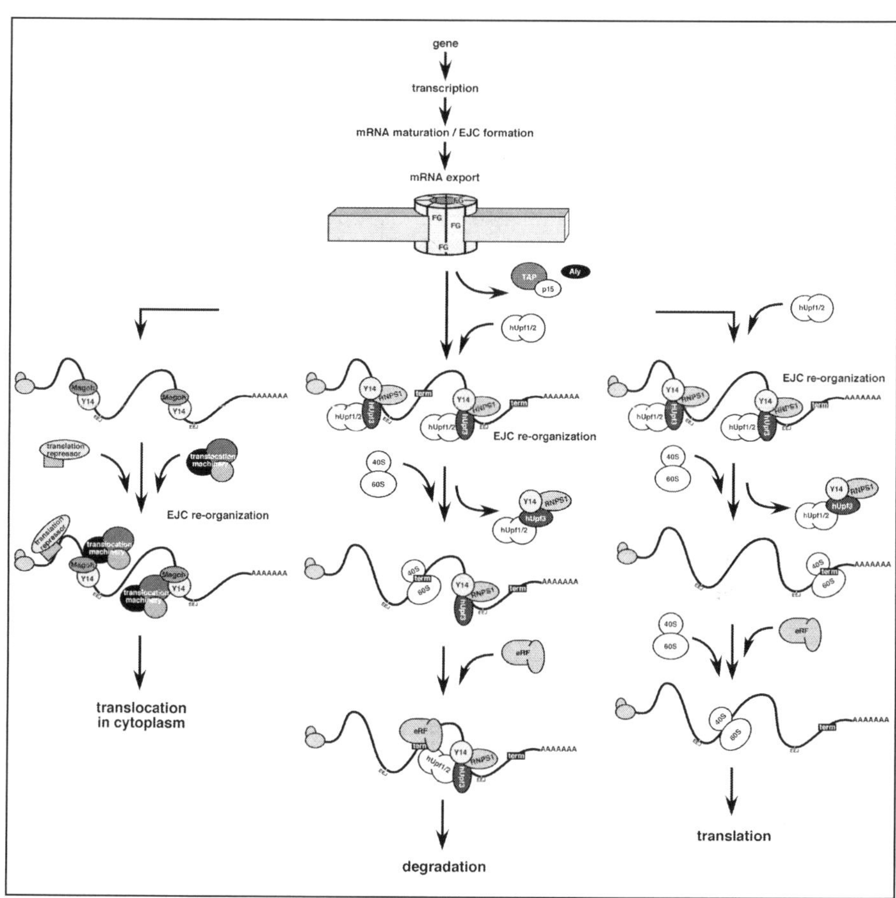

Figure 2. The nuclear history of mRNA determines the fate of mRNA in the cytoplasm. The fate of mRNA in the cytoplasm may be imprinted as protein complex during nuclear processing events. Nonsense-mediated mRNA decay (NMD) is a system for leading mRNAs carrying nonsense termination codons into a degradation pathway. The exon-exon junction complex (EJC) organized during splicing events acts as molecular memory of splicing and a NMD factor hUpf3 deposited into the EJC may have a major role. After being exported to cytoplasm, cytoplasmic NMD components hUpf1/2 may be recruited to the complex by function of hUpf3. Normally, these reorganized 'marks' should be checked and displaced by actively translating ribosomes in the cytoplasm and the mRNA can be used for following rounds of translation (translation). However, if the ribosome is released before reaching the legitimate termination codon by nonsense mutation, the remaining EJC may become a 'RNA decay' tag to prevent from synthesizing immature peptides (degradation). EJC can also function as a transit pass for being transferred to the final destination in the cytoplasm. For example, oogenesis requires strictly regulated timing and localization of gene expression. After nuclear export, transcripts of Drosophila genes osker and bicoid, both essential for oogenesis, are further transported to posterior pole without being translated during the transport. In this case, EJC components Mago Nashi and Y14 are critical, probably for deposition of the translocation machinery onto mRNA (translocation in cytoplasm).

developed to satisfy the requirements of both diversity and accuracy. Although we now have an elegant model of the export system, several unanswered questions still persist. First, the mechanism of export of intronless mRNA is still not clear. Although the TAP system may be used for this class of mRNAs, distinct factors may be involved in the recruitment of export factors to the mRNP. Another question is the extent to which the RanGTPase system is involved in cellular mRNA export. It is conceivable that such a critical cellular function as mRNA export may require a back up system for emergencies, such in cases of environmental stress. In addition, some mRNAs may employ the RanGTPase system to find a niche for their export, using specific adapters that can regulate the export efficiency or gene expression, as is the case for *c-fos* mRNA. The answers to these questions should be provided by extensive studies on the mRNA export combined with those on other physiological functions and the findings by these research might give us another different view of the export process.

Acknowledgments

We thank Jennifer Hood-DeGrenier for critical reading of the manuscript and all the members of Siomi laboratory for helpful discussions. We apologize to researchers whose original papers we did not cite due to the space limitations. T.T. and H.S. are supported by the Japan Society for the Promotion of Science. M.C.S. is supported by the The Mitsubishi Foundation.

References

1. Arts G-J, Fornerod M, Mattaj IW. Identification of a nuclear export receptor for tRNA. Curr Biol 1998a; 8:305-314.
2. Arts G-J, Kuersten S, Romby P et al. The role of exportin-t in selective nuclear export of mature tRNAs. EMBO J 1998b; 17:7430-7441.
3. Askjaer P, Bachi A, Wilm M et al. RanGTP-regulated interactions of CRM1 with nucleoporins and a shuttling DEAD-box helicase. Mol Cell Biol 1999; 19:6276-6285.
4. Askjaer P, Jensen TH, Nilsson J et al. The specificity of the CRM1-Rev nuclear export signal interaction is mediated by RanGTP. J Biol Chem 1998; 273:33414-33422.
5. Bachi A, Braun IC, Rodrigues JP et al. The C-terminal domain of TAP interacts with the nuclear pore complex and promotes export of specific CTE-bearing RNA substrates. RNA 2000; 6:136-158.
6. Bailer SM, Balduf C, Katahira J et al. Nup116p associates with the Nup82p-Nsp1p-Nup159p nucleoporin complex. J Biol Chem 2000; 275:23540-23548.
7. Bailer SM, Siniossoglou S, Podtelejnikov A et al. Nup116p and Nup100p are interchangeable through a conserved motif which constitutes a docking site for the mRNA transport factor Gle2p. EMBO J 1998; 17:1107-1119.
8. Bear J, Tan W, Zolotukhin AS et al. Identification of novel import and export signals of human TAP, the protein that binds to the constitutive transport element of the type D retrovirus mRNAs. Mol Cell Biol 1999; 19:6306-6317.
9. Ben-Efraim I, Gerace L. Gradient of increasing affinity of importin β for nucleoporins along the pathway of nuclear import. J Cell Biol 2001; 152:411-418.
10. Braun IC, Herold A, Rode M et al. Overexpression of TAP/p15 heterodimers bypasses nuclear retention and stimulates nuclear mRNA export. J Biol Chem 2001; 276:20536-20543.
11. Braun IC, Rohrbach E, Schmitt C et al. TAP binds to the constitutive transport element (CTE) through a novel RNA-binding motif that is sufficient to promote CTE-dependent RNA export from the nucleus. EMBO J 1999; 18:1953-1965.
12. Brennan CM, Gallouzi I-E, Steitz JA. Protein ligands to HuR modulate its interaction with target mRNAs in vivo. J Cell Biol 2000; 151:1-14.
13. Brown JA, Bharathi A, Whalen AG et al. A mutation in the Schizosaccharomyces pombe rae1 gene causes defects in poly(A)+ RNA export and in the cytoskeleton. J Biol Chem 1995; 270:7411-7419.
14. Bruhn L, Munnerlyn A, Grosschedl R. ALY, a context-dependent coactivator of LEF-1 and AML-1, is required for TCRalpha enhancer function. Genes Dev 1997; 11:640-653.
15. Bullock TL, Clarkson WD, Kent HM et al. The 1.6 Å resolution crystal structure of nuclear transport factor 2 (NTF2). J Mol Biol 1996; 260:422-431.
16. Clouse NK, Luo M-j, Zhou Z et al. A Ran-independent pathway for export of spliced mRNA. Nat Cell Biol 2001; 3:97-99.

17. Conti E, Izaurralde E. Nucleocytoplasmic transport enters the atomic age. Curr Opin Cell Biol 2001; 13:310-319.
18. Cui Y, Gonzalez CI, Kinzy TG et al. Mutations in the MOF2/SUI1 gene affect both translation and nonsense-mediated mRNA decay. RNA 1999; 5:794-804.
19. Culbertson MR. RNA surveillance. Unforeseen consequences for gene expression, inherited genetic disorders and cancer. Trends Genet 1999; 15:74-80.
20. Cullen BR. Nuclear RNA export pathways. Mol Cell Biol 2000; 20:4181-4187.
21. Daneholt B. Assembly and transport of a premessenger RNP particle. Proc Natl Acad Sci USA 2001; 98:7012-7017.
22. Fabre E, Hurt E. Yeast genetics to dissect the nuclear pore complex and nucleocytoplasmic trafficking. Annu Rev Genet 1997; 31:277-313.
23. Fan XC, Steitz JA. HNS, a nuclear-cytoplasmic shuttling sequence in HuR. Proc Natl Acad Sci USA 1998; 95:15293-15298.
24. Fleckner J, Zhang M, Valcarcel J et al. U2AF65 recruits a novel human DEAD box protein required for the U2 snRNP-branchpoint interaction. Genes Dev 1997; 11:1864-1872.
25. Fornerod M, Ohno M, Yoshida M et al. CRM1 is an export receptor for leucine-rich nuclear export signals. Cell 1997; 90:1051-1060.
26. Fribourg S, Braun IC, Izaurralde E et al. Structural basis for the recognition of a nucleoporin FG repeat by the NTF2-like domain of the TAP/p15 mRNA nuclear export factor. Mol Cell 2001; 8:645-656.
27. Gadal O, Strauss D, Kessl J et al. Nuclear export of 60s ribosomal subunits depends on Xpo1p and requires a nuclear export sequence-containing factor, Nmd3p, that associates with the large subunit protein Rpl10p. Mol Cell Biol 2001; 21:3405-3415.
28. Gallouzi IE, Steitz JA. Delineation of mRNA export pathways by the use of cell-permeable peptides. Science 2001; 294:1895-1901.
29. Gonzalez CI, Ruiz-Echevarria MJ, Vasudevan S et al. The yeast hnRNP-like protein Hrp1/Nab4 marks a transcript for nonsense-mediated mRNA decay. Mol. Cell 2000; 5:489-499.
30. Görlich D, Kutay U. Transport between the cell nucleus and the cytoplasm. Annu Rev Cell Dev Biol 1999; 15:607-660.
31. Grüter P, Tabernero C, von Kobbe C et al. TAP, the human homolog of Mex67p, mediates CTE-dependent RNA export from the nucleus. Mol Cell 1998; 1:649-659.
32. Guzik BW, Levesque L, Prasad S et al. NXT1 (p15) is a crucial cellular cofactor in TAP-dependent export of intron-containing RNA in mammalian cells. Mol Cell Biol 2001; 21:2545-2554.
33. Hachet O, Ephrussi A. Drosophila Y14 shuttles to the posterior of the oocyte and is required for oskar mRNA transport. Curr Biol 2001; 11:1666-1674.
34. Herold, A., Klymenko, T. and Izaurralde, E. NXF1/p15 heterodimers are essential for mRNA nuclear export in Drosophila. RNA 2001; 7:1768.
35. Hirose Y, Manley JL. RNA polymerase II and the integration of nuclear events. Genes Dev 2000; 14:1415-1429.
36. Ho AK, Shen TX, Ryan KJ et al. Assembly and preferential localization of Nup116p on the cytoplasmic face of the nuclear pore complex by Interaction with Nup82p. Mol Cell Biol 2000a; 20:5736-5748.
37. Ho JH, Kallstrom G, Johnson AW. Nmd3p is a Crm1p-dependent adapter protein for nuclear export of the large ribosomal subunit. J Cell Biol 2000b; 151:1057-1066.
38. Hodge CA, Colot HV, Stafford P et al. Rat8p/Dbp5p is a shuttling transport factor that interacts with Rat7p/Nup159p and Gle1p and suppresses the mRNA export defect of XPO1-1 cells. EMBO J 1999; 18:5778-5788.
39. Huang Y, Steitz JA. Splicing factors SRp20 and 9G8 promote the nucleocytoplasmic export of mRNA. Mol Cell 2001; 7:899-905.
40. Hurt E, Hannus S, Schmelzl B et al. A novel in vivo assay reveals inhibition of ribosomal nuclear export in Ran-cycle and nucleoporin mutants. J Cell Biol 1999; 144:389-401.
41. Izaurralde E, Jarmolowski A, Beisel C et al . A role for the M9 transport signal of hnRNP A1 in mRNA nuclear export. J Cell Biol 1997; 137:27-35.
42. Jensen TH, Boulay J, Rosbash M et al. The DECD box putative ATPase Sub2p is an early mRNA export factor. Curr Biol 2001a ; 11:1711-1715.

43. Jensen TH, Patricio K, McCarthy T et al. A block to mRNA nuclear export in S. cerevisiae leads to hyperadenylation of transcripts that accumulate at the site of transcription. Mol Cell 2001b; 7:887-898.

44. Kadowaki T, Hitomi M, Chen S et al. Nuclear mRNA accumulation causes nucleolar fragmentation in yeast mtr2 mutant. Mol Biol Cell 1994; 5:1253-1263.

45. Kang Y, Cullen BR. The human Tap protein is a nuclear mRNA export factor that contains novel RNA-binding and nucleocytoplasmic transport sequences. Genes Dev 1999; 13:1126-1139.

46. Katahira J, Straesser K, Saiwaki T et al. Complex formation between Tap and p15 affects binding to FG-repeat nucleoporins and nucleocytoplasmic shuttling. J Biol Chem 2002; 277:9242-9246.

47. Katahira J, Sträßer K, Podtelejnikov A et al. The Mex67p-mediated nuclear mRNA export pathway is conserved from yeast to human. EMBO J 1999; 18:2593-2609.

48. Kataoka N, Diem MD, Kim VN et al. Magoh, a human homolog of Drosophila mago nashi protein, is a component of the splicing-dependent exon-exon junction complex. EMBO J 2001; 20:6424-6433.

49. Kataoka N, Yong J, Kim VN et al. Pre-mRNA splicing imprints mRNA in the nucleus with a novel RNA-binding protein that persists in the cytoplasm. Mol Cell 2000; 6:673-682.

50. Kehlenbach RH, Dickmanns A, Kehlenbach A et al. A role for RanBP1 in the release of CRM1 from the nuclear pore complex in a terminal step of nuclear export. J Cell Biol 1999; 145:645-657.

51. Kim VN, Kataoka N, Dreyfuss G. Role of the nonsense-mediated decay factor hUpf3 in the splicing-dependent exon-exon junction complex. Science 2001; 293:1832-1836.

52. Kraemer D, Blobel G. mRNA binding protein mrnp 41 localizes to both nucleus and cytoplasm. Proc Natl Acad Sci USA 1997; 94:9119-9124.

53. Krecic AM, Swanson MS. hnRNP complexes: Composition, structure, and function. Curr Opin Cell Biol 1999; 11:367-371.

54. Kutay U, Lipowsky G, Izaurralde E et al. Identification of a tRNA-specific nuclear export receptor. Mol Cell 1998; 1:359-369.

55. Le Hir H, Gatfield D, Braun IC et al. The protein Mago provides a link between splicing and mRNA localization. EMBO R 2001a; 2:1119-1124.

56. Le Hir H, Gatfield D, Izaurralde E et al. The exon-exon junction complex provides a binding platform for factors involved in mRNA export and nonsense-mediated mRNA decay. EMBO J 2001b; 20:4987-4997.

57. Le Hir H, Izaurralde E, Maquat LE et al. The spliceosome deposits multiple proteins 20-24 nucleotides upstream of mRNA exon-exon junctions. EMBO J 2000; 19:6860-6869.

58. Lei EP, Krebber H, Silver PA. Messenger RNAs are recruited for nuclear export during transcription. Genes Dev 2001; 15:1771-1782.

59. Lei EP, Silver PA. Protein and RNA export from the nucleus. Dev Cell 2002; 2:261-72.

60. Lévesque L, Guzik B, Guan T et al. RNA export mediated by Tap involves NXT1-dependent interactions with the nuclear pore complex. J Biol Chem 2001; 276:44953-44962.

61. Lopez PJ, Seraphin B. YIDB: The Yeast Intron DataBase. Nucleic Acids Res 2000; 28:85-86.

62. Luo M.-J, Zhou Z, Magni K et al. Pre-mRNA splicing and mRNA export linked by direct interactions between UAP56 and Aly. Nature 2001; 413:644-647.

63. Lykke-Andersen J, Shu M-D, Steitz JA. Human Upf proteins target an mRNA for nonsense-mediated decay when bound downstream of a termination codon. Cell 2000; 103:1121-1131.

64. Lykke-Andersen J, Shu M-D, Steitz JA. Communication of the position of exon-exon junctions to the mRNA surveillance machinery by the protein RNPS1. Science 2001; 293:1836-1839.

65. Maniatis T, Reed R. An extensive network of coupling among gene expression machines. Nature 2002; 416:499-506.

66. Mattaj IW, Englmeier L. Nucleocytoplasmic transport: The soluble phase. Annu Rev Biochem 1998; 67:265-306.

67. Mingot J-M, Kostka S, Kraft R et al. Importin 13: A novel mediator of nuclear import and export. EMBO J 2001; 20:3685-3694.

68. Mohr SE, Dillon ST, Boswell RE. The RNA-binding protein Tsunagi interacts with Mago Nashi to establish polarity and localize oskar mRNA during Drosophila oogenesis. Genes Dev 2001; 15:2886-2899.

69. Moy TI, Silver PA. Nuclear export of the small ribosomal subunit requires the Ran-GTPase cycle and certain nucleoporins. Genes Dev 1999; 13:2118-2133.
70. Murphy R, Watkins JL, Wente SR. GLE2, a Saccharomyces cerevisiae homologue of the Schizosaccharomyces pombe export factor RAE1, is required for nuclear pore complex structure and function. Mol Biol Cell 1996; 7:1921-1937.
71. Murphy R, Wente SR. An RNA-export mediator with an essential nuclear export signal. Nature 1996; 383:357-360.
72. Nakielny S, Dreyfuss G. Transport of proteins and RNAs in and out of the nucleus. Cell 1999; 99:677-690.
73. Neville M, Rosbash M. The NES-Crm1p export pathway is not a major mRNA export route in Saccharomyces cerevisiae. EMBO J 1999; 18:3746-3756.
74. Newmark P, Boswell R. The mago nashi locus encodes an essential product required for germ plasm assembly in Drosophila. Development 1994; 120:1303-1313.
75. Ohno M, Segref A, Bachi A et al. PHAX, a mediator of U snRNA nuclear export whose activity is regulated by phosphorylation. Cell 2000; 101:187-198.
76. Paschal BM, Gerace L. Identification of NTF2, a cytosolic factor for nuclear import that interacts with nuclear pore complex protein p62. J Cell Biol 1995; 129:925-937.
77. Pollard VW, Michael WM, Nakielny S et al. A novel receptor-mediated nuclear protein import pathway. Cell 1996; 86:985-994.
78. Pritchard CEJ, Fornerod M, Kasper LH et al. RAE1 is a shuttling mRNA export factor that binds to a GLEBS-like NUP98 motif at the nuclear pore complex through multiple domains. J Cell Biol 1999; 145:237-254.
79. Reed R, Hurt E. A conserved mRNA export machinery coupled to pre-mRNA splicing. Cell 2002; 108:523-531.
80. Reed R, Magni K. A new view of mRNA export: Separating the wheat from the chaff. Nat Cell Biol 2001; 3:E201-4.
81. Ribbeck K, Görlich D. Kinetic analysis of translocation through nuclear pore complexes. EMBO J 2001; 20:1320-1330.
82. Rodrigues JP, Rode M, Gatfield D et al. REF proteins mediate the export of spliced and unspliced mRNAs from the nucleus. Proc Natl Acad Sci USA 2001; 98:1030-1035.
83. Rout MP, Aitchison JD, Suprapto A et al. The yeast nuclear pore complex: Composition, architecture, and transport mechanism. J Cell Biol 2000; 148:635-652.
84. Ruiz-Echevarria MJ, Gonzalez CI, Peltz SW. Identifying the right stop: Determining how the surveillance complex recognizes and degrades an aberrant mRNA. EMBO J 1998; 17:575-589.
85. Santos-Rosa H, Moreno H, Simos G et al. Nuclear mRNA export requires complex formation between Mex67p and Mtr2p at the nuclear pores. Mol Cell Biol 1998; 18:6826-6838.
86. Schmitt C, von Kobbe C, Bachi A et al. Dbp5, a DEAD-box protein required for mRNA export, is recruited to the cytoplasmic fibrils of nuclear pore complex via a conserved interaction with CAN/Nup159p. EMBO J 1999; 18:4332-4347.
87. Schmitt I, Gerace L. In vitro analysis of nuclear transport mediated by the C-terminal shuttle domain of Tap. J Biol Chem 2001; 276:42355-42363.
88. Seedorf, M, Damelin M, Kahana J et al. Interactions between a nuclear transporter and a subset of nuclear pore complex proteins depend on Ran GTPase. Mol Cell Biol 1999; 19:1547-1557.
89. Segref A, Sharma K, Doye V et al. Mex67p, a novel factor for nuclear mRNA export, binds to both poly(A)⁺ RNA and nuclear pores. EMBO J 1997; 16:3256-3271.
90. Siniossoglou S, Lutzmann M, Santos-Rosa H et al. Structure and assembly of the Nup84p complex. J Cell Biol 2000; 149:41-54.
91. Siomi MC, Eder PS, Kataoka N et al. Transportin-mediated nuclear import of heterogeneous nuclear RNP proteins. J Cell Biol 1997; 138:1181-1192.
92. Stoffler D, Fahrenkrog B, Aebi U. The nuclear pore complex: From molecular architecture to functional dynamics. Curr Opin Cell Biol 1999; 11:391-401.
93. Strahm Y, Fahrenkrog B, Zenklusen D et al. The RNA export factor Gle1p is located on the cytoplasmic fibrils of the NPC and physically interacts with the FG-nucleoporin Rip1p, the DEAD-box protein Rat8p/Dbp5p and a new protein Ymr255p. EMBO J 1999; 18:5761-5777.

94. Sträßer K, Hurt E. Yra1p, a conserved nuclear RNA-binding protein, interacts directly with Mex67p and is required for mRNA export. EMBO J 2000; 19:410-420.

95. Sträßer K, Hurt E. Splicing factor Sub2p is required for nuclear mRNA export through its interaction with Yra1p. Nature 2001; 413:648-652.

96. Stutz F, Bachi A, Doerks T et al. REF, an evolutionary conserved family of hnRNP-like proteins, interacts with TAP/Mex67p and participates in mRNA nuclear export. RNA 2000; 6:638-650.

97. Tan W, Zolotukhin AS, Bear J et al. The mRNA export in Caenorhabditis elegans is mediated by Ce-NXF-1, an ortholog of human TAP/NXF and Saccharomyces cerevisiae Mex67p. RNA 2000; 6:1762-1772.

98. Visa N, Alzhanova-Ericsson AT, Sun X et al. A pre-mRNA-binding protein accompanies the RNA from the gene through the nuclear pores and into polysomes. Cell 1996; 84:253-264.

99. Wiegand HL, Coburn GA, Zeng Y et al. Formation of Tap/NXT1 heterodimers activates Tap-dependent nuclear mRNA export by enhancing recruitment to nuclear pore complexes. Mol Cell Biol 2002; 22:245-256.

100. Wilkie GS, Zimyanin V, Kirby R et al. Small bristles, the Drosophila ortholog of NXF-1, is essential for mRNA export throughout development. RNA 2001; 7:1781-1792.

101. Yoshida K, Blobel G. The karyopherin Kap142p/Msn5p mediates nuclear import and nuclear export of different cargo proteins. J Cell Biol 2001; 152:729-740.

102. Zhao X-F, Nowak NJ, Shows TB et al. MAGOH interacts with a novel RNA-binding protein. Genomics 2000; 63:145-148.

103. Zhou Z, Luo M-J, Straesser K et al. The protein Aly links pre-messenger-RNA splicing to nuclear export in metazoans. Nature 2000; 407:401-405.

104. Zolotukhin AS, Tan W, Bear J et al. U2AF participates in the binding of TAP (NXF1) to mRNA. J Biol Chem 2002; 277:3935-3942.

105. Mayeda A, Badolato J, Kobayashi R et al. Purification and characterization of human RNPS1: a general activator of pre-mRNA splicing. EMBO J 1999; 18:4560-4570.

Nuclear Import and Export of Mammalian Viruses

Michael Bukrinsky

Viruses are intracellular parasites that commandeer cellular processes, such as RNA processing or protein synthesis, to perform virus-specific functions. For this purpose, many viral proteins shuttle between the nuclear and cytoplasmic compartments, even when the viral genome is replicated in the cytoplasm. This shuttling process is usually regulated by classical nuclear import and export signals (NLSs and NESs, respectively), which are also found in many cellular proteins and are described elsewhere in this book. In this Chapter, we will focus on viruses that replicate in the nucleus, and in particular on the mechanisms by which they transport their genomes into and out of the nucleus.

While replication in the host cell's nucleus provides clear benefits for the virus, such as ready access to cellular transcription and splicing apparatus, it imposes the barrier of the nuclear envelope that viruses have to overcome. During the early stages of infection, they have to transport their genome from the site of penetration to the nucleus and then through the nuclear membrane, while at the late stage they need to export newly made genomes or assembled viral capsids out of the nucleus to the site of virus assembly. With some notable exceptions, viruses accomplish these tasks by mimicking nuclear import or export signals used by cellular proteins, which allows them to hijack the nuclear transport machinery evolved to transport cellular proteins or RNPs in and out of the nucleus.

Nuclear envelope is disassembled during mitosis, thus providing the viruses with an opportunity for unobstructed nuclear entry and exit. Some viruses, such as most retroviruses, critically depend on mitosis for nuclear entry.[61] However, because mitosis constitutes only a small part of the cell cycle, this dependency greatly diminishes the efficiency of viral infection. Therefore, many viruses with high replicative capacity (which often correlates with high pathogenicity), including HIV and other members of the *Lentivirus* genus of the *Retroviridae* family, evolved mechanisms that ensure efficient transport of their genome into the interphase nucleus.

Viral nuclear import in most cases occurs through the nuclear pore and relies on the cellular nuclear import machinery. Given the diffusional barrier imposed by the cytoplasm to the movement of macromolecules, the step of nuclear targeting is unlikely to proceed via free diffusion. Instead, viruses use some form of active transport by engaging actin or microtubulin cytoskeleton to deliver the viral nucleoprotein complex to the nuclear membrane.[10,70,72] The next step is actual translocation of the viral genome (together with viral proteins important for initiation of the replication cycle) through the nuclear pore complex. This step is relatively unified between virus families and relies in most instances on interaction with cellular soluble import factors. Because of restrictions imposed by the nuclear pore on the size of the passing molecules, complex (and poorly characterized) steps of uncoating and capsid rearrangement are required before entry of the viral nucleoprotein complex can occur through the nuclear pore. Different virus families display diverse uncoating programs involving interactions between viral and cellular molecules. Uncoating either precedes or coincides with migration of the viral nucleoprotein complex towards the nucleus.

Nuclear envelope presents a barrier also during exit of viruses from the nucleus. One simple solution of this problem is lysis of the nucleus along with the cell. This mechanism is used by nonenveloped viruses, such as adeno- or papovaviruses (e.g., SV40), which mature inside the nucleus. This strategy, however, cannot be employed by enveloped viruses, which need an intact cell membrane for assembly. Enveloped viruses with small genomes (retroviruses, hepadnaviruses, orthomyxoviruses) exit through the nuclear pore, while viruses with large genomes that assemble their capsid in the nucleus, such as herpesviruses, exit the nucleus via budding through the nuclear envelope.

Despite certain differences between the virus families in the mechanisms of nuclear entry and exit, there are many common features that reflect interaction of the viral nucleoprotein complex with the cellular nuclear transport machinery. In this Chapter, we will review sequential steps of viral nuclear import and export using human immunodeficiency virus type 1 (HIV-1) as an example. Intense research by the international scientific community into the fundamental processes of HIV-1 replication has yielded knowledge that in many aspects equals or exceeds that of many other viruses. Other viruses will be used to illustrate alternative mechanisms or events that are not well characterized for HIV-1.

Transport to the Nuclear Envelope

As a member of the *Retroviridae* family, HIV-1 copies its RNA genome into a double-stranded DNA molecule that is subsequently integrated into the host chromosomal DNA. This process is mainly carried out in the cytoplasm of infected cells in the context of the viral reverse transcription complex (RTC).[74] These complexes contain viral genomic RNA associated with several proteins, including reverse transcriptase (RT), integrase (IN), matrix protein (MA), and viral protein R (Vpr).[12,49] While the function of RT and IN is well defined, the role of two other protein components of the RTC (MA and Vpr) is less clear. One likely possibility is that they facilitate nuclear import of the PIC (see below). The final product of reverse transcription reaction (termed now pre-integration complex or PIC), a blunt-ended linear duplex DNA associated with several viral proteins, is then transported to and through the nuclear pore. A specific feature of reverse transcription of HIV (and other lentiviruses as well) appears to influence subsequent nuclear import steps. The synthesis of plus-strand cDNA of these viruses is discontinuous: in addition to normal initiation at the polypurine tract proximal to the U3 region of the LTR, it can be reinitiated at the additional polypurine tract in the middle of genome. As a result, unintegrated HIV-1 DNA has a central plus-strand overlap 99 nucleotides long. This central DNA flap was shown recently to be necessary for the nuclear import of the PIC.[83] While the mechanism by which the DNA flap promotes nuclear uptake of HIV DNA is unclear, it is unlikely to perform nuclear import of the HIV PIC on its own. Most likely, it synergizes with the nuclear import function of HIV proteins (see below) by providing the optimal conformation to the PIC necessary for its translocation through the nuclear pore.

Viral reverse transcription complexes were found to associate with the host cell cytoskeleton, and in particular, with actin filaments.[10] This association appears to be mediated by the HIV-1 matrix protein (MA), which is also a component of the PIC.[12,49] It is logical to surmise, therefore, that the PICs' movement towards the nucleus might be mediated by actin cables. Consistent with this hypothesis, disruption of actin microfilaments or inhibition of myosin-mediated movement along actin microfilaments significantly impaired viral infectivity.[10] If the role of actin microfilaments in nuclear targeting of the HIV-1 PIC is confirmed by future studies, it will set HIV-1 apart from several other virus families which employ microtubules to reach the nucleus.

One such virus family is adenoviruses. In contrast to retroviruses, which enter target cells by fusing at neutral pH with cell plasma membrane, adenoviruses enter by receptor-mediated endocytosis followed by a low-pH-dependent fusion that releases the viral capsid to the cytosol (reviewed in ref. 51). After a series of disassembly events, the viral capsid associates with micro-tubules (MT), which mediate its transport toward the nucleus.[80] The MT network of an inter-phase cell is a dynamic polarized structure. The stable minus ends are localized to the MT organizing center located at a perinuclear position, while the dynamic plus ends extend to-wards the cell periphery.[13] Cytosolic adenovirus particles engage in both minus- and plus-end-directed motilities along the microtubules,[73] but obviously only the minus-end-directed move-ment delivers the capsid to the nucleus. The transport balance is tipped towards the nucleus by activation of PKA and p38/MAPK, which are triggered by the incoming adenovirus and boost microtubule-dependent minus-end-directed motilities of the cytosolic capsid.[72]

Interactions at the Nuclear Pore

As mentioned above, the HIV-1 PIC carries several proteins which presumably determine its karyophilic properties. The MA protein was the first viral protein found to participate in the process of HIV-1 nuclear import.[11,76] Its role also turned out to be the most controversial. Work by our group[12] and by Nadler and colleagues[50] demonstrated that a basic region in the MA protein encompassing amino acids 25-33, G^{25}KKKYKLKH functions as an NLS when conjugated to BSA. Compared to the NLS of SV40 large T antigen, this MA NLS was weak, requiring the presence of multiple peptides per BSA molecule to achieve partial nuclear local-ization. Another basic region in the C-terminal part of the MA protein, N^{109}KSKKKA, was found to be an even weaker NLS, though still capable of targeting the BSA-NLS conjugate to the nucleus.[50,36] In addition to the inherent weakness of MA NLSs, a report by Dupont and colleagues[15] identified a potent nuclear export signal (NES) in MA, which determines the cytoplasmic localization of the protein observed by others in classical nuclear import assays.[26] Therefore, it is not surprising that several groups questioned the role of MA in HIV-1 nuclear import.[26,27] When the NES was inactivated, MA localized exclusively to the nucleus, thus confirming the karyophilic properties of this protein.

Genetic experiments supported the role of MA in HIV-1 nuclear import. Mutations in-troduced into the MA NLSs significantly attenuated HIV-1 replication in non-proliferating cells, and the defect appeared to localize to the step of nuclear import.[11,36,76] The presence of multiple copies (1,000) of MA in the HIV-1 PIC[32] may compensate to some extent for the weakness of the MA NLS, as shown for other weak NLSs.[16] In addition, other proteins within the PIC (see below) may help MA to transport the PIC into the nucleus.

Another HIV protein implicated in the process of PIC nuclear import is integrase (IN). Gallay and colleagues[30] demonstrated that IN associates with karyopherin α and can target a fusion GST-IN protein into the nucleus of microinjected COS cells. They also showed that mutations in the IN gene, combined with mutations in Vpr and the MA NLS, completely eliminated nuclear import of HIV-1 PICs in HeLa cells. Together with reports demonstrating that IN appears to be the only factor required for the ability of lentiviral vectors to transduce non-proliferating cells,[22,86] these results clearly indicate that IN might be involved in regula-tion of HIV-1 nuclear import, at least in certain cell types. Less clear is the mechanism by which IN participates in this process. While early report implicated basic-type NLS in nuclear import activity of IN,[30] later work refuted this hypothesis.[56] It appears therefore that IN nuclear import activity might be mediated by a non-conventional NLS.

The third viral protein implicated in nuclear import of the HIV-1 PIC is Vpr.[19,39,76] A study by Sherman and colleagues[67] identified two non-classical NLSs that mediate nuclear import of the Vpr-containing fusion protein. The first signal is the leucine-rich bipartite NLS located in the amino-terminal half of the protein ([22]LLEEL[26] and [64]LQQLL[68]), while the second one is the arginine-rich bipartite NLS in the carboxyl-terminal half ([73]RIGCR[77] and [85]RQRRAR[90]). Interestingly, the second half of the leucine-rich NLS overlapped with the nuclear export signal (NES). Mutation of leucine in position 67 specifically disrupted nuclear export of the GFP-PK-Vpr fusion protein without affecting its import. The cellular binding partners for these NLSs remain to be determined. However, several studies demonstrated that Vpr specifically interacts with the NLS receptor karyopherin α.[60,75] The interaction occurs outside of the NLS-binding site on the karyopherin α molecule[1] and leads to enhanced affinity of NLS-karyopherin α binding.[60] Therefore, Vpr might function to enhance otherwise weak karyophilic activity of HIV-1 NLSs. In addition, Vpr directly binds to nucleoporines[25,60] thus resembling activity of karyopherin β. The mechanism of Vpr's activity in HIV-1 nuclear import has been hotly debated in the literature. Three hypotheses (not necessarily mutually exclusive) for the mode of action of Vpr have been proposed: (i), Vpr targets the HIV-1 PIC to the nucleus via a distinct, karyopherin α-independent pathway;[31,44] (ii), Vpr-mediated import requires karyopherin α, but not β ;[75] and (iii), Vpr modifies cellular karyopherin α/β-dependent import machinery.[59,60] It remains to be determined whether karyopherins α and β are necessary for HIV-1 nuclear import, but it seems unlikely that Vpr alone can mediate nuclear import of such a large complex as the HIV-1 PIC. Instead, a cooperative activity of three karyophilic proteins within the HIV-1 PIC (MA, IN, and Vpr) can be envisioned. Based on described properties of these proteins, the following sequence of events can be proposed (Fig. 1).

1. Interaction of the PIC with karyopherin α. This interaction is mediated by MA and IN and is stimulated by Vpr.
2. Binding of karyopherin β. This step is similar to that occurring during nuclear import of other cargoes and is mediated by the karyopherin β-binding domain (IBB) in the karyopherin α molecule.
3. Binding to nucleoporins at the nuclear pore. This step is governed by interactions of karyopherin β and Vpr with nucleoporins.
4. Additional uncoating of the PIC and migration through the NPC.

Vpr's ability to bind certain nucleoporins might be critical at this last step, when interaction between PIC and transport factors is weakened by uncoating. Evidence for additional uncoating accompanying nuclear import comes from temporal analysis of the HIV-1 reverse transcription complexes, which demonstrated gradual reduction of sedimentation coefficient associated with the loss of RT and MA.[21] This alteration in PIC composition coincided with the PIC migration from the cytoplasm to the nucleus. This model is also consistent with previous finding that complexes composed only of HIV-1 cDNA and the IN protein are fully integration-competent.[20]

A similar mechanism of karyopherin α/β-dependent nuclear docking and entry is employed by most viruses, including influenza viruses. These viruses have a segmented RNA genome consisting of eight distinct RNA molecules individually packaged into complexes with the NP protein and an RNA-dependent RNA polymerase complex (PA, PB1, and PB2). After binding to the cell surface, virus particles are delivered into the cytoplasm by endocytosis and low-pH fusion.[40] Low pH is required not only for release of viral nucleocapsids from the endosome, but also for their disassembly into eight individual nucleoproteins (reviewed in ref. 79). All four proteins of each nucleoprotein complex (NP, PA, PB1, and PB2) carry NLSs that are recognized by karyopherin α and most likely direct the nuclear import process via karyopherins α/β-dependent pathway.[53] Involvement of multiple karyophilic proteins resembles the situation with HIV-1 and might reflect general requirement for multiple NLSs during transport of viral nucleoprotein complexes.

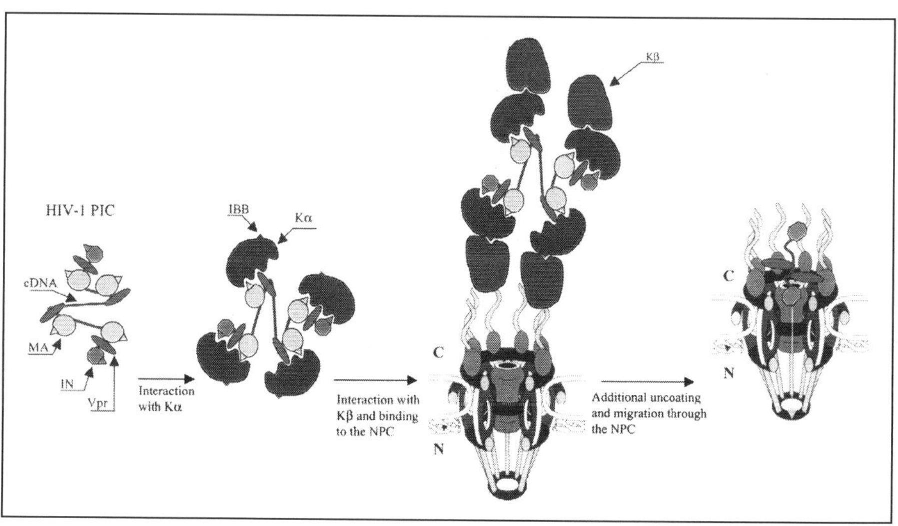

Figure 1. Mechanisms of HIV-1 nuclear import. The cartoon on the left depicts the HIV-1 preintegration complex (PIC) and shows only proteins with known karyophilic activity: integrase (IN), matrix antigen (MA), and Vpr. The PIC is derived from the reverse transcription complex (RTC) and proteins involved in RTC organization, such as nucleocapsid (NC) or reverse transcriptase (RT), may also contribute to PIC nuclear import through conformational effects. Triangles on IN and MA depict NLSs. Karyopherin α (Kα) binds to NLSs on MA and IN; this interaction is assisted by Vpr, which also binds to Kα but outside of the NLS-binding region (Agostini et al, 2000). The NLSs of MA and IN may occupy two NLS-binding sites identified in karyopherin α (Herold et al, 1998). The blue triangle on Kα depicts the karyopherin (importin) β-binding domain (IBB) crucial for interaction with karyopherin β (Kβ) (Weis et al, 1996). Initial attachment to the NPC occurs via Kβ-dependent binding. C and N depict the cytoplasmic and nucleoplasmic sides, respectively, of the nuclear membrane. Additional uncoating at the nuclear pore may lead to the loss of most proteins except IN and, probably, Vpr, which might mediate binding to nucleoporins during migration through the pore. The yellow arrow shows the direction of import.

Herpesviruses employ a slightly different strategy for nuclear entry. Herpesvirus capsids move towards the nucleus along microtubules and accumulate at the nuclear envelope in association with the nuclear pore complex.[70] The virus next releases its DNA, which is translocated through the nuclear pore into the nucleus leaving an empty capsid behind. Whether this translocation involves soluble transport factors or relies totally on viral proteins is unclear. Electron microscopy revealed alterations in the capsid pentons at the vertices where DNA extruded,[52] suggesting that DNA might be 'injected' into the nucleus.

Export through the Nuclear Pore

The ways by which viral genomes leave the nucleus reflect differences in assembly strategies and are in general less well understood than the entry mechanisms. HIV-1 virions are assembled at the plasma membrane, and the viral genomic RNA has to be delivered there from the nucleus. Since HIV-1 genomic RNA is unspliced, it cannot be exported by the cellular mRNA nuclear export machinery, which deals with intron-less RNAs. To circumvent this problem, HIV-1 encodes for a specific nuclear export factor, Rev. This protein contains a leucine-based NES which interacts directly with the cellular exportin CRM1.[23] Rev also binds to the viral RNA via the structured Rev-response element (RRE), thus linking it to the export machinery (reviewed in ref. 57). Due to its location within the Env-encoding sequence of the genome, RRE is present on all unspliced and incompletely spliced viral RNAs, thus ensuring their nuclear export. In addition to Rev, eukaryotic initiation factor 5A (eIF-5A) was shown to be essential for HIV-1 RNA export in microinjected *Xenopus laevis* oocytes (Hofmann et al, 2001). In vitro binding studies demonstrated that eIF-5A was required for efficient interaction of Rev-NES with CRM1/exportin1 and that eIF-5A interacted with the nucleoporins CAN/nup214, nup153, nup98, and nup62.[42] eIF-5A was also shown to directly interact with CRM1, to accumulate at nuclear pore-associated intranuclear filaments in normal cells, and to shuttle between the nucleus and the cytoplasm (Rosorius et al, 1999).[63] Therefore, eIF-5A appears to be a component of the cellular nuclear export machinery. Identification of the ribosomal protein L5, originally described as a factor involved in the assembly of 5S rRNA into ribosomes and later shown to participate in nuclear export of non-ribosomal-associated 5S rRNA,[35,71] as a cellular interaction partner of eIF-5A[66] suggested that eIF-5A might be a part of the nuclear export pathway for 5S rRNA. Interestingly, experiments with dominant-negative form of eIF-5A or microinjection of anti-eIF-5A antibodies demonstrated that eIF-5A is specifically required for Rev-dependent nuclear export and not for export governed by the NES of protein kinase inhibitor, which also uses export receptor CRM1.[17] The role of eIF-5A might be to transport the viral ribonucleoprotein complex from specific subnuclear compartments to the NPC, where CRM1 is localized in association with CAN/Nup214 (Fig. 2).[24] While eIF-5A does not interact directly with Rev, it can bind to peptide mimics corresponding to Rev NES or to RRE-Rev complexes.[6,65] These findings suggest that RNA binding may trigger a conformational change in Rev required for eIF-5A interaction.

Since all retroviruses need to export intron-containing genomic RNA, they encode one or the other form of *cis*-acting elements, termed constitutive transport elements (CTEs). In contrast to lentiviral Rev proteins, which work in *trans* by binding to RRE-containing RNAs, CTEs of other retroviruses, such as Mazon-Pfizer monkey virus (MPMV), are RNA elements (present, in case of MPMV, in an untranslated region near the 3' end of the genome) that can activate nuclear export of incompletely spliced mRNAs when present in *cis*.[8] The cellular nuclear export factor, TAP, was found to bind specifically to the CTE of MPMV and to rescue CTE function when expressed in the otherwise nonpermissive quail cell line QCI-3.[7,34,46] While TAP is not homologous to transportin/karyopherin β proteins (CRM1 belongs to this class), it can also bind to nucleoporins, including CAN/Nup214,[5] and via this mechanism may target viral ribonucleoproteins to the nuclear pore. Alternatively, TAP might serve as an adapter between CTE and an unknown exportin, just as HIV-1 Rev serves as an adapter between RRE and CRM1. This hypothesis is supported in part by the observation that the essential TAP nucleocytoplasmic shuttle domain can be functionally replaced by the HIV-1 Rev NES.[46] Despite the fact that both TAP- and CRM1-mediated export pathways proceed through the nuclear pore and may involve same nucleoporins (e.g., CAN/Nup214), they do not overlap, as each one can be specifically inhibited without affecting the other.[46] Therefore, at least two distinct pathways, TAP- and CRM-dependent, can be used for export of retrovirus genomes. The CRM1-dependent mechanism appears to be used by many viruses, as export of such different virus families as paramyxoviruses and hepadnaviruses can be blocked by agents targeting CRM1 (leptomycin B or peptides corresponding to the HIV-1 Rev NES).[18,64]

In the case of influenza virus, assembly of viral capsids (NP-encapsidated viral genomic RNA complexed with RNA-dependent RNA polymerase) occurs in the nucleus. Three viral polypeptides have been implicated in nuclear export of these complexes: M1, NS2, and NP itself. Viral RNP export was defective in M1-mutant viruses or when M1 was blocked by microinjection of antibodies.[47] How M1 regulates nuclear export of influenza genome is unclear but two possible mechanisms can be envisioned. One deals with the proposed activity of M1 to displace vRNPs from the nuclear matrix.[84] Indeed, vRNPs were found to tightly associate with the nuclear matrix[9] which may impede their export. The second potential mechanism

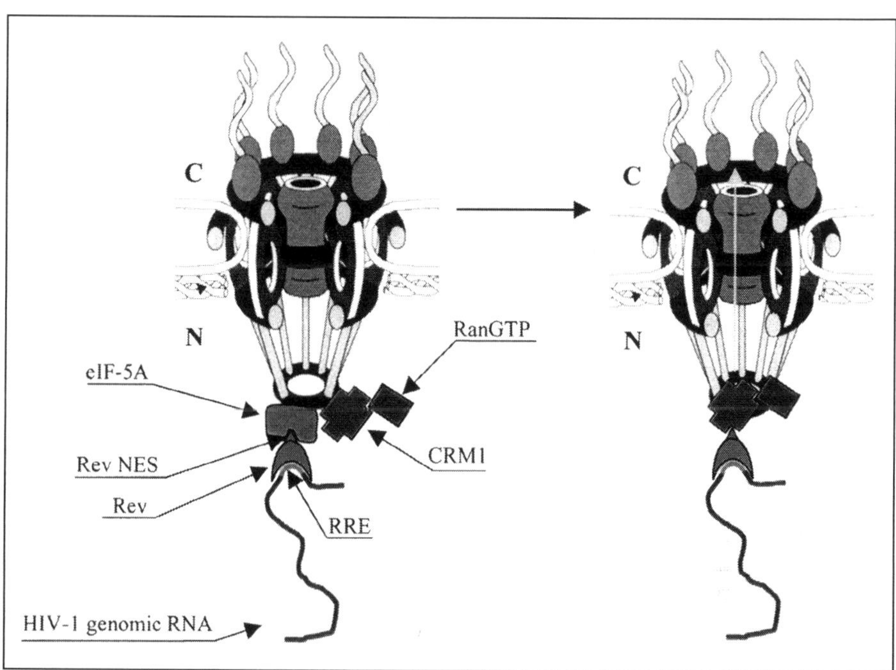

Figure 2. Mechanisms of HIV-1 nuclear export. Intron-containing HIV-1 genomic RNA is delivered to the inner nuclear membrane by eIF-5A, which interacts with the Rev NES when Rev is bound to the Rev-response element (RRE) (an integral part of the unspliced RNA). At the nuclear pore, Rev NES is recognized by CRM1 (this interaction is greatly stimulated by RanGTP) which mediates further export of the viral RNP. N and C depict nucleoplasmic and cytoplasmic sides, respectively, of the nuclear membrane. The yellow arrow shows the direction of export.

of nuclear export activity of the M1 protein is based on its interaction with the viral NS2 protein.[54] NS2 (also termed nuclear export protein, or NEP) contains a leucine-based NES and can also interact with nucleoporin-related molecules termed RIP or Rab,[29] which are involved in nuclear export of leucine-based NESs (reviewed in ref. 82). Therefore, NS2 may act as an adapter molecule that links M1-RNP complexes to the NPC.

Support for the role of NP in influenza vRNP nuclear export comes from analysis of the effects of leptomycin B (LMB), a drug that specifically inactivates CRM1-dependent export pathway, on subcellular distribution of viral proteins and RNP.[18] This study has found that LMB treatment caused nuclear retention of RNP and NP, while distribution of M1 and NS2 proteins was not changed. Although no NES has been yet identified in NP, this protein can shuttle between nucleus and cytoplasm and was shown to interact with CRM1 in vitro.[18] It appears possible, therefore, that nuclear export of influenza RNP is mediated by at least two redundant pathways, M1/NS2- and NP-dependent. While the latter has been shown to be CRM1-dependent,[18] factors involved in the first are not fully characterized. The finding that LMB does not affect subcellular distribution of M1 and NS2 and that inhibition of the CRM1 export pathway failed to completely block RNP export and virus assembly suggest that this pathway, while playing a predominant role, is not the only one by which influenza virus RNP is exported from the nucleus.

For large DNA viruses, such as herpesviruses, that assemble their capsids in the nucleus, transport through the NPC is not physically possible. While the mechanism of capsid assembly and attachment is not fully established, evidence obtained by electron microscopy indicates that capsid assembly occurs at the nuclear membrane and viral DNA is incorporated into preformed capsids (reviewed in ref. 62). To exit the nucleus, DNA-containing capsids become enveloped at the inner lamellae of the nuclear membrane.[3] In the space between the inner and outer nuclear membranes, the enveloped capsids are incorporated into transport vesicles derived from the outer nuclear membranes and are transported into the cytoplasm. The UL11 protein appears to regulate both initial budding events and subsequent transport of capsids.[2] Deletion of UL11 resulted in accumulation of mature cores at the inner lamellae, an increase in unenveloped virions within the cytoplasm, and a decrease of extracellular virions.[3] UL11 is a myristoylated protein that can interact with membranes, including nuclear membrane. However, this protein does not bind to virus capsids,[2] indicating that it is not directly involved in binding of capsids to the inner membrane. Instead, UL11 may alter properties of the inner nuclear membrane at the site of capsid attachment and thus facilitate budding. In the cytoplasm, virus loses the envelope and releases the nucleocapsids.[69] This step is likely controlled by the UL20 gene product, as HSV-1 mutants lacking this gene accumulate in the perinuclear space.[4] The site where final envelopment occurs is uncertain, with *trans*-Golgi network and late endosomal compartment being possible candidates.[38,85]

Conclusions

The common theme in nuclear transport of viral genomes appears to be multiplicity and often redundancy of the pathways engaged. Given the critical role of the nuclear transport steps in the viral life cycle, this characteristic does not come as a surprise. Since viruses exploit the cellular nuclear transport machinery, they are excellent tools to analyze basic nuclear transport mechanisms. Indeed, studies of HIV-1 Rev lead to discovery of nuclear export mechanisms that regulate subcellular localization and activity of many cellular protein.[28,48,78] Another example is HIV-1 Vpr, whose activity during early steps of viral replication can be replaced by heat-shock protein 70 (Hsp70) (our unpublished observation). This finding suggests that nuclear import activity of Vpr is similar to that of Hsp70, consistent with our recent observation that Hsp70 competes with Vpr for binding to karyopherin α and that Hsp70 can rescue nuclear import of Vpr-deficient HIV-1 in vitro.[1] Therefore, Hsp70 may participate in the physiological nuclear import process by the same mechanism that Vpr utilizes to facilitate HIV-1 nuclear import: binding of Hsp70 to karyopherin α increases affinity of interaction between this NLS receptor and proteins carrying weak NLSs, thus facilitating their nuclear translocation. This hypothesis is supported by the finding that ectopic expression of Hsp70 rescued the defective function of a mutant (weakened) SV-40 large T antigen NLS.[45] Consistent with the role of Hsp70 as a component of the nuclear import machinery, nuclear import of proteins containing basic-type NLSs was inhibited by the microinjection of anti-Hsp70 antibodies,[43] and depletion of Hsp70 from cytosolic extracts prevented import of karyophiles in an in vitro nuclear import model.[55,68,81] Hsp70 may accompany the import complex all the way to the nucleus, thus accounting for its nuclear accumulation coincident with the import of NLS substrate.[55] Further analysis of nuclear import and export pathways used by the viruses will certainly help us understand how these steps are performed in uninfected cells.

Since nuclear import and export of the viral genome are critical for productive infection by the viruses that replicate in the nucleus, these steps present an attractive target for anti-viral therapeutics. Unfortunately, due to the lack of detailed knowledge about the mechanisms of viral nuclear transport, the work on compounds targeting these steps is in its early stages. A big potential problem facing development of such compounds is similarity between the viral and cellular nuclear transport signals, which raises a concern about possible drug toxicity. One way to avoid toxicity is to design compounds that would be selectively targeted to the viral nucleoprotein complex. Such compounds would be much more likely to inactivate targeting signals on viral than on cellular proteins. This approach was used to design a class of arylene bis(methyl ketone) compounds that target basic-type NLSs.[14] These small molecules, via their carbonyl moieties, can form Schiff-base adducts with the lysine residues abundant in basic-type NLSs. The specificity of the compounds for HIV appears to be derived from their ability to associate with the viral RT via the pyrimidine side chain on the compounds.[58] Binding to RT seems to stabilize the otherwise reversible Schiff base adducts between the compounds and lysine residues in the NLS, since elimination of the pyrimidine side-chain dramatically reduced the inhibitory activity of the compounds. Specificity of the compounds is reflected in their low cytotoxicity and high anti-HIV activity, making them viable drug candidates.[33,37]

Arylene bis(methyl ketone) compounds provide a proof of concept that nuclear transport is a possible target for drug discovery. Better characterization of the signals and proteins involved in the regulation of the viral nuclear transport will certainly lead to new drugs against such pathogenic viruses as HIV, influenza, and herpesvirus.

References

1. Agostini I, Popov S, Li J et al. Heat-shock protein 70 can replace viral protein R of HIV-1 during nuclear import of the viral pre-integration complex. Exp Cell Res 2000; 259:398-403.
2. Baines JD, Jacob RJ, Simmerman L et al. The herpes simplex virus 1 UL11 proteins are associated with cytoplasmic and nuclear membranes and with nuclear bodies of infected cells. J Virol 1995; 69:825-833.
3. Baines JD, Roizman B. The UL11 gene of herpes simplex virus 1 encodes a function that facilitates nucleocapsid envelopment and egress from cells. J Virol. 1992; 66:5168-5174.
4. Baines JD, Ward PL, Campadelli-Fiume G et al. The UL20 gene of herpes simplex virus 1 encodes a function necessary for viral egress. J Virol. 1991; 65:6414-6424.
5. Bear J, Tan W, Zolotukhi AS et al. Identification of novel import and export signals of human TAP, the protein that binds to the constitutive transport element of the type D retrovirus mRNAs. Mol Cell Biol 1999; 19:6306-6317.
6. Bevec D, Jaksche H, Oft M et al. Inhibition of HIV-1 replication in lymphocytes by mutants of the Rev cofactor eIF-5A. Science 1996; 271:1858-1860.
7. Braun IC, Rohrbach E, Schmitt C et al. TAP binds to the constitutive transport element (CTE) through a novel RNA-binding motif that is sufficient to promote CTE-dependent RNA export from the nucleus. EMBO J 1999; 18:1953-1965.
8. Bray M, Prasad S, Dubay JW et al. A small element from the Mason-Pfizer monkey virus genome makes human immunodeficiency virus type 1 expression and replication Rev-independent. Proc Natl Acad Sci USA 1994; 91:1256-1260.
9. Bui M, Wills EG, Helenius A et al. Role of the influenza virus M1 protein in nuclear export of viral ribonucleoproteins. J Virol 2000; 74:1781-1786.
10. Bukrinskaya A, Brichacek B, Mann A et al. Establishment of a functional human immunodeficiency virus type 1 (HIV- 1) reverse transcription complex involves the cytoskeleton. J Exp Med 1998; 188:2113-2125.
11. Bukrinsky MI, Haggerty S, Dempsey MP et al. A nuclear localization signal within HIV-1 matrix protein that governs infection of non-dividing cells. Nature 1993; 365:666-669.
12. Bukrinsky MI, Sharova N, McDonald TL et al. Association of integrase, matrix, and reverse transcriptase antigens of human immunodeficiency virus type 1 with viral nucleic acids following acute infection. Proc Natl Acad Sci USA 1993; 90:6125-6129.
13. Desai A, Mitchison TJ. Microtubule polymerization dynamics. Annu Re. Cell Dev Biol 1997; 13:83-117.
14. Dubrovsky L, Ulrich P, Nuovo GJ et al. Nuclear localization signal of HIV-1 as a novel target for therapeutic intervention. Mol Med 1995; 1:217-230.
15. Dupont S, Sharova N, DeHoratius C et al. A novel nuclear export activity in HIV-1 matrix protein required for viral replication. Nature 1999; 402:681-685.
16. Dworetzky SI, Lanford RE, Feldherr CM. The effects of variations in the number and sequence of targeting signals on nuclear uptake. J Cell Biol 1998; 107:1279-1287.

17. Elfgang C, Rosorius O, Hofer L et al. Evidence for specific nucleocytoplasmic transport pathways used by leucine-rich nuclear export signals. Proc Natl Acad Sci USA 1999; 96:6229-6234.
18. Elton D, Simpson-Holley M, Archer K et al. Interaction of the influenza virus nucleoprotein with the cellular CRM1-mediated nuclear export pathway. J Virol 2001; 75:408-419.
19. Emerman M, Bukrinsky M, Stevenson M. HIV-1 infection of non-dividing cells. Nature 1994; 369:107-108.
20. Farnet CM, Haseltine WA. Determination of viral proteins present in the human immunodeficiency virus type 1 preintegration complex. J Virol 1991; 65:1910-1915.
21. Fassati A, Goff SP. Characterization of intracellular reverse transcription complexes of human immunodeficiency virus type 1. J Virol 2001; 75:3626-3635.
22. Follenzi A, Ailles LE, Bakovic S et al. Gene transfer by lentiviral vectors is limited by nuclear translocation and rescued by HIV-1 pol sequences. Nat Genet 2000; 25:217-222.
23. Fornerod M, Ohno M, Yoshida M et al. CRM1 is an export receptor for leucine-rich nuclear export signals. Cell 1997; 90:1051-1060.
24. Fornerod M, van Deursen J, van Baal S et al. The human homologue of yeast CRM1 is in a dynamic subcomplex with CAN/Nup214 and a novel nuclear pore component Nup88. EMBO J 1997; 16:807-816.
25. Fouchier RA, Meyer BE, Simon JH et al. Interaction of the human immunodeficiency virus type 1 Vpr protein with the nuclear pore complex. J Virol 1998; 72:6004-6013.
26. Fouchier RA, Meyer BE, Simon JH et al. HIV-1 infection of non-dividing cells: Evidence that the amino-terminal basic region of the viral matrix protein is important for Gag processing but not for post-entry nuclear import. EMBO J 1997; 16:4531-4539.
27. Freed EO, Englund G, Martin MA. Role of the basic domain of human immunodeficiency virus type 1 matrix in macrophage infection. J Virol 1995; 69:3949-3954.
28. Fritz CC, Green MR. HIV Rev uses a conserved cellular protein export pathway for the nucleocytoplasmic transport of viral RNAs. Curr Biol 1996; 6:848-854.
29. Fritz CC, Zapp ML, Green MR. A human nucleoporin-like protein that specifically interacts with HIV Rev Nature 1995; 376:530-533.
30. Gallay P, Hope T, Chin D et al. HIV-1 infection of nondividing cells through the recognition of integrase by the importin/karyopherin pathway. Proc Natl Acad Sci USA 1997; 94:9825-9830.
31. Gallay P, Stitt V, Mundy C et al. Role of the karyopherin pathway in human immunodeficiency virus type 1 nuclear import. J Virol 1996; 70:1027-1032.
32. Gallay P, Swingler S, Song J et al. HIV nuclear import is governed by the phosphotyrosine-mediated binding of matrix to the core domain of integrase. Cell 1995; 83:569-576.
33. Glushakova S, Dubrovsky L, Grivel JC et al. Small molecule inhibitor of HIV-1 nuclear import suppresses HIV-1 replication in human lymphoid tissue ex vivo: A potential addition to current anti-HIV drug repertoire. Antiviral Res 2000; 47:89-95.
34. Gruter P, Tabernero C, von Kobbe C et al. TAP, the human homolog of Mex67p, mediates CTE-dependent RNA export from the nucleus. Mol Cell 1998; 1:649-659.
35. Guddat U, Bakken AH, Pieler T. Protein-mediated nuclear export of RNA: 5S rRNA containing small RNPs in xenopus oocytes. Cell 1998; 60:619-628.
36. Haffar OK, Popov S, Dubrovsky L et al. Two nuclear localization signals in the HIV-1 matrix protein regulate nuclear import of the HIV-1 pre-integration complex. J Mol Biol 2000; 299:359-368.
37. Haffar OK, Smithgall MD, Popov S et al. CNI-H0294, a nuclear importation inhibitor of the human immunodeficiency virus type 1 genome, abrogates virus replication in infected activated peripheral blood mononuclear cells. Antimicrob Agents Chemother 1998; 42:1133-1138.
38. Harley CA, Dasgupta A, Wilson DW. Characterization of herpes simplex virus-containing organelles by subcellular fractionation: Role for organelle acidification in assembly of infectious particles. J Virol 2001; 75:1236-1251.
39. Heinzinger NK, Bukrinsky MI, Haggerty SA et al. The Vpr protein of human immunodeficiency virus type 1 influences nuclear localization of viral nucleic acids in nondividing host cells. Proc Natl Acad Sci USA 1994; 91:7311-7315.
40. Hernandez LD, Hoffman LR, Wolfsberg TG et al. Virus-cell and cell-cell fusion. Annu Rev Cell Dev Biol 1996; 12:627-661.

41. Herold A, Truant R, Wiegand H et al. Determination of the functional domain organization of the importin alpha nuclear import factor. J Cell Biol 1998; 143:309-318.

42. Hofmann W, Reichart B, Ewald A et al. Cofactor requirements for nuclear export of Rev response element (RRE)- and constitutive transport element (CTE)-containing retroviral RNAs. An unexpected role for actin. J Cell Biol 2001; 152:895-910.

43. Imamoto N, Matsuoka Y, Kurihara T et al. Antibodies against 70-kD heat shock cognate protein inhibit mediated nuclear import of karyophilic proteins. J Cell Biol 1992; 119:1047-1061.

44. Jenkins Y, McEntee M, Weis K et al. Characterization of HIV-1 vpr nuclear import: Analysis of signals and pathways. J Cell Biol 1998; 143:875-885.

45. Jeoung DI, Chen S, Windsor J et al. Human major HSP70 protein complements the localization and functional defects of cytoplasmic mutant SV40 T antigen in Swiss 3T3 mouse fibroblast cells. Genes Dev 1991; 5:2235-2244.

46. Kang Y, Cullen BR. The human Tap protein is a nuclear mRNA export factor that contains novel RNA-binding and nucleocytoplasmic transport sequences. Genes Dev 1999; 13:1126-1139.

47. Martin K, Helenius A. Nuclear transport of influenza virus ribonucleoproteins: The viral matrix protein (M1) promotes export and inhibits import. Cell 1991; 67:117-130.

48. Meyer BE, Meinkoth JL, Malim MH. Nuclear transport of human immunodeficiency virus type 1, visna virus, and equine infectious anemia virus Rev proteins: Identification of a family of transferable nuclear export signals. J Virol 1996; 70:2350-2359.

49. Miller MD, Farnet CM, Bushman FD. Human immunodeficiency virus type 1 preintegration complexes: Studies of organization and composition. J Virol 1997; 71:5382-5390.

50. Nadler SG, Tritschler D, Haffar OK et al. Differential expression and sequence-specific interaction of karyopherin alpha with nuclear localization sequences. J Biol Chem 1997; 272:4310-4315.

51. Nemerow GR. Cell receptors involved in adenovirus entry. Virology 2000; 274:1-4.

52. Newcomb WW, Brown JC. Induced extrusion of DNA from the capsid of herpes simplex virus type 1. J Virol 1994; 68:433-440.

53. O'Neill RE, Jaskunas R, Blobel G et al. Nuclear import of influenza virus RNA can be mediated by viral nucleoprotein and transport factors required for protein import. J Biol Chem 155; 270:22701-22704.

54. O'Neill RE, Talon J, Palese P. The influenza virus NEP (NS2 protein) mediates the nuclear export of viral ribonucleoproteins. EMBO J 1998; 17:288-296.

55. Okuno Y, Imamoto N, Yoneda Y. 70-kDa heat-shock cognate protein colocalizes with karyophilic proteins into the nucleus during their transport in vitro. Exp Cell Res 1993; 206:134-142.

56. Petit C, Schwartz O, Mammano F. The karyophilic properties of human immunodeficiency virus type 1 integrase are not required for nuclear import of proviral DNA. J Virol 2000; 74:7119-7126.

57. Pollard VW, Malim MH. The HIV-1 Rev protein. Annu Rev Microbiol 1998; 52:491-532.

58. Popov S, Dubrovsky L, Lee MA et al. Critical role of reverse transcriptase in the inhibitory mechanism of CNI-H0294 on HIV-1 nuclear translocation. Proc Natl Acad Sci USA 1996; 93:11859-11864.

59. Popov S, Rexach M, Ratner L et al. Viral protein R regulates docking of the HIV-1 preintegration complex to the nuclear pore complex. J Biol Chem 1998; 273:13347-13352.

60. Popov S, Rexach M, Zybarth G et al. Viral protein R regulates nuclear import of the HIV-1 preintegration complex. EMBO J 1998; 17:909-917.

61. Roe T, Reynolds TC, Yu G et al. Integration of murine leukemia virus DNA depends on mitosis. EMBO J 1993; 12:2099-2108.

62. Roizman B, Sears AY. Herpes simplex viruses and their replication. In: Fields BN, Knipe DM, Howley PM, eds. Fields Virology. Philadelphia: Lippincott-Raven, 1996:2231-2295.

63. Rosorius O, Reichart B, Kratzer F et al. Nuclear pore localization and nucleocytoplasmic transport of eIF-5A: Evidence for direct interaction with the export receptor CRM1. J Cell Sci 1999; 112:2369-2380.

64. Roth J, Dobbelstein M. Export of hepatitis B virus RNA on a Rev-like pathway: Inhibition by the regenerating liver inhibitory factor IkappaB alpha. J Virol 1997; 71:8933-8939.

65. Ruhl M, Himmelspach M, Bahr GM et al. Eukaryotic initiation factor 5A is a cellular target of the human immunodeficiency virus type 1 Rev activation domain mediating trans-activation. J Cell Biol 1993; 123:1309-1320.

66. Schatz O, Oft M, Dascher C et al. Interaction of the HIV-1 rev cofactor eukaryotic initiation factor 5A with ribosomal protein L5. Proc Natl Acad Sci USA 1998; 95:1607-1612.

67. Sherman MP, de Noronha CM, Heusch MI et al. Nucleocytoplasmic shuttling by human immunodeficiency virus type 1 Vpr. J Virol 2001; 75:1522-1532.

68. Shi Y, Thomas JO. The transport of proteins into the nucleus requires the 70-kilodalton heat shock protein or its cytosolic cognate. Mol Cell Biol 1992; 12:2186-2192.

69. Skepper JN, Whiteley A, Browne H et al. Herpes simplex virus nucleocapsids mature to progeny virions by an envelopment —> deenvelopment —> reenvelopment pathway. J Virol 2001; 75:5697-5702.

70. Sodeik B, Ebersold MW, Helenius A. Microtubule-mediated transport of incoming herpes simplex virus 1 capsids to the nucleus. J Cell Biol 1997; 136:1007-1021.

71. Steitz JA, Berg C, Hendrick JP et al. A 5S rRNA/L5 complex is a precursor to ribosome assembly in mammalian cells. J Cell Biol 1988; 106:545-556.

72. Suomalainen M, Nakano MY, Boucke K et al. Adenovirus-activated PKA and p38/MAPK pathways boost microtubule-mediated nuclear targeting of virus. EMBO J 2001; 20:1310-1319.

73. Suomalainen M, Nakano MY, Keller S et al. Microtubule-dependent plus- and minus end-directed motilities are competing processes for nuclear targeting of adenovirus. J Cell Biol 1999; 144:657-672.

74. Telesnitsky A, Goff SP. Reverse transcriptase and the generation of retroviral DNA. In: Coffin JM, Highes SH, Varmus HE, eds. Retroviruses. Cold Spring Harbor: Cold Spring Harbor Press, 1997:121-160.

75. Vodicka MA, Koepp DM, Silver PA et al. HIV-1 Vpr interacts with the nuclear transport pathway to promote macrophage infection. Genes Dev 1998; 12:175-185.

76. von Schwedler U, Kornbluth RS, Trono D. The nuclear localization signal of the matrix protein of human immunodeficiency virus type 1 allows the establishment of infection in macrophages and quiescent T lymphocytes. Proc Natl Acad Sci USA 1994; 91:6992-6996.

77. Weis K, Ryder U, Lamond AI. The conserved amino-terminal domain of hSRP1 alpha is essential for nuclear protein import. EMBO J 1996; 15:1818-1825.

78. Wen W, Meinkoth JL, Tsien RY et al. Identification of a signal for rapid export of proteins from the nucleus. Cell 1995; 82:463-473.

79. Whittaker GR, Helenius A. Nuclear import and export of viruses and virus genomes. Virology 1998; 246:1-23.

80. Whittaker GR, Kann M, Helenius A. Viral entry into the nucleus. Annu Rev Cell Dev Biol 2000; 16:627-651.

81. Yang J, DeFranco DB. Differential roles of heat shock protein 70 in the in vitro nuclear import of glucocorticoid receptor and simian virus 40 large tumor antigen. Mol Cell Biol 1994; 14:5088-5098.

82. Yoneda Y. Nucleocytoplasmic protein traffic and its significance to cell function. Genes Cells 2000; 5:777-787.

83. Zennou V, Petit C, Guetard D et al. HIV-1 genome nuclear import is mediated by a central DNA flap. Cell 2000; 101:173-185.

84. Zhirnov OP, Klenk HD. Histones as a target for influenza virus matrix protein M1. Virology 1997; 235:302-310.

85. Zhu Z, Gershon MD, Hao Y et al. Envelopment of varicella-zoster virus: targeting of viral glycoproteins to the trans-Golgi network. J Virol 1995; 69:7951-7959.

86. Zufferey R, Nagy D, Mandel RJ et al. Multiply attenuated lentiviral vector achieves efficient gene delivery in vivo. Nat Biotechnol 1997; 15:871-875.

CHAPTER 12

Nuclear Import of DNA

David A. Dean and Kerimi E. Gokay

Nuclear trafficking of macromolecules usually brings to mind the nuclear import of NLS-containing proteins and certain RNAs and the export of NES-containing proteins and mRNAs. One macromolecule whose nuclear import is often overlooked and under-appreciated is exogenously administered DNA. While extrachromosomal DNA may not be a "normal" species in the cell, its nuclear localization is integral to the life cycles of many pathogens and necessary for the success of transfections in the laboratory and gene therapy in the clinic. Moreover, the movement of DNA from the cytoplasm to the nucleus remains one of the major barriers to efficient gene transfer and expression (Fig. 1). Without localization of DNA to the nucleus, no transcription, replication, integration, maintenance, or "gene therapy" can take place. Surprisingly, there has been relatively little attention directed toward either discovering or exploiting the mechanisms used by the cell to direct DNA to the nucleus, despite its importance in gene therapy. The discussion that follows will highlight our working knowledge of the mechanisms of DNA nuclear import in both non-viral and viral systems.

The Nuclear Envelope Is a Barrier to Gene Delivery

That the nuclear envelope presents a major barrier to gene transfer and viral infections was realized over 20 years ago in seminal experiments by Capecchi and others in which plasmids that had been microinjected into the cytoplasm were found to be virtually incapable of directing gene expression while those injected into the nucleus were highly proficient for gene expression.[8] Using similar microinjection strategies, Graessman demonstrated that when 1000 to 2000 copies of a plasmid were injected into the cytoplasm, less than 3% of the expression was seen as compared to cells injected in the nucleus with the same number of plasmids.[33] Other experiments in a variety of mammalian cell types,[63,84] as well as in *Xenopus* oocytes[104] has confirmed that plasmids injected into the cytoplasm are much less capable of directing gene expression than those injected into the nucleus. The same is true for many viral genomes: direct nuclear injection, instead of cytoplasmic injection, of the DNA genome from SV40 or a reverse-transcribed retroviral genome resulted in 10 to 100-fold more infectious virus particles in a given time.[32,49,64]

During mitosis, the nuclear envelope breaks down, eliminating a major barrier to gene transfer. If plasmids or viral DNA genomes are present in the cytoplasm, they have unencumbered access to the nuclear compartment during this stage of the cell cycle. By contrast, in non-dividing cells, the nuclear envelope provides a substantial barrier to the DNA (see above). Indeed, it has been demonstrated that retroviruses cannot productively infect non-dividing cells due to the fact that their reverse-transcribed viral genomes (rtDNA) cannot traverse the nuclear envelope to gain access to the nucleus. However, when the cells undergo mitosis, rtDNA can localize to the nucleus, integrate, and lead to new rounds of replication.[54,62] Similarly, it is greatly appreciated by researchers the world over that non-dividing cells (growth-arrested cells, confluent cells, primary

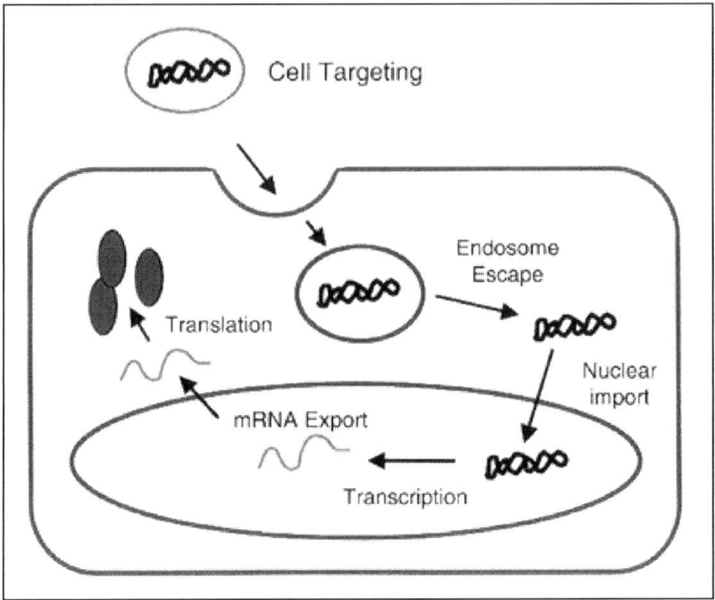

Figure 1. Cellular barriers to gene delivery. Extracellular DNA, delivered to cells in either viral particles, liposomes, or other vehicles, must traverse the plasma, endosomal, and nuclear membranes before any transcription, replication, or integration can occur.

cells, etc.) are exceedingly difficult to transfect. Again, the main reason for this is that the nuclear envelope restricts access of exogenous DNA to the site of transcription. In at least a number of the cells of an actively dividing population, DNA is able to enter the nucleus and express itself. Using primary human airway epithelial cells, it was shown that actively dividing cells, identified by BrdU incorporation, were ten times more likely to express a transferred gene product than BrdU-negative cells.[27] In a more recent study, synchronized cells transfected by a variety of techniques in the G2 or G2-M stage expressed 50- to 300-fold more gene product than those transfected in G1.[6] Thus, nuclear import of DNA is crucial to gene transfer reactions.

During transfections, large numbers of plasmids are delivered into the cytoplasm of cells, but only a fraction of these plasmids make it to the nucleus for gene expression. In one recent study, HeLa and CV1 cells (1-2 x 10^5 cells) were transfected with fluorescently labeled plasmids (1.25 µg) using liposomes, and the amount of DNA in the cytoplasm and nucleus was quantified by flow cytometry.[43] Within 2 hours, approximately 2000 copies of plasmid were found to be intracellular in either cell type, and by 24 hours, the number had increased only slightly. These numbers are in close agreement with previous studies.[14,85] However, when the amount of DNA in the nucleus was measured, more than 60% of the plasmids were in the nuclei of HeLa cells, versus 30% for CV1 cells.[43] One interpretation of these results is that different cell types may be more or less efficient at nuclear targeting of DNA than others. Alternatively, differences in cell division rates between the cells or differences in rates of cytoplasmic degradation of plasmids[18,52] could also account for the differences. Nonetheless, relatively few plasmids enter the nuclei of dividing cells, let alone those of non-dividing cells. In studies from our laboratory, it took approximately 30 to 100-times more plasmid injected into the cytoplasm of dividing cells to observe the same level of gene expression as compared to nuclear-injected DNAs, again illustrating that nuclear import is a relatively inefficient process, even when the nuclear envelope breaks down.[20]

Nuclear Import of DNAs in Non-Dividing Cells

Based on the data above, it could appear that it is necessary to use dividing cells for successful transfections and certain viral infections (e.g., retroviruses). The question then arises, does DNA ever enter the nuclei of non-dividing cells? Based on a number of criteria and experimental data, the answer is yes. Indeed, a number of viruses, including HIV, SV40, adenovirus, herpes simplex virus, and hepatitis B virus and bacterial pathogens such as *Agrobacterium tumefaciens* have developed very efficient ways to infect non-dividing cells (for recent reviews, see refs. 16, 48, 91, 108). Furthermore, plasmids have been successfully delivered to a variety of tissues in animals and humans that are made up largely of cells that either do not divide or do so very slowly. Studies over the past 5 to 10 years have begun to elucidate some of the pathways used by viruses and plasmids to enter the nuclei of non-dividing cells. The common theme of these various mechanisms is that they all employ proteins complexed to the DNA for nuclear transport. The next section will highlight some of the better studied non-viral, viral, and bacterial systems.

Plasmid Nuclear Import

Nuclear import of plasmids in the absence of cell division was first demonstrated by Jon Wolff and colleagues.[23] Plasmids were microinjected into the cytoplasm of cultured myotubes and nuclear localization was followed either directly using biotin-labeled plasmids, or indirectly by gene expression from the plasmid. They showed that plasmids were able to localize to the nucleus in these quiescent cells in a dose-dependent manner in an energy-requiring process. Co-injection of the lectin WGA that blocks signal-mediated transport through the NPC inhibited gene expression, leading the authors to conclude that the DNA entered the nuclei through the NPC. However, the import they detected was not efficient. Using biotin or gold-labeled plasmids, they were unable to detect nuclear DNA after cytoplasmic injection until 20 hours post-injection. Again, this illustrates the inefficiency of DNA nuclear uptake.

In a similar set of experiments, but using epithelial cells, our laboratory also showed that plasmids were able to enter the nuclei of non-dividing cells using an NPC-mediated pathway. Synchronized, confluent (i.e., contact-inhibited), or aphidicolin-arrested cells, as well as asynchronous populations have all been used to demonstrate that plasmids can enter the nuclei in the absence of cell division. To confirm that the cells did not divide over the course of our experiments, fluorescent markers that were too large to translocate across the NPC were co-injected with the DNA; if the marker localized to the nucleus, it indicated that the cells had either divided during the course of the experiment or that the DNA had been injected into the nucleus accidentally. Our initial experiments were performed in African green monkey kidney epithelial cells using protein-free, purified SV40 DNA (5,243 bp) which was detected post-injection by in situ hybridization.[19] The advantage of this approach is that it is direct and does not rely on the transcriptional activity or post-transcriptional processes that must occur in order to detect the expression of a gene product as an indicator of DNA nuclear import. When SV40 DNA (about 5,000 copies per cell) was injected in to the cytoplasm, the DNA could be detected diffusely within the cytoplasm immediately following injection. Within 2 to 4 hours of injection, much of the in situ signal was localized in the perinuclear region, perhaps suggesting that the DNA was accumulating at the nuclear envelope awaiting import. Finally, by 6 to 8 hours post-injection, the majority of the DNA was localized to the nucleus. Once in the nucleus, the DNA accumulates in distinct regions of the nucleus that co-localize with proteins involved in transcription and splicing (e.g., SC-35), indicating that the DNA is functional for transcription. This time course of SV40 DNA nuclear accumulation is in agreement with that observed by Nakanishi et al, who followed nuclear import by measuring large T-antigen expression from the cytoplasmically injected SV40 DNA or virions.[64]

Nuclear transport of SV40 DNA was shown to utilize the NPC for its translocation based on the inhibitory effects of the lectin WGA or an antibody (mAb414) that has been shown to block the central pore of the NPC.[17] Energy depletion also inhibited nuclear localization of the DNA as did low temperatures.[19] These results demonstrate that plasmids enter the nucleus using the same nuclear pore complexes as do NLS-containing proteins and small RNAs. It was also demonstrated that nuclear import of the DNA was not affected by inhibition of translation using either puromycin or cycloheximide.[19] However, when RNA polymerase II-mediated transcription was inhibited using several different drugs, nuclear import of SV40 DNA was halted. A similar link between transcription and nuclear transport has also been observed for the import of the A1 hnRNA binding protein.[70,71] The A1 protein shuttles into and out of the nucleus using the importin homologue transportin. When RNA polymerase II activity is inhibited, A1 is not re-imported into the nuclei of heterokaryons, while other nuclear proteins are transported. Thus, transcription is required for nuclear import of A1 (as well as plasmids), although how remains unclear. One possibility for the transcription-dependent plasmid nuclear import is that proteins similar to A1 in their dependence on active transcription for nuclear import, bind to plasmids in the cytoplasm and facilitate their nuclear uptake. However, at present, this remains to be determined.

Although protein-free SV40 DNA is readily taken up by nuclei of non-dividing cells, many other plasmids are not (Fig. 2). When plasmids including pBR322, pUC19, and pGL3basic (Promega, Madison, WI) are injected into the cytoplasm, they do not localize to the nucleus within 8 to 12 hours of injection.[19,87] However, these plasmids do gain access to the nuclei when the cells divide.[19,20,105] The simplest interpretation of these results is that plasmid nuclear import is sequence-specific and that the SV40 genome contains a sequence that can mediate nuclear uptake. Indeed, when as little as 50 bp of the SV40 enhancer region is cloned into any of the other bacterial plasmids that normally remain cytoplasmic, they are targeted to the nucleus with the same kinetics as the entire 5.2 Kbp SV40 genome.[19,20] Furthermore, the intranuclear localization of the plasmids are the same: they remain absent from nucleoli and have a punctate staining pattern in the nucleoplasm.

The SV40 enhancer contains binding sites for a number of well-defined, general transcription factors (Fig. 3).[25,26,92] Among others, AP1 (fos and jun), AP2, AP3, NF-κB, Oct-1, TEF-I, and TEF-II have been shown to bind to this 72 bp region of DNA. Like other nuclear proteins, these transcription factors are synthesized in the cytoplasm and contain NLSs for their nuclear import. Under normal circumstances in the cell, once translated, these proteins are transported into the nucleus where they recognize their binding sites within chromosomal promoters and enhancers and bind to the DNA to carry out their roles in regulation. However, if exogenous DNA is present in the cytoplasm (as in the case with transfections or gene therapy), the transcription factors can bind to their binding sites on this DNA to form a protein-DNA complex. One or more of the transcription factors bound to the DNA may have an NLS that is exposed at the surface of the complex such that it can interact with the importin machinery for nuclear import. Thus, the cytoplasmic plasmid containing the SV40 enhancer will be transported into the nucleus by a "piggy-back" mechanism, similar to the mechanism of the HIV preintegration complex (Fig. 4; see below).

One caveat for the above model for sequence-specific DNA nuclear import is that at least one or more of the transcription factors that bind must possess an NLS that is away from the DNA binding domain and free to interact with the importin proteins. Thus, zinc-finger transcription factors such as SP1 would not be able to mediate DNA nuclear import because the NLS is embedded within the DNA–binding zinc-finger domain; once bound to the DNA, the NLS would be buried at the protein-DNA interface. Similarly, the NLS of the c-jun component of the AP1 transcription factor would also be unable to bind to the importins because its NLS is immediately adjacent to the DNA binding region of the protein (Fig. 5A). By contrast,

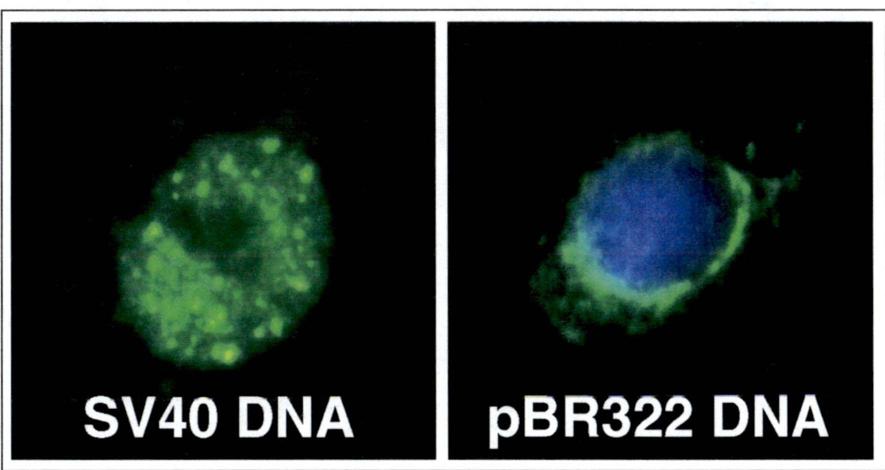

Figure 2. Sequence-specific nuclear import of plasmids. Growth-arrested CV1 cells were cytoplasmically microinjected with either pBR322 or SV40 DNA (5000 copies per cell) as described.[19] Eight hours later the location of the DNA was visualized by in situ hybridization (green signal). Cellular DNA is stained with DAP1 (blue signal).

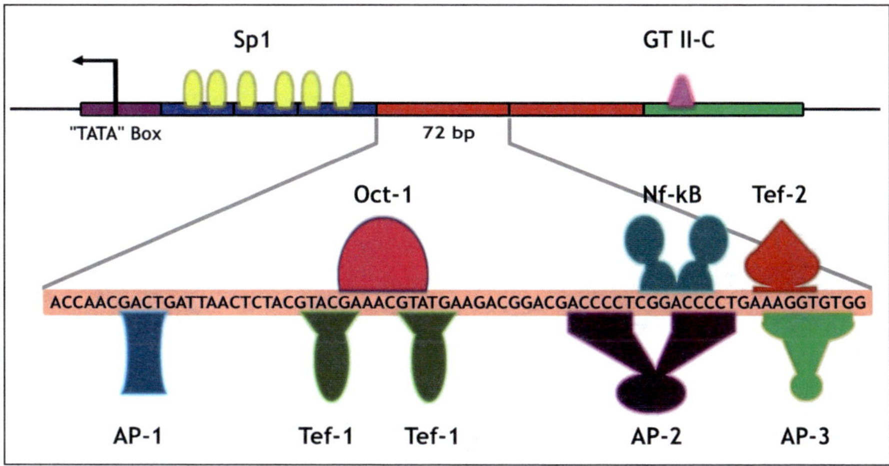

Figure 3. Cartoon of the SV40 origin region. A portion of the SV40 origin and early/late promoter region, containing two copies of the 72 bp enhancer repeats is shown with identified binding sites for transcription factors. As can be seen, multiple transcription factors have been identified to bind to this region (not all proteins are listed).

NF-κB could potentially be the adapter protein between DNA and the import machinery because its DNA-binding domain and NLS are structurally separated (Fig. 5B).

Based on the above model for plasmid nuclear import, it could follow that any eukaryotic promoter or enhancer sequence could act as a scaffold for transcription factor binding and as a nuclear targeting sequence. However, this does not appear to be the case. Several very strong viral promoters have been tested for their ability to promote nuclear import of DNA, and found not to have import activity. The immediate early promoter and enhancer from CMV

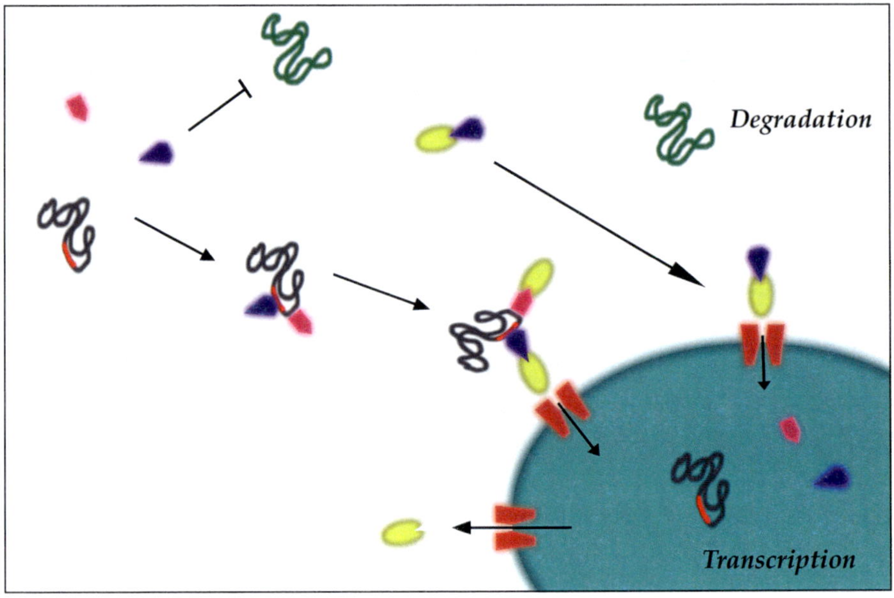

Figure 4. Model for SV40 enhancer-mediated sequence-specific plasmid nuclear import. Newly synthesized NLS-containing transcription factors bind to the SV40 enhancer in the cytoplasm to form a protein-DNA complex. These transcription factors would normally bind to the importins after translation for their nuclear import. Importin family members can recognize the DNA-bound NLSs to facilitate nuclear import of the complex. Because the transcription factors bound by this DNA sequence are ubiquitously expressed, SV40 DNA localizes to the nuclei of all cell types. Plasmids lacking the SV40 sequence cannot form DNA-protein complexes and are thus not imported, remaining in the cytoplasm where they are degraded.

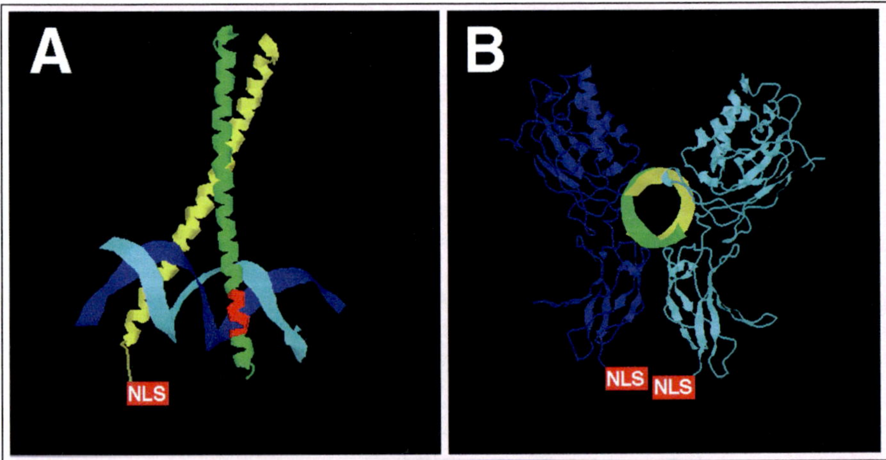

Figure 5. Transcription factor X-ray crystal structures. A) AP1. The crystal structure of AP1 is shown with fos and jun shown binding to DNA. The NLSs of the proteins are shown in red. B) NF-κB. The p65 homodimer is shown with the NLS indicated in red. The DNA helix would bind within the central hole and come out of the page at 90°. Both structures are from the Brookhaven National Laboratory Protein Structure database and were manipulated with RasMol 2.6.

(CMV$_{iep}$), the Rous sarcoma virus LTR, the Moloney murine leukemia virus LTR, and the herpes simplex virus thymidylate kinase (TK) promoter all are incapable of causing a plasmid to localize to the nucleus in the absence of cell division.[20,33,87] Although each of these promoters and enhancers contain multiple binding sites for a variety of transcription factors, they do not function as DNA nuclear targeting sequences (Fig. 6). The simplest explanation would be that the SV40 enhancer binds to one or more proteins that none of the others bind to and that these unique proteins are responsible for import. The only identified transcription factor that binds to the SV40 enhancer and not the other sequences is AP2. However, AP2 has been shown to bind to another promoter that is not targeted to the nuclei of all cells,[87] so it is not the responsible protein. The more likely explanation for why the SV40 sequence is unique is that it is the overall organization and structure of the transcription factor-DNA complex that is important and that more than one transcription factor is needed for nuclear localization. To support this hypothesis, our laboratory has examined the nuclear import activity of plasmids containing mutations that disrupt most of the individual binding sites within the 72 bp enhancer repeat[25] and found that disruption of any individual binding site kills import activity (D.A.D., unpublished observations). Thus, it is likely that the three-dimensional structure and organization is necessary for DNA nuclear localization.

Several other sequences have been identified that act as DNA nuclear targeting sequences.[87,103] The common feature of these sequences is that, unlike the SV40 enhancer which acts in all cell types, they act only in specific cell types. Based on our model for plasmid nuclear localization, it is possible that binding of cell-specific transcription factors to DNA could result in cell-specific nuclear import in those cells in which the transcription factors are expressed; in cells that do not express the proteins, no DNA-protein complexes could be formed and no nuclear import would occur (Fig. 7). Indeed, the smooth muscle gamma actin (SMGA) promoter, which binds to a set of transcription factors found only in smooth muscle cells,[5,9,50] is transported into the nuclei of smooth muscle cells, but not the nuclei of endothelial cells, fibroblasts, or epithelial cells.[87] The minimal sequence of the SMGA promoter necessary for nuclear transport encompassed the first ~400 bp of the promoter from −404 to +25. This region contains binding sites for both general (C/EBP and AP2) and smooth muscle specific transcription factors (SRF and Nkx3).[5,9,50] When a plasmid containing the 400 bp SMGA promoter was injected into the cytoplasm of SRF-expressing stable transfectants of non-smooth muscle cells (CV1 cells), some of the plasmids were able to localize to the nuclei of the cells, whereas they remained completely cytoplasmic in the parent CV1 cells.[87] However, the level of nuclear import was less than in smooth muscle cells, suggesting that additional transcription factors other than SRF alone are needed for maximal nuclear import of the DNA.

More recently, it has been reported that the incorporation of NF-κB binding sites alone in a plasmid can increase the nuclear localization of the plasmid in HeLa cells in a regulated fashion.[61] In this study, a series of 5 consensus binding sites for NF-κB were cloned into the pGL3-control plasmid (Promega, Madison, WI) and the plasmid was transfected or microinjected into the cytoplasm of cells. Nuclear import was followed indirectly by gene expression or directly by labeling the DNA with a fluorescently-labeled peptide nucleic acid (PNA). In transfection studies, the NF-κB site-containing plasmid expressed about the same as the parent plasmid under normal circumstances, but expressed gene product much more robustly in the presence of the NF-κB activator TNF-α. Further, more of the NF-κB plasmid appeared to localize to the nucleus in the presence of TNF-α than did the parent plasmid. Because TNF-α is known to promote the nuclear translocation of NF-κB, the authors' conclusion was that the activated transcription factor bound to its sites on the plasmid and facilitated transport, as in our model. However, it should be pointed out that the plasmids used in this study also contained the SV40 enhancer. Consequently, it is unclear whether inclusion of NF-κB binding sites alone would support nuclear import of the DNA.

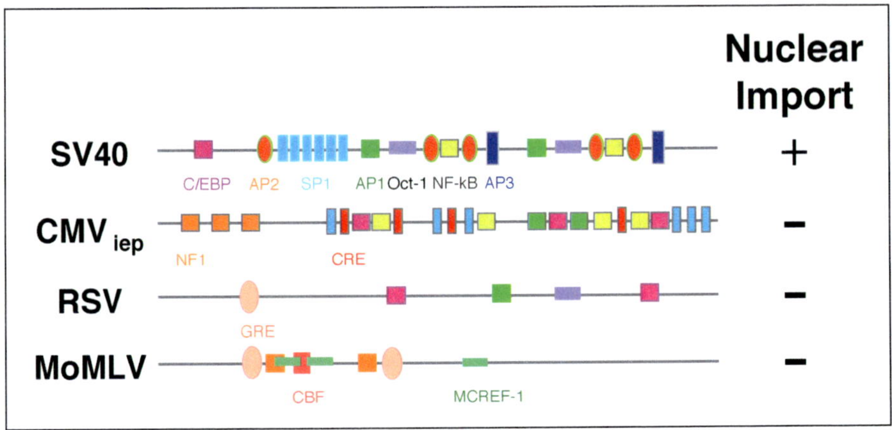

Figure 6. Cartoon of transcription factor binding sites in viral promoters. A number of viral promoters are shown with some of the transcription factors known to bind to them.

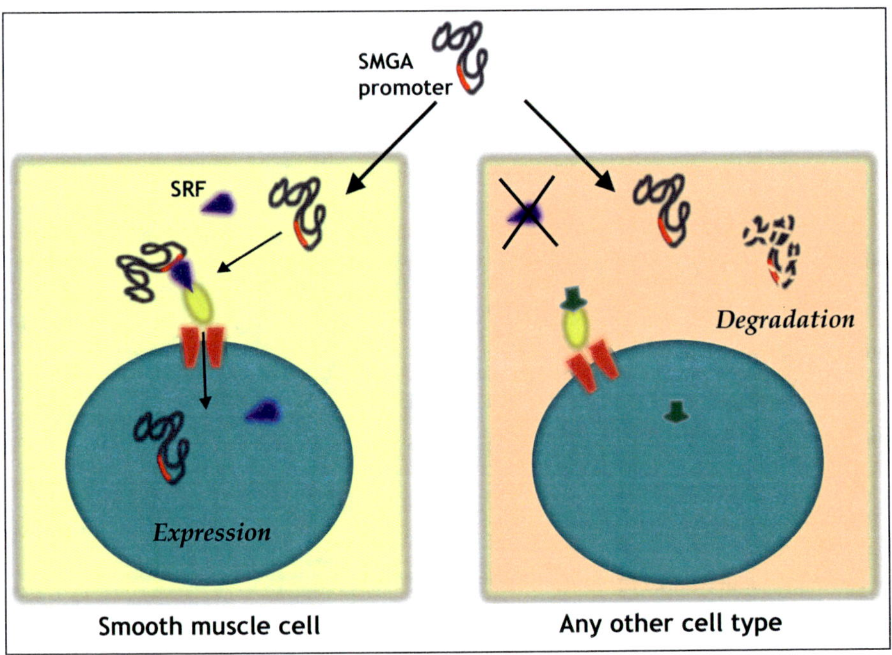

Figure 7. Smooth muscle-specific plasmid nuclear import. Smooth muscle-specific transcription factors, including SRF among others, can bind to their target sites within the SMGA promoter carried on a plasmid and serve to transport the DNA to the nucleus, via interactions with the NLS-mediated protein import machinery. Since these factors are not expressed in other cell types, no nuclear import will occur in non-smooth muscle cells.

Nuclear Import of Plasmids in Cell-Free Systems

To characterize the mechanisms of DNA nuclear import in more detail, and to identify the proteins involved, several groups have utilized digitonin-permeabilized cells. Wolff and colleagues demonstrated that linear fragments of labeled double-stranded DNA can localize to the nuclei of permeabilized cells in a reaction that is inhibited by agents that block the NPC and by energy depletion.[36] DNA nuclear import was also saturable, but was not competed by excess NLS-containing proteins, suggesting that the DNA is entering the nucleus through a pathway distinct from that of classic NLS-containing proteins. In contrast to their,[23] and other's,[19,20] findings that in microinjected cells 5 to 15 Kbp plasmids can be imported into the nucleus, they found that nuclear uptake of the DNA in permeabilized cells was size-dependent. DNA fragments less than 1000 bp in length were able to accumulate in the nuclei, but longer fragments remained excluded from the nuclei. Furthermore, DNA nuclear import was inhibited by the addition of cytoplasmic extracts.[36] This is in contrast to what has been seen for the nuclear import of NLS-containing proteins and U snRNAs. Perhaps the reasons for these differences was that the DNA fragments used were highly conjugated to fluorescent tags. Thus, the properties of the DNA may be vastly different from native DNA. By contrast to the results with linear fragments of DNA, when the Wolff group used plasmids, they found that nuclear uptake in permeabilized cells was dependent on cytoplasmic extracts only when the DNA was covalently cross-linked to a number of NLS peptides.[79] However, these NLS-conjugated plasmids failed to localize to the nuclei of microinjected cells as determined by fluorescence microscopy, but did show modest increases in gene expression compared to unmodified plasmids when cytoplasmically microinjected into the cells.[79]

In a separate set of experiments, using permeabilized cells and intact plasmids that were fluorescently tagged with a triplex-forming PNA, our laboratory demonstrated that plasmid nuclear uptake was time-, energy-, and temperature-dependent and utilized the NPC for translocation.[93] Furthermore, nuclear entry of the plasmids, as well as NLS-containing proteins was absolutely dependent on the addition of cytoplasmic extracts. When using purified importinα, importinβ, and Ran, plasmids were excluded from the nuclei unless nuclear extract was also provided. The probable reason for this is that the nuclear extracts are providing a source of transcription factors. Because importinα, importinβ, and Ran do not bind to DNA directly, DNA-binding, NLS-containing proteins must also be added to act as adapters between the DNA and the importins. As seen in microinjected cells, DNA nuclear entry was sequence specific: a 4.2 kbp plasmid lacking the SV40 enhancer was excluded from the nuclei while an isogenic plasmid containing the SV40 sequence localized to the nucleus efficiently. Plasmids up to 14 kbp were efficiently imported into the nuclei within 4 hours, although larger plasmids were not tested. Thus, like most other nuclear localizing macromolecules, plasmids were imported into the nuclei using the same mechanisms.

Nuclear import of plasmids has also been studied in synthetic nuclei reconstituted from phage λ DNA and interphase extracts from *Xenopus laevis*.[77] Nuclear import of NLS-containing proteins has been extensively characterized using this system.[28,65-68] To study DNA nuclear uptake, 48.5 kbp linear λ DNA was labeled along its length with Cy3 and added to reconstituted nuclei. In its protein-free, extended state, naked double stranded λ DNA is ~2 nm in diameter and about 15 μm in length. However, once added to cellular extracts, the DNA would become covered with proteins and condensed to take on a different hydrodynamic shape. Like NLS-mediated uptake, import of λ DNA was inhibited by WGA, suggesting that the DNA enters the nucleus via the NPC. However, although ATP depletion and low temperatures effectively inhibited BSA-NLS import, the treatments had no effect on DNA nuclear uptake. Because of this it is likely that this linear DNA is entering the nuclei through a pathway different from that seen in intact cells. To study how the DNA moved through the NPC, a 3 μm polystyrene sphere was attached to one end of the DNA and the DNA was then added to reconstituted nuclei. Within several minutes, laser tweezers were used to immobilize DNA-bead complexes adjacent to nuclei and the bead was pulled away from the nucleus. When some of the beads were pulled, the nuclei followed, indicating that the DNA was bound to the nucleus. In other cases, the bead could be pulled a defined distance from the nucleus (although less than the total persistence length of the DNA) before it recoiled toward the envelope. Measurements were taken at 2 second intervals to determine the kinetics of uptake. Import of the DNA occurred with an initial rate of 28 nm/sec, with or without ATP present. Further, the presence of an NLS at the end of the DNA distal to the bead (i.e., at the NPC-binding end) caused no increase in nuclear import rates. These results suggest that such linear DNA enters the nucleus by a mechanism distinct from that of NLS-mediated proteins. Because energy is not required, it is unlikely that molecular motors are involved, but rather a model is favored whereby the DNA is "ratcheted" into the nucleus in a serpentine fashion. Similar models have been postulated for the nuclear import of ssDNA (e.g., T-DNA from *Agrobacterium tumefaciens*) or peptide translocation into the endoplasmic reticulum.[60,81] Such a mechanism may also be involved in plasmid nuclear import for the translocation of the bulk of the DNA; the NLS-transcription factor complex may be necessary for initial localization and binding to the NPC, but the remainder of the DNA could be pulled in by a similar mechanism.

Alternative Pathways for Plasmid Nuclear Uptake

It is well documented that two of the strongest promoters for in vivo expression are the CMV immediate early promoter and enhancer and the RSV LTR promoter.[40,58,95,102] When plasmids carrying these promoters are injected into the post-mitotic myotubes in skeletal muscle from rodents or primates, robust gene expression is observed, even as quickly as 15 minutes post-injection.[22] Because these cells are non-dividing, the DNA is somehow gaining access to the intact nucleus, although how remains to be determined. In Dowty's experiments, a plasmid using the RSV LTR to drive luciferase expression (i.e., no SV40 sequence) was efficiently imported into the nuclei of myotubes by a pathway using the NPC.[23] A common thread to all of the above studies is that they employed skeletal muscle. One possible explanation is that skeletal muscle represents a tissue unlike any others in the body. Another finding that suggests that this at least partially could be the case is that skeletal muscle is known to express plasmid-encoded genes for extended periods of time, in some cases over the lifetime of the experimental animal,[94] whereas all other cells in the body tend to express plasmid-encoded genes on the order of days to weeks. While differences in patterns of gene expression are routinely seen in muscle versus other tissues, this may not be the sole reason. Another alternative is that plasmid nuclear import may not be absolutely sequence-dependent, but rather, the presence of certain sequences such as the SV40 enhancer increase the rate of nuclear targeting. As such, promoters like the CMV_{iep}, which we have shown does not localize to the nucleus within 8-12 hours,[19,20] may drive nuclear import at reduced rates. Simple kinetics could then account for the gene expression that is seen within minutes of plasmid injections into muscle: the more plasmid delivered to the cytoplasm, the more likely at least some will be driven into the nucleus. By microinjecting single myotubes in vivo, it has been shown that 10^6 plasmids (lacking SV40 sequences) are needed for gene expression.[86] Once in the cytoplasm, any plasmid will be quickly coated with proteins. It is more than possible that at least some of these DNA-binding proteins will contain NLSs, creating a complex of DNA weakly bound to a few nuclear-localizing proteins. When compared to the SV40 sequence which binds multiple transcription factors with exposed NLSs, import of the other DNAs could appear much slower, and in the time frame of our experiments, not at all.

Several experiments from Peter Traub's group have suggested that there may exist another pathway for DNA nuclear import. The intermediate filament protein vimentin is a cytoplasmic protein, but when bound to single- or double-stranded oligonucleotides, it migrates to the nucleus.[38] Similarly, when supercoiled plasmids, regardless of sequence, were bound to vimentin, through the cryptic DNA-binding activity of this protein, they caused the rapid movement of the protein into the nucleus.[39] Although the experiment was not performed, these results suggest that supercoiled DNAs may have the ability to rapidly migrate into the nucleus. However, in our hands, this is not the case.[18-20,87,93] Thus, these results suggest that the nuclear import of supercoiled DNA/vimentin complexes may be due to the complex rather than either of the parts.

Viral Nuclear Import

A number of DNA and RNA viruses must target their genomes to the nucleus in order to complete their life cycles. As for plasmid nuclear import, in all cases, the viral genome is targeted to the nucleus through the interactions of the nucleic acid with one or more viral proteins (Fig. 8). Several excellent reviews have covered the nuclear import strategies used by viruses, and for extensive discussions, the reader is directed to these.[16,48,91] To summarize, strategies for the nuclear import of viral genomes can be categorized into three groups: (1) targeting of the viral capsid to the nuclear envelope and "dumping" of the DNA into the nucleus, (2) the use viral capsid proteins to target the DNA to the nucleus, and (3) the use of core proteins to target the DNA into the nucleus.

Herpes simplex virus (HSV) enters the cell through the interaction of a group of envelope glycoproteins with proteins on the cell surface. Once internalized, the viral capsid, surrounded by a set of tegument proteins, is targeted toward the nucleus via the cytoskeleton. Fluorescently-labeled HSV capsids have been shown to target to the NPC using the microtubule-associated motor protein, dynein; depolymerization of the microtubule network abolished intracellular movement of the capsids, but not cell entry.[82,101] Once the capsid reaches the nuclear envelope, the core "docks" and uncoats at the NPC. Experiments using isolated rat liver nuclei have shown that HSV capsids can bind to the NPC only when cytoplasmic extracts or importin β and RAN and an energy source are provided.[69] This implies that NLS-containing proteins within the tegument and/or capsid are also required for NPC targeting. Immediately following NPC binding, capsid-associated DNA becomes DNaseI-sensitive, mimicking the in vivo uncoating at the NPC.[69] Somewhat similar to HSV, adenovirus enters the cell, although by escape from endosomes, and also targets to the nucleus using the cytoskeleton. Based on studies using fluorescently-labeled viruses and real-time video microscopy, microtubules have also been implicated in the nuclear targeting of adenovirus,[34,78] although other studies have not confirmed this.[31] Like HSV, the nuclear import machinery is needed for nuclear entry of adenoviral DNA, but hsc70 appears also required, perhaps to aid in the disassembly of the capsid before nuclear entry of the genome.[78] Hepatitis B virus also targets its capsid to the NPC in a process dependent on phosphorylation of the core protein and use of its NLS.[46,47]

Other viruses, including SV40 and baculovirus, target their genomes into the nucleus using capsid proteins that accompany the DNA into the nucleus.[88] The best studied of these is SV40. When SV40 virions were microinjected into the cytoplasm of SV40 permissive cells, the injected virion proteins localized to the nucleus and were immediately followed by large T-antigen expression, indicating that the viral genome had also entered the nucleus.[12] Virion (or more likely virion protein-DNA complex) entry into the nucleus and subsequent T-antigen expression could be blocked by co-injection of WGA or antibodies against NPC proteins, or by energy depletion, all treatments known to inhibit translocation through the NPC. The NPC also appears to be used for the nuclear targeting of incoming SV40 virions during an infection since cytoplasmic microinjection of anti-Vp3 antibodies blocked T-antigen expression during cell surface-mediated infection.[100] This study also showed electron dense particles, interpreted to be intact or partially dissociated virion particles, in transit through the NPC.[100] However, whether these were indeed intact viral particles was not convincingly proven. While it has been largely assumed that uncoating of the papovaviruses occurs exclusively in the nucleus,[21] there is a great amount of evidence demonstrating that uncoating can occur in the cytoplasm. For example, the

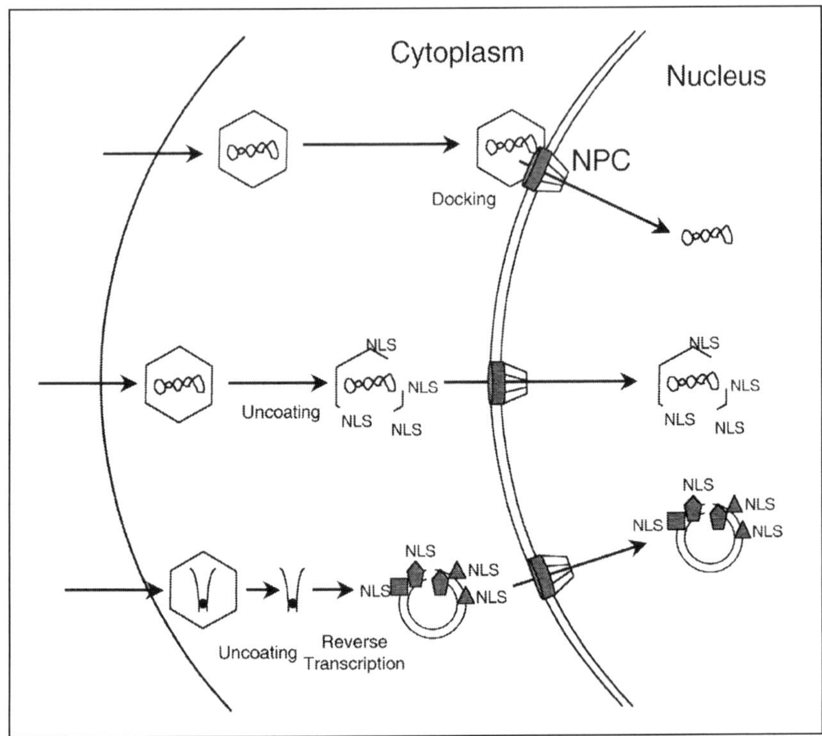

Figure 8. Major nuclear import strategies used by viruses. The three major routes of nuclear import for viruses are shown. Viruses including HSV and adenovirus, target their capsids to the NPC and "dump" their genomes into the nucleus (top). Other viruses including SV40 and baculovirus uncoat at least partially in the cytoplasm and their genomes migrate into the nucleus in association with their NLS-containing, DNA-binding capsid proteins (middle). Finally, viruses exemplified by HIV use core proteins to target the genomes into the nucleus (bottom).

DNA in over 50% of infecting virions recovered from cytoplasm were nuclease-sensitive within 30 minutes of infection, indicating that they had at least partially uncoated.[1] Thus, it is more probable that the virus partially uncoats in the cytoplasm and that the viral DNA remains associated with many of the capsid proteins and then enters the nucleus as a protein-DNA complex. The SV40 capsid consists of 360 copies of the major capsid protein and 72 copies of the minor two proteins. All three viral structural proteins contain well defined nuclear localization signals (NLSs) that direct the proteins to the nucleus.[11,30,97,98] Furthermore, all three proteins bind DNA.[10,13,83] Consequently, even if the virus uncoats losing 90% of its capsid proteins, the genome will still be complexed with almost 50 NLS-containing proteins. Since the incoming viral genome is already coated with a large number of NLS-containing proteins, nuclear import is much more rapid than that seen when purified SV40 DNA is injected.[19,64]

A third common theme used by viruses to target their genomes to the nucleus is best exemplified by HIV. Like oncoretroviruses, HIV is internalized into the cytoplasm and the virus uncoats. The incoming viral RNA genome is then reverse transcribed in the cytoplasm into a double stranded DNA molecule of approximately 14 kbp. A set of viral proteins that are carried within the virus particle remain associated with the viral genome during reverse transcription. This complex is referred to as the preintegration complex, and is made up of at least 4 proteins: nucleocapsid (NC), matrix (MA), integrase (IN), and Vpr. Both matrix and integrase possess NLSs that have been proposed to play a role in the nuclear import of HIV. Vpr also plays a role in nuclear import, although its role is less well defined. When the NLS of either matrix or integrase is mutated or Vpr is deleted, the infectivity of the resulting virus is greatly diminished.[7,41,73,89,90] While the matrix protein appears to function by donating its NLS for DNA nuclear import, experiments suggest that Vpr may both promote association of importinα with the matrix NLS and act directly as an importinβ-like protein to facilitate nuclear import of the complex.[44,72,73] Thus, the model for HIV preintegration complex nuclear targeting is that the reverse-transcribed viral DNA is transported into the nucleus through the interactions of its associated NLS-containing proteins, a mechanism similar to that seen for plasmids and other viral genomes.

Nuclear Import of Single-Stranded DNA

All of the preceding discussion has dealt with the nuclear import of double-stranded DNA, either in its linear or circular forms. Single-stranded DNAs are also transported to the nuclei of cells by very similar mechanisms. The most extensively studied ssDNA in terms of nuclear localization is that of the plant pathogen *Agrobacterium tumefaciens*. *A. tumefaciens* transforms host cells by transferring a segment of a single-stranded DNA, referred to as the T-DNA (transferred DNA), derived from its tumor-inducing (Ti) plasmid (for recent reviews, see refs. 51, 80 and 108). Nuclear import of T-DNA complexes into host cell nuclei is very fast and efficient, suggesting an active nuclear transport process. Furthermore, T-DNA import has been shown to be inhibited in presence of competitor molecules such as BSA-NLS or agents that block the NPC.[106,107,109] Like viruses, the T-DNA transfer process utilizes specialized bacterial virulence proteins provided by the pathogen for nuclear transport.

It has been shown that for effective T-DNA transfer, virulence proteins are not only necessary for the assembly of the nucleoprotein complex, but also for its subsequent nuclear import in the host cell. Briefly, the virulence proteins VirD1 and VirD2 are shown to be required for the excision of the T-DNA from its adjacent sequences within the Ti plasmid. Subsequently, virulence protein E2 (VirE2) is recruited to the complex by its single-stranded DNA-binding properties and is thought to exert its influence on T-DNA transfer efficiency by coating and compacting the excised DNA sequence necessary both to preserve the T-DNA integrity and to facilitate its nuclear import.[76,107] All three proteins contain well-defined NLSs. Subsequent to transfer of the T-DNA complex into the host cell, VirD2 through its nuclear localization signal (NLS) is thought to guide the entire complex in to the nucleus.[74,75] However, for import of short stretches of single stranded DNA, VirD2 alone has been shown to be sufficient.

Agrobacterium-mediated T-DNA transfer to plant cells is the only known example for interkingdom DNA transfer and is widely used for plant transformation. However, the interest in this field has broadened by a recent report showing that reconstituted complexes consisting of the bacterial virulence proteins VirD2, VirE2, and single-stranded DNA can mediate nuclear import of nucleic acids in permeabilized mammalian cells in vitro and in *Xenopus* oocytes.[35,106] Furthermore, in mammalian cells, a VirD2 mutant lacking its C terminal nuclear localization signal was deficient in import of the ssDNA-protein complexes into nuclei and import of DNA-protein complexes was dependent on importinα, Ran, and an energy source, suggesting a classical importin-mediated nuclear import pathway.[106]

Nuclear Import of Oligonucleotides

It is well established that, like double-stranded DNA, oligonucleotide polymers can be used to specifically express, or even more importantly, modulate the expression of a gene. Short stretches of single stranded DNA and RNA molecules have been shown to exert complementary antisense effects or enzymatic activities to modulate and in some cases to completely abolish expression of selective host genes. The ability to specifically target a single gene has brought antisense oligonucleotides to the fore front of rational drug design with great potential for research as well as clinical applications.[24,45,53,59]

Antisense oligonucleotides are synthetic polymers of oligonucleotides usually with built-in structural modifications, such as a phosphorothioate backbone, to modulate their stability, cellular uptake, and/or toxicity. Antisense oligonucleotides are generally designed to bind through Watson-Crick base pairing to a unique complementary RNA sequence and this hybridization selectively interferes with nuclear processing, transport and/or cytoplasmic translation of the target molecule as well as its degradation in an RNase H-dependent manner.[15] Although the intended site of action in the majority of cases is cytoplasmic, it is important to emphasize that in many cell types and using different delivery methods, oligonucleotides have been shown to localize to the nucleus and in some cases this nuclear localization has been shown to correlate with an increase in antisense activity.[3,56,57]

Any oligonucleotide polymer used to specifically express or modulate the expression of a gene, must first reach the cytosol to be able to exert its biological effect. A variety of studies have shown that oligonucleotide uptake by cells is time-, temperature-, energy-, concentration-, cell type- and size-dependent, and can be affected by chemical modifications. Kinetic analysis of oligonucleotide binding and uptake has generally pointed to a bimodal process where receptor mediated uptake predominates only at low micromolar concentrations whereas fluid-phase endocytosis prevails at higher concentrations.[2] Abundant evidence has also been presented demonstrating the existence of cell surface receptors that bind oligonucleotides.[2,37,55,99] Taken together these data also support the notion that a receptor-mediated but passive uptake, referred to as adsorptive endocytosis, may be the prevailing mechanism of oligonucleotide uptake.[96] Cationic liposomes have also been used extensively for the successful delivery of oligonucleotides to cells.[42] However, the success rate for liposome mediated antisense oligonucleotide delivery varies depending on the cell type, particular liposome formulation used, and cell culture conditions utilized. Nevertheless, using a cationic liposome assisted delivery and the C6 glioma cell line as an in vitro model, Islam and colleagues have shown a 10-12 fold increase in cellular uptake of both biotin-labeled and radiolabeled oligonucleotides.[42] Furthermore, in the absence of liposomes, internalized oligonucleotides were shown to be sequestered mostly within endosomal and lysosomal vesicles. In the presence of cationic lipids, distribution of internalized oligonucleotides showed an enhanced penetration of the cytosol and increased accumulation within the nucleus consistent with an increased endosomal release.[42] Finally, electroporation has also been used to attain high intracellular concentrations of oligonucleotide in a large proportion of viable cells. Even more interestingly, following electroporation transfected oligonucleotides rapidly localized to the nucleus and remained predominantly nuclear for at least 2 days.[4] Taken together, these results demonstrate that the route of cellular uptake can have profound effects on the intracellular distribution of nucleic acid.

More and more compelling evidence implies that antisense DNA may exert their effects in the nucleus. For example, uptake studies with ^{35}S-labeled phosphorothioate oligonucleotides have shown that following cellular uptake labeled oligonucleotides were found in significant amounts in the nucleus.[2] In contrast to other studies, nuclear uptake was not blocked by WGA, suggesting a diffusion driven process, rather than an importin-mediated one. However, using oligonucleotides conjugated to a photoactivatable, radiolabeled crosslinker, it has been shown that, even following uptake, internalized oligonucleotides remain largely associated with proteins. Although several candidate proteins were detected, internalized oligonucleotides were predominantly found to be associated with a 75 kD protein that appears to be membrane-associated.[29] Therefore, it has been suggested that the majority of intracellular oligonucleotides remain associated in vesicles with the same protein to which they bind on the cell surface and only a small percentage of non-protein-bound cytosolic oligonucleotide can be detected. Additionally, the majority of free cytosolic oligonucleotides readily accumulated in the nuclei where they were shown to associate with a new set of nuclear oligonucleotide binding proteins.[29]

More recent evidence suggests that phosphorothioate oligonucleotides utilize the classic NPC-mediated pathway for nuclear localization.[57] Further, at least a fraction of the nuclear pool of oligonucleotides are in dynamic balance with the cytoplasmic pool and are able to continuously shuttle between the nucleus and the cytoplasm while maintaining their antisense activity.[57] The shuttling of phosphorothioate oligonucleotides was shown to be an active transport process and sensitive to treatment with WGA indicating an NPC-mediated translocation process. Nucleocytoplasmic shuttling was only moderately affected by disruption of the Ran/RCC1 system, suggesting a transport process similar to that of U snRNA, tRNA and mRNA export.[57] However, oligonucleotides without a phosphorothioate backbone chemistry do not share these characteristics and are only weakly restricted in their migration upon chilling, ATP depletion and WGA treatment.[2] Therefore, whether a particular oligonucleotide is going to be subject to active nuclear transport or rather moved by diffusion may indeed depend on its affinity for protein binding partners.

In another study of nuclear targeting and retention of oligonucleotides, using high resolution imaging and fluorescently tagged phosphorothioate oligonucleotides, Lorenz et al have shown that under conditions expected to display optimal antisense activity, phosphorothioate oligonucleotides predominantly localize to the cell nucleus (but not to the nucleoli). Under these conditions, nuclear oligonucleotides accumulated in a number of bright spherical foci referred to as "phosphorothioate bodies".[56] Furthermore, these nuclear foci formed in both transfected as well as microinjected cells, suggesting that formation of nuclear phosphorothioate bodies was independent of the oligodeoxynucleotide delivery method used. Also, formation of these nuclear foci did not correlate with antisense activity or the primary sequence of oligonucleotide used. Ultrastructurally, these nuclear foci have been shown to correspond to electron-dense structures of 150-300 nm diameter and resemble nuclear bodies that were described to occur at a lower frequency in non-treated control cells. These foci developed in living cells within minutes after the introduction of the oligonucleotides and they were relatively stable entities that underwent noticeable reorganization only during mitosis. These findings support the notion that, following uptake and endosomal release, phosphorothioate oligonucleotides associate with a new set of nuclear proteins including the nuclear matrix.[56]

Conclusions

Although the nuclear import of DNA may not be a normal event in the cell, mechanisms do exist for its transport. Some of these have evolved over a billion years, as viruses and other pathogens have perfected ways to invade the host, while others appear to be fortuitous piracy, as in the case of the SV40 enhancer which binds to proteins on their way to the nucleus. Regardless, the mechanism is the same: NLS-containing proteins, either provided by the host or pathogen, bind to the DNA and target it to the nucleus. The goal of all gene therapy approaches is to target enough DNA to the nuclei of cells to obtain sufficient expression for a therapeutic effect. As is well accepted, one of the major barriers to this goal is the nuclear envelope and our relative inability to target substantial amounts of DNA to the nucleus. By characterizing and understanding the mechanisms of DNA nuclear import we can begin to exploit these pathways to increase the nuclear targeting of genes for transfection, transgenic plant production, and ultimately, gene therapy.

References

1. Barbanti-Brodano G, Swetly P, Koprowski H. Early events in the infection of permissive cells with simian virus 40: Adsorption, penetration, and uncoating. J Virol 1970; 6:78-86.
2. Beltinger C, Saragovi HU, Smith RM et al. Binding, uptake, and intracellular trafficking of phosphorothioate-modified oligodeoxynucleotides. J Clin Invest 1995; 95:1814-1823.
3. Bennett M, Pinol-Roma S, Staknis D et al. Differential binding of heterogeneous nuclear ribonucleoproteins to mRNA precursors prior to spliceosome assembly in vitro. Mol Cell Biol 1992; 12:3165-75.

4. Bergan R, Connell Y, Fahmy B et al. Electroporation enhances c-myc antisense oligodeoxynucleotide efficacy. Nucleic Acids Res 1993; 21:3567-3573.
5. Browning CL, Culberson DE, Aragon IV et al. The developmentally regulated expression of serum response factor plays a key role in the control of smooth muscle-specific genes. Dev Biol 1998; 194:18-37.
6. Brunner S, Sauer T, Carotta S et al. Cell cycle dependence of gene transfer by lipoplex, polyplex and recombinant adenovirus. Gene Therapy 2000; 7:401-407.
7. Bukrinsky MI and Haffar OK. HIV-1 nuclear import: Matrix protein is back on center stage, this time together with Vpr. Mol Med 1998; 4:138-143.
8. Capecchi MR. High efficiency transformation by direct microinjection of DNA into cultured mammalian cells. Cell 1980; 22:479-88.
9. Carson JA, Fillmore RA, Schwartz RJ et al. The smooth muscle gamma-actin gene promoter is a molecular target for the mouse bagpipe homologue, mNkx3-1, and serum response factor. J Biol Chem 2000; 275:39061-39072.
10. Chang D, Cai X, Consigli RA. Characterization of the DNA binding properties of polyomavirus capsid proteins. J Virol 1993; 67:6327-6331.
11. Clever J, Kasamatsu H. Simian virus 40 Vp2/3 small structural proteins harbor their own nuclear transport signal. Virology 1991; 181:78-90.
12. Clever J, Yamada M, Kasamatsu H. Import of simian virus 40 virions through nuclear pore complexes. Proc Natl Acad Sci USA 1991; 88:7333-7337.
13. Clever JL, Dean DA, Kasamatsu H. Identification of a DNA-binding domain within the simian virus 40 capsid proteins Vp2 and Vp3. J Biol Chem 1993; 268:20877-20883.
14. Coonrod A, Li FQ, Horwitz M. On the mechanism of DNA transfection: efficient gene transfer without viruses. Gene Ther 1997; 4:1313-1321.
15. Crooke ST. Molecular mechanisms of action of antisense drugs. Biochim Biophys Acta 1999; 1489:31-44.
16. Cullen BR. Journey to the center of the cell. Cell 2001; 105:697-700.
17. Davis LI, Blobel G. Identification and characterization of a nuclear pore complex protein. Cell 1986; 45:699-709.
18. Dean BS, Byrd JN Jr, Dean DA. Nuclear targeting of plasmid DNA in human corneal cells. Cur. Eye Res. 1999; 19:66-75.
19. Dean DA. Import of plasmid DNA into the nucleus is sequence specific. Exp. Cell Res. 1997; 230:293-302.
20. Dean DA, Dean BS, Muller S et al. Sequence requirements for plasmid nuclear entry. Exp Cell Res 1999; 253:713-722.
21. Diacumakos EG, Gershey EL. Uncoating and gene expression of simain virus 40 in CV-1 cell nuclei inoculated by microinjection. J Virol 1977; 24:903-906.
22. Doh SG, Vahlsing HL, Hartikka J et al. Spatial-temporal patterns of gene expression in mouse skeletal muscle after injection of lacZ plasmid DNA. Gene Ther 1997; 4:648-663.
23. Dowty ME, Williams P, Zhang G et al. Plasmid DNA entry into postmitotic nuclei of primary rat myotubes. Proc Natl Acad Sci USA 1995; 92:4572-4576.
24. Dvorchik BH. The disposition (ADME) of antisense oligonucleotides. Curr Opin Mol Ther 2000; 2:253-257.
25. Dynan WS, Chervitz SA. Characterization of a minimal simian virus 40 late promoter: enhancer elements in the 72-base-pair repeat not required. J Virol 1989; 63:1420-1427.
26. Dynan WS, Tjian R. The promoter-specific transcription factor Sp1 binds to upstream sequences in the SV40 early promoter. Cell 1983; 35:79-87.
27. Fasbender A, Zabner J, Zeiher BG et al. A low rate of cell proliferation and reduced DNA uptake limit cationic lipid-meidated gene tranfer to primary cultures of ciliated human airway epithelia. Gene Therapy 1997; 4:1173-1180.
28. Finlay DR, Newmeyer DD, Price TM et al. Inhibition of in vitro nuclear transport by a lectin that binds to nuclear pores. J Cell Biol 1987; 104:189-200.
29. Geselowitz DA, Neckers LM. Analysis of oligonucleotide binding, internalization, and intracellular trafficking utilizing a novel radiolabeled crosslinker. Antisense Res Dev 1992; 2:17-25.

30. Gharakhanian E, Takahashi J, Kasamatsu H. The carboxyl 35 amino acids of SV40 Vp3 are essential for its nuclear accumulation. Virology 1987; 157:440-448.

31. Glotzer JB, Michou A-I, Baker A et al. Microtubule-independent motility and nuclear targeting of adenoviruses with fluorescently labeled genomes. J Virol 2001; 75:2421-2434.

32. Graessman M and Graessman A. Regulation of SV40 gene expression. Adv Cancer Res 1981; 35:111-149.

33. Graessman M, Menne J, Liebler M et al. Helper activity for gene expression, a novel function of the SV40 enhancer. Nucleic Acids Res 1989; 17:6603-6612.

34. Greber UF, Suomalainen M, Stidwill RP et al. The role of the nuclear pore complex in adenovirus DNA entry. Embo J 1997; 16:5998-6007.

35. Guralnick B, Thomsen G, Citovsky V. Transport of DNA into the nuclei of Xenopus oocytes by a modified VirE2 protein of Agrobacterium. Plant Cell 1996; 8:363-373.

36. Hagstrom JE, Ludtke JJ, Bassik MC et al. Nuclear import of DNA in digitonin-permeabilized cells. J Cell Sci 1997; 110:2323-2331.

37. Hanss B, Leal-Pinto E, Bruggeman LA et al. Identification and characterization of a cell membrane nucleic acid channel. Proc Natl Acad Sci USA 1998; 95:1921-1926.

38. Hartig R, Shoeman RL, Janetzko A et al. Active nuclear import of single-stranded oligonucleotides and their complexes with non-karyophilic macromolecules. Biol Cell 1998; 90:407-426.

39. Hartig R, Shoeman RL, Janetzko A et al. DNA-mediated transport of the intermediate filament protein vimentin into the nucleus of cultured cells. J Cell Sci 1998; 111:3573-3584.

40. Hartikka J, Sawdey M, Cornefert-Jensen F et al. An improved plasmid DNA expression vector for direct injection into skeletal muscle. Hum Gene Ther 1996; 7:1205-1217.

41. Heinzinger NK, Bukrinsky MI, Haggerty SA et al. The Vpr protein of human immunodeficiency virus type 1 influences nuclear localization of viral nucleic acids in nondividing host cells. Proc Natl Acad Sci USA 1994; 91:7311-7315.

42. Islam A, Handley SL, Thompson KS et al. Studies on uptake, sub-cellular trafficking and efflux of antisense oligodeoxynucleotides in glioma cells using self-assembling cationic lipoplexes as delivery systems. J Drug Target 2000; 7:373-382.

43. James MB, Giorgio TD. Nuclear-associated plasmid, but not cell-associated plasmid, is correlated with transgene expression in cultured mammalian cells. Mol Ther 2000; 1:339-346.

44. Jenkins Y, McEntee M, Weis K et al. Characterization of HIV-1 vpr nuclear import: analysis of signals and pathways. J Cell Biol 1998; 143:875-885.

45. Juliano RL, Yoo H. Aspects of the transport and delivery of antisense oligonucleotides. Curr Opin Mol Ther 2000; 2:297-303.

46. Kann M, Bischof A, Gerlich WH. In vitro model for the nuclear transport of the hepadnavirus genome. J Virol 1997; 71:1310-1316.

47. Kann M, Sodeik B, Vlachou A et al. Phosphorylation-dependent binding of hepatitis B virus core particles to the nuclear pore complex. J Cell Biol 1999; 145:45-55.

48. Kasamatsu H, Nakanishi A. How do animal DNA viruses get to the nucleus? Annu Rev Microbiol 1998; 52:627-686.

49. Kopchick JJ, Ju G, Skalka AM et al. Biological activity of cloned retroviral DNA in microinjected cells. Proc Natl Acad Sci USA 1981; 78:4383-4387.

50. Kovacs AM, Zimmer WE. Cell specific transcription of the smooth muscle g-actin gene requires both positive and negative acting cis-elements. Gene Exp 1998; 7:115-129.

51. Lartey R, Citovsky V. Nucleic acid transport in plant-pathogen interactions. Genet Eng 1997; 19:201-214.

52. Lechardeur D, Sohn K-J, Haardt M et al. Metabolic instability of plasmid DNA in the cytosol: A potential barrier to gene transfer. Gene Ther 1999; 6:482–497.

53. Levin AA. A review of the issues in the pharmacokinetics and toxicology of phosphorothioate antisense oligonucleotides. Biochim Biophys Acta 1999; 1489:69-84.

54. Lewis PF, Emerman M. Passage through mitosis is required for oncoretroviruses but not for the human immunodeficiency virus. J Virol 1994; 68:510-516.

55. Loke SL, Stein CA, Zhang XH et al. Characterization of oligonucleotide transport into living cells. Proc Natl Acad Sci USA 1989; 86:3474-3478.

56. Lorenz P, Baker BF, Bennett CF et al. Phosphorothioate antisense oligonucleotides induce the formation of nuclear bodies. Mol Biol Cell 1998; 9:1007-1023.

57. Lorenz P, Misteli T, Baker BF et al. Nucleocytoplasmic shuttling: a novel in vivo property of antisense phosphorothioate oligodeoxynucleotides. Nucleic Acids Res 2000; 28:582-592.
58. Manthorpe M, Cornefert-Jensen F, Hartikka J et al. Gene therapy by intramuscular injection of plasmid DNA: Studies on firefly luciferase gene expression in mice. Hum Gene Ther 1993; 4:419-431.
59. Marwick C. First "antisense" drug will treat CMV retinitis. Jama 1998; 280:871.
60. Matlack KE, Misselwitz B, Plath K et al. BiP acts as a molecular ratchet during posttranslational transport of prepro-alpha factor across the ER membrane. Cell 1999; 97:553-564.
61. Mesika A, Grigoreva I, Zohar M et al. A regulated, NFkappaB-assisted import of plasmid DNA into mammalian cell nuclei. Mol Ther 2001; 3:653-657.
62. Miller DG, Adam MA, Miller AD. Gene transfer by retrovirus vectors occurs only in cells that are actively replicating at the time of infection. Mol Cell Biol 1990; 10:4239-4242.
63. Mirzayans R, Remy AA, Malcom PC. Differential expression and stability of foreign genes introduced into human fibroblasts by nuclear versus cytoplasmic microinjection. Mutation Res 1992; 281:115-122.
64. Nakanishi A, Clever J, Yamada M et al. Association with capsid proteins promotes nuclear targeting of simian virus 40. Proc Natl Acad Sci USA 1996; 93:96-100.
65. Newmeyer DD, Finlay DR, Forbes DJ. In vitro transport of a fluorescent nuclear protein and exclusion of non-nuclear proteins. J Cell Biol 1986; 103:2091-2102.
66. Newmeyer DD, Forbes DJ. Nuclear import can be separated into distinct steps in vitro: Nuclear pore binding and translocation. Cell 1988; 52:641-653.
67. Newmeyer DD, Forbes DJ. An N-ethylmaleimide-sensitive cytosolic factor necessary for nuclear protein import: requirement in signal-mediated binding to the nuclear pore. J Cell Biol 1990; 110:547-557.
68. Newmeyer DD, Lucocq JM, Bürglin TR et al. Assembly in vitro of nuclei active in nuclear protein transport: ATP is required for nucleoplasmin accumulation. EMBO J 1986; 5:501-510.
69. Ojala PM, Sodeik B, Ebersold MW et al. Herpes simplex virus type 1 entry into host cells: reconstitution of capsid binding and uncoating at the nuclear pore complex in vitro. Mol Cell Biol 2000; 20:4922-4931.
70. Piñol-Roma S and Dreyfuss G. Transcription-dependent and transcription-independent nuclear transport of hnRNP proteins. Science 1991; 253:312-314.
71. Piñol-Roma S and Dreyfuss G. Shuttling of pre-mRNA binding proteins between nucleus and cytoplasm. Nature 1992; 355:730-732.
72. Popov S, Rexach M, Ratner L et al. Viral protein R regulates docking of the HIV-1 preintegration complex to the nuclear pore complex. J Biol Chem 1998; 273:13347-13352.
73. Popov S, Rexach M, Zybarth G et al. Viral protein R regulates nuclear import of the HIV-1 pre-integration complex. Embo J 1998; 17:909-917.
74. Relic B, Andjelkovic M, Rossi L et al. Interaction of the DNA modifying proteins VirD1 and VirD2 of Agrobacterium tumefaciens: analysis by subcellular localization in mammalian cells. Proc Natl Acad Sci USA 1998; 95:9105-9110.
75. Rossi L, Hohn B, Tinland B. The VirD2 protein of Agrobacterium tumefaciens carries nuclear localization signals important for transfer of T-DNA to plant. Mol Gen Genet 1993; 239:345-353.
76. Rossi L, Hohn B and Tinland B. Integration of complete transferred DNA units is dependent on the activity of virulence E2 protein of Agrobacterium tumefaciens. Proc Natl Acad Sci USA 1996; 93:126-130.
77. Salman H, Zbaida D, Rabin Y et al. Kinetics and mechanism of DNA uptake into the cell nucleus. Proc Natl Acad Sci USA 2001; 98:7247-7252.
78. Saphire ACS, Guan T, Schirmer EC et al. Nuclear import of adenovirus DNA in vitro involves the nuclear protein import pathway and hsc70*. J Biol Chem 2000; 275:4298–4304.
79. Sebestyén MG, Ludtke JL, Bassik MC et al. DNA vector chemistry: the covalent attachment of signal peptides to plasmid DNA. Nature Biotech. 1998; 16:80-85.
80. Sheng J and Citovsky V. Agrobacterium-plant cell DNA transport: have virulence proteins, will travel. Plant Cell 1996; 8:1699-1710.
81. Simon SM, Peskin CS and Oster GF. What drives the translocation of proteins? Proc Natl Acad Sci USA 1992; 89:3770-3774.
82. Sodeik B, Ebersold MW, Helenius A. Microtubule-mediated transport of incoming herpes simplex virus 1 capsids to the nucleus. J Cell Biol 1997; 136:1007-1021.

83. Soussi T. DNA-binding properties of the major structural protein of simian virus 40. J Virol 1986; 59:740-742.

84. Thornburn AM, Alberts AS. Efficient expression of miniprep plasmid DNA after needle micro-injection into somatic cells. Biotechniques 1993; 14:356-358.

85. Tseng W, Haselton F, Giogio T. Transfection by cationic liposomes using simultaneous single cell measurements of plasmid delivery and transgene expression. J Biol Chem 1997; 272:25641-25647.

86. Utvik JK, Nja A, Gundersen K. DNA injection into single cells of intact mice. Hum Gene Ther 1999; 10:291-300.

87. Vacik J, Dean BS, Zimmer WE et al. Cell-specific nuclear import of plasmid DNA. Gene Ther 1999; 6:1006-1014.

88. van Loo ND, Fortunati E, Ehlert E et al. Baculovirus infection of nondividing mammalian cells: mechanisms of entry and nuclear transport of capsids. J Virol 2001; 75:961-970.

89. Vodicka MA, Koepp DM, Silver PA et al. HIV-1 Vpr interacts with the nuclear transport pathway to promote macrophage infection. Genes Dev 1998; 12:175-185.

90. von Schwedler U, Kornbluth RS and Trono D. The nuclear localization signal of the matrix protein of human immunodeficiency virus type 1 allows the establishment of infection in macrophages and quiescent T lymphocytes. Proc Natl Acad Sci USA 1994; 91:6992-6996.

91. Whittaker GR, Helenius A. Nuclear import and export of viruses and virus genomes. Virology 1998; 246:1-23.

92. Wildeman AG. Regulation of SV40 early gene expression. Biochem Cell Biol 1988; 66:567-577.

93. Wilson GL, Dean BS, Wang G et al. Nuclear import of plasmid DNA in digitonin-permeabilized cells requires both cytoplasmic factors and specific DNA sequences. J Biol Chem 1999; 274:22025-22032.

94. Wolff JA, Ludtke JJ, Acsadi G et al. Long-term persistence of plasmid DNA and foreign gene expression in mouse muscle. Hum Mol Genetics 1992; 1:363-369.

95. Wolff JA, Malone RW, Williams P et al. Direct gene transfer into mouse muscle in vivo. Science 1990; 247:1465-1468.

96. Wu-Pong S, Weiss TL, Hunt CA. Antisense c-myc oligonucleotide cellular uptake and activity. Antisense Res Dev 1994; 4:155-163.

97. Wychowski C, Benichou D, Girard M. A domain of SV40 capsid polypeptide Vp1 that specifies migration into the cell nucleus. EMBO J 1986; 5:2569-2576.

98. Wychowski C, Benichou D, Girard M. The intranuclear localization of simian virus 40 polypeptides Vp2 and Vp3 depends on a specific amino acid sequence. J Virol 1987; 61:3862-3869.

99. Yakubov LA, Deeva EA, Zarytova VF et al. Mechanism of oligonucleotide uptake by cells: involvement of specific receptors? Proc Natl Acad Sci USA 1989; 86:6454-6458.

100. Yamada M, Kasamatsu H. Role of nuclear pore complex in simian virus 40 nuclear targeting. J Virol 1993; 67:119-130.

101. Ye G-J, Vaughan KT, Vallee RB et al. The herpes simplex virus 1 U_L34 protein interacts with a cytoplasmic dynein intermediate chain and targets nuclear membrane. J Virol 2000; 74:1355–1363.

102. Yew NS, Wysokenski DM, Wang KX et al. Optimization of plasmid vectors for high-level expression in lung epithelial cells. Hum Gene Ther 1997; 8:575-584.

103. Young JL, Byrd JN, Wyatt CR et al. Endothelial cell-specific plasmid nuclear import. Mol Biol Cell 1999; 10S:443a.

104. Zabner J, Fasbender AJ, Moninger T et al. Cellular and molecular barriers to gene transfer by a cationic lipid. J Biol Chem 1995; 270:18997-19007.

105. Zelphati O, Liang X, Hobart P et al. Gene chemistry: functionally and conformationally intact fluorescent plasmid DNA. Hum Gene Ther 1999; 10:15-24.

106. Ziemienowicz A, Gorlich D, Lanka E et al. Import of DNA into mammalian nuclei by proteins originating from a plant pathogenic bacterium. Proc Natl Acad Sci USA 1999; 96:3729-3733.

107. Ziemienowicz A, Merkle T, Schoumacher F et al. Import of Agrobacterium T-DNA into Plant Nuclei. Two distinct functions of vird2 and vire2 proteins. Plant Cell 2001; 13:369-384.

108. Zupan J, Muth TR, Draper O et al. The transfer of DNA from Agrobacterium tumefaciens into plants: a feast of fundamental insights. Plant J 2000; 23:11-28.

109. Zupan JR, Citovsky V. Zambryski P. Agrobacterium VirE2 protein mediates nuclear uptake of single-stranded DNA in plant cells. Proc Natl Acad Sci USA 1996; 93:2392-2397.

Research Methodologies for the Investigation of Cell Nucleus

Jose Omar Bustamante

It is impossible to detail all of the techniques used in the nuclear transport field within the allotted space. Since some methods (EM, cell and molecular biology, etc.) are dealt with in other Chapters, here I will focus on the approaches that are less familiar to the cell and molecular biologist (e.g., patch-clamp and atomic force microscopy). I believe that some of the observations I make are significant when interpreting experimental measurements.

Introduction

Any particle traveling directly from cytosol to nucleus, or vice versa, must go through the nuclear pores. The pores connect the inner and outer nuclear membranes (INM and ONM, respectively) of the nuclear envelope (NE). When the pores have no macromolecule traversing inside them, they are filled with the nucleocytoplasmic liquid (probably a mixture of both cytosol and nucleosol). Their denomination as pores, however, does not convey their supramolecular nature of 50 to 130 megadaltons (yeast and vertebrates, respectively). To emphasize their supramolecular architecture, nuclear pores are often referred to as nuclear pore complexes (NPCs). This explanation is necessary to understand the complexities of this structure when compared to much less massive membrane pores such as plasmalemmal ion channels (e.g., Na^+-, K^+-, Ca^{2+}- and Cl^- channels).

One more point deserves clarification for those who have just crossed into either cell biology or physiology. Nuclear transport investigators, mostly cell/molecular biologists, understand that the NPC has a channel inside it (i.e., one that can be filled with liquid). Physiologists, instead, are used to the concept that a channel (i.e., ion channel) is a membrane protein (or several proteins) with a pore inside. And, while physiologists think that (ion) channel gating means the fast statistical opening of the channel gates (a switch-like mechanism), nuclear transport researchers use the term NPC gating in the sense of a relatively sluggish multi-step translocation of macromolecules along the liquid channel of the NPC.

All the many methods used to study nucleocytoplasmic transport are based on physical and chemical principles. Accordingly, they may be classified as biochemical approaches if the purpose is to determine the role of particular molecular species in intracellular/intranuclear cascades. Alternatively, they may be identified as biophysical approaches if the purpose is to identify the structures involved in a particular phenomenon. Methods can also be viewed as purely biophysical if the aim is to determine physical properties of a particular structure or its function. Clearly, the more the approaches, the stronger the case will be made for a particular conclusion. However, quantity not always means quality for it is clear that one must use good judgement in order to reach meaningful data.

Nuclear Import and Export in Plants and Animals, edited by Tzvi Tzfira and Vitaly Citovsky.
©2005 Eurekah.com and Kluwer Academic / Plenum Publishers.

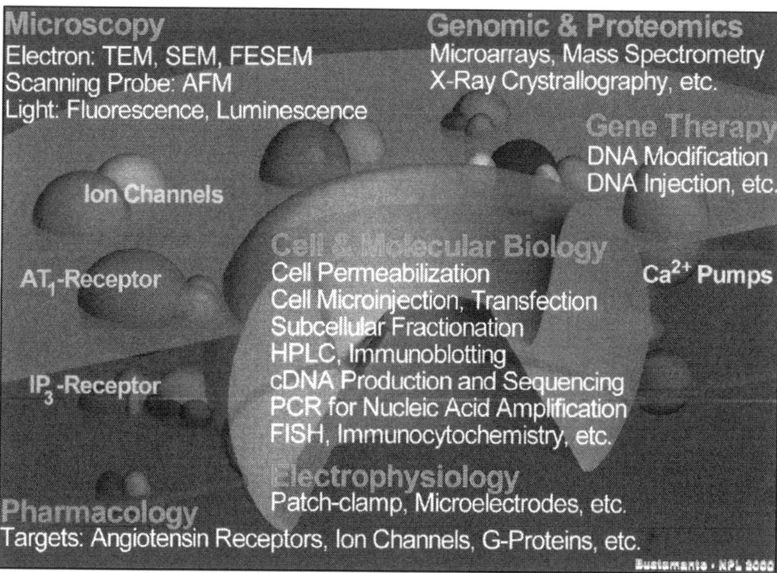

Figure 1. Nuclear transport methods. The foreground groups methods according to various areas. The background represents a single NPC connected to receptors and surrounded by various structural elements, including ion channels, pumps, etc.

In what follows, I shall try to describe most of the methods currently used to study nuclear transport. In Figure 1 I attempt to summarize them. Rather than trying to classify each method within a particular field (e.g., biochemistry), I simply describe the method, for it may appear a pure exercise of semantics to apply such a classification. Reviews on the subject have appeared[1] and I encourage the reader not only to look up reviews but also to data mine the literature with Internet search engines. Most of the references related to this Chapter can be obtained from freely accessible databases. The major database used in the preparation of this Chapter is PubMed.[2] Since this book will be accessible from the Internet, most of the references are given with their Internet link for easy access. Some of the links lead to the free access of the full text article. Due to space limitations, I have also left out some classical and less critical references.

Methods

Electron Microscopy (EM)

As shown elsewhere in this book, EM has been successful in the study of the structure and function of NEs and NPCs.[3-6] Both transmission and scanning EM (TEM and SEM, respectively) have been useful in identifying the physical structures responsible for specific functions of NPCs as well as in analyzing various transport mechanisms. Both, however, must be applied with caution. For example, in either TEM or SEM, the presence of macromolecular transport requirements, prior and/or during fixation of the sample, appears to be an important determinant of whether or not a central plug is observed.[4] Thus, whether the plug is intrinsic to the NPC has been subject of debate.[4]

TEM images are formed by the absorption of electrons during their passage through the sample. The absorptive properties of matter are referred to as its electron-density. The more electron-dense the material, the more the absorption and the darker it will appear in the image. As with other imaging techniques, pre-treatment and enhancement of the image must be made because, like many biological materials, the nucleus is not rigid and because it has low electron density. In EM, this is accomplished initially with chemicals. For this reason, the sample must be treated with reagents (e.g., glutaraldehyde, osmium tetroxide, etc.) that confer it rigidity against the high electrical fields used. The treatment also enhances the electron density contrast. Since these chemical treatments may alter the sample (e.g., shrinking), a cryogenic EM approach (cryo-EM) is sometimes used.

The procedure used for sample preparation (e.g., time of fixation, immunolabeling, time of exposure to transport substrates, etc.) is critical for obtaining images that depict realistic features of NPCs and their transport mechanisms. When used with electron-dense labels (e.g., nanogold particles with antibodies against specific NPC proteins), TEM becomes a useful tool in determining the localization of NPC protein components.

In contrast to TEM, which gives a sort of x-ray of the sample, SEM is a topological technique, for its images are formed by scattering and reflection of the electrons beamed onto the surface of the sample. A shortcoming of SEM is the lower resolving power relative to TEM. For this reason, the instrument of preference for topological EM studies is the field emission SEM (FESEM). The resolving power of this method is far superior to that of conventional SEM.[5,6] The resolution in all EM instruments will depend on their particular design and procedure for sample preparation. In general, the resolution for TEM is better than 0.5 nm and for FESEM, no better than 0.5 nm (e.g., ref. 7—EM instruments with which I have worked). Although the application of EM at atmospheric pressure is being studied[7], to date all EM protocols for nuclear transport include sample fixation (chemically or cryogenically). For this reason, there is no way to follow up a particular process for a specific NPC. That is, the statistics of the NPC population must be used to make inferences on a particular process.

Since EM quantitative data is often used to calculate NPC channel ion permeability and conductance, I shall take the opportunity to briefly discuss the importance of error analysis for both instrument and measurement. Firstly, statistics cannot improve the resolution of the instrument. If the instrument resolution is 1 nm, and the statistics gives a value of, say, 10 ± 0.1 nm (mean±SD), this does not mean that the determination has such accuracy. Secondly, if the value is not measured directly but, instead, derived from a mathematical equation, the error of the determination increases. This is because errors never subtract each other. That is, each and every error adds up to each other.

The error of a measurement is generally derived through a principle that is paramount to experimental physics: the L'Hospital rule.[8] One case is that of the calculation of the volume of a nuclear pore. The volume may later be used for the calculation of NPC channel conductance. The NPC is approximated to a cylinder as shown in Figure 2. Say that we calculate the volume, V, of the NPC channel from the values of the NPC length, l, and radius, r. The volume is given by:

$$V = \pi r^2 \times l \qquad\qquad (1)$$

The error in V, ΔV, is calculated by taking the derivative of the logarithm of the formula (so as to have the relative error, $\Delta V/V$):

$$\ln (V) = \ln (\pi r^2) + \ln (l) \qquad\qquad (2)$$

thus,

$$(\Delta V/V) = 2\pi (\Delta r/r) + (\Delta l/l) \qquad\qquad (3)$$

Thus, for example, if $r=5$ nm and $l=5$ nm, and Δr and Δl of 1 nm (instrument resolution), the resulting relative error, is:

$$\Delta V/V = 2\pi (1/5) + (1/5)$$

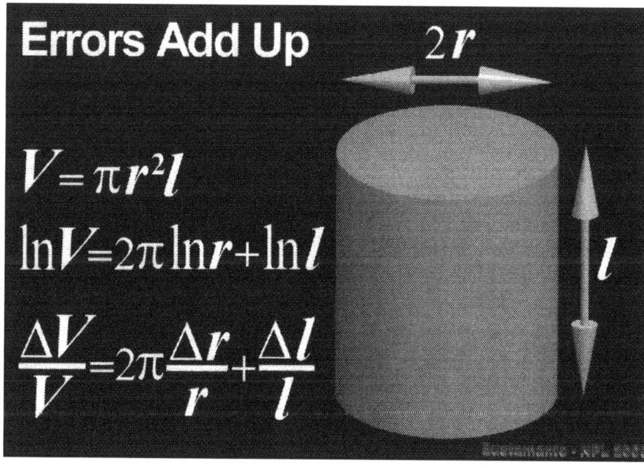

Figure 2. Error estimation for quantities that are derived from formulas. The example is given for the calculation of the volume, *V*, of a cylinder (e.g., NPC) with the measured values for its radius and length, *r* and *l*, respectively. Errors never cancel each other. Instead, they add up.

or

$$\Delta V / V = 1.45$$

that is,

$$(\Delta V / V)\,\% = 145\%$$

Thus, when trying to evaluate measurements such as ionic conductance in terms of formulas that use NPC channel volume, one must take into account the errors introduced by indirect measurement (in addition to the intrinsic resolution of the instrument). From (3) it is obvious that, for the same instrumentation error (e.g., Δr), smaller values of the variables (e.g., *r*) give greater relative and percentile errors.

Atomic Force Microscopy (AFM)

AFM, one of the several types of scanning probe microscopy (SPM), offers the possibility of measuring NE and NPC topology in unfixed preparations.[3,9] Although the very name and principles of operation of AFM imply atomic resolution, when applied to biological specimens, AFM resolution suffers considerably. The principle of operation of AFM is that of the classical phonograph. The tip of the AFM probe, placed at the end of a cantilever, scans the surface of the sample while keeping the distance between the tip and the sample at a safe atomic distance controlled by a piezoelectric feedback mechanism. As the AFM tip is not infinitesimal, its dimensions restrict the depth at which the probe can advance into the sample, thus preventing the identification of structures that traverse the thickness of the NE. The more so in living samples exposed to physiological solutions. Sample preparation and recording conditions are as important to AFM as they are to EM. For example, the plug imaged by two research groups,[10,11] was not present in the images made by another group when the sample was provided with nuclear transport substrates.[12] This led to the interpretation that the NPC plug or transporter is actually a macromolecular cargo caught in transit inside the NPC channel.[12] Likewise, the recent AFM observation of putative peripheral channels,[10] reported a decade ago in a study using image reconstruction with cryo-EM data[13], has yet to be confirmed. I shall refer to this issue in the section devoted to electrophysiology. As this Chapter deals with nuclear transport methods, it can not be emphasized enough that one must recognize the limitation of the technique being used. For AFM, one limitation is the penetrating distance of the recording tip. This limitation is compounded by the tempting computer-based smoothing of surface roughness which renders the image nice and shining (see, for example, ref. 10). Another limitation is that, since the surface is neither flat nor horizontal, the feedback gain (which controls the speed with which the AFM tip readjusts its distance to the sample) must be set at a level that will allow manual correction before it hits the sample. This restricts the speed with which the image can be acquired: a problem that may be relevant to transport of macromolecules.[14] Despite these limitations, AFM has great potentials yet to be applied to the characterization of NE inner or outer surfaces. For example, the tip can be painted with monoclonal antibodies to specific NPC molecules.

Light Microscopy: Fluorescence and Luminescence

Light microscopy (transmission or otherwise) is one of the most used techniques in cell biology. Standard, transmission light microscopy can be used in living cells to determine the passage of colored particles (e.g., vital stains) into the nucleus.[15] However, this approach is not a popular one because of its limited resolution. When used with fluorescence accessories (i.e., fluorescence microscopy equipped with stabilized light source, dichroic filters and sensors), light microscopy becomes a powerful tool, one that has been well exploited in the field of nuclear transport. This is due to the increased sensitivity and resolution/distinction of individual reporter particles. The introduction of intensified CCDs and photomultipliers for laser-scanning confocal microscopy (i.e., fluorescence microscopy—confocal and otherwise) render this technique a highly sensitive one for determining the distributions of important particles such as second messengers and transcription factors. When proper measures are taken to avoid sample damage by the excitation beam, epifluorescence (excitation and emission beams passing through the same objective) can provide long-lasting recording of living cellular events. The principle of operation is simple. A reporter molecule, the fluorochrome that binds to a particular target with a known binding constant (e.g., Ca^{2+}—see ref. 16), is introduced into the system at a particular concentration. The fluorochrome emits light at a wavelength (say green) different from that with which is excited (say blue). Since the energy of emission can not be greater than that of excitation, the wavelength of the emitted light is always longer than that of the excitation light. Fluorochromes are selected for their well-defined separation between the emission and excitation peaks. This separation is used in the design of the separating optics—usually packed into a cube known as a dichroic (i.e., two-color) filter. The separating optics may be so designed as to allow more than one fluorochrome to be observed simultaneously (see, for example, ref. 16). Clearly, more than one fluorochrome can be introduced into the system as long as their properties do not interfere with each other. There are many commercial probes available for nuclear transport research (e.g., ref. 16). In one example, inert dextran and dendrimer molecules (fluorescently labeled) are used to monitor the diameter of the NPC channel.[17] In another, highly fluorescent macromolecules (e.g., B-phycoerythrin) conjugated to nuclear localization signal (NLS) are used in the determination of the macromolecular transport capacity of NPCs.[18] Several fluorochromes may be combined in tandem or used in what is known as fluorescence resonance energy transfer or FRET (see, for example, ref. 16).

The recent availability of cloning vectors for the production of the stable green fluorescence protein (GFP)[19,20] has greatly facilitated the advancement in the nuclear transport field. DNA vectors for translocating particles such as androgen receptor[21] or 60S ribosomal subunit proteins,[23] can be constructed to produce molecules fused with GFP or its enhanced fluorescent mutant: EGFP. There are now commercial vectors[19] for the study of EGFP-fused factors such as p53, NFκB, CREB, PKA, PKC, etc. And the fluorescence is not limited to green, for EGFP mutants are now readily available[19] in yellow, blue, red, etc. Thus, cell transfection with these vectors will result in cells that produce factors intrinsic to the cell, yet of different fluorescent colors.[19] A recent breakthrough, known as fluorescent timer, can be used to monitor both activation and down-regulation of target promoters at the whole-organism scale.[22] The fluorochrome is a mutant of the red fluorescence protein drFP583. The mutant (E5) changes its fluorescence over time: from green to red. Since the rate of color conversion of E5 is independent of its concentration it can be used to trace time-dependent expression.[22]

If confocal microscopy is needed, for example to eliminate doubts regarding the potential contribution of the NE cisterna, but no laser-scanning confocal microscope is available, then there is the option of using computer-based deconvolution. The software sorts out the data out of focus according to a pattern contained in what is called the point spread function. The software is based on that introduced to eliminate the blur in astronomical images. Note that, since the depth of focus is reduced with large magnifications of the objective, the larger the magnification of the objective the less the blur the image will have.

One problem of fluorescence microscopy, and related techniques, is that some of the probes used have been shown to be compartment-dependent. For example, Fluo-3, commonly used as a Ca^{2+} indicator, was shown to be pH-sensitive.[24] This, along with the prevalent idea that NPCs are permanently open to monoatomic ions such as Ca^{2+}, has led to the neglect by some investigators of the observations of nucleocytoplasmic gradients. To overcome some of the problems posed by compartmentalized fluorescence probes, luminiscence microscopy has been used. In this approach, the light resulting from a chemical reaction is used. Luminescence microscopy supports the view that nucleocytoplasmic gradients of Ca^{2+} are real.[25]

One characteristic of many fluorochromes is that of photobleaching upon repeated or relatively prolonged exposure. Clearly, the higher the sensitivity the detector has (whether a photomultiplier or intensified CCD), the lower excitation level needed and the smaller the photobleaching. Photobleaching reduces the observation time but the phenomenon can be reduced with certain reagents (e.g., ref. 16). The phenomenon, however, can be exploited to study the kinetics of recovery of the fluorescence signal. Another problem of fluorescence microscopy is that it can not describe the microscopic function of the single NPC channel. Recent efforts to overcome this limitation through mathematical equations have been published.[26,27] Fluorescence microscopy is also a useful tool for the assessment of nuclear transport kinetics, even at the single molecule level.[28,29] Such measurements have attained levels of accuracy of near 30 nm and 18 ms.[28] The greater accuracy revealed time-dependent diffusion coefficients, long-distance displacement and immobile particles.[28] Therefore, the increased resolution may uncover processes that were not predicted on the basis of classical fluorescence microscopy. Due to space limitations, I will not cover FISH (fluorescence in situ hybridization) and FACS (fluorescence activated cell sorting), techniques which are important but that have not been major contributors to the nuclear transport field. Note that a definite advantage of fluorescence and luminescence microscopy is that it can be coupled to recording techniques such as AFM and patch-clamp.

Cell and Molecular Biology

The majority of contributors to this book are cell/molecular biologists. And this is so because it has been the field of cell/molecular biology that has given the greatest contribution to our current understanding of nuclear transport.[30] The first hurdle that cell biologists overcame was the problem of simplifying the study of nuclear transport in living cells. The hurdle was posed by the difficulty in controlling the intracellular environment without inserting into the cells a potentially damaging pipette. Most studies, therefore, were conducted with microinjection or with isolated nuclei bathed in cell lysate. Both of these approaches were either too time consuming, cumbersome or could pose questions about artifactual interference of the procedures with the nuclear transport mechanisms. The solution was simple enough. The cell surface membrane had to be permeabilized. Plasmalemmal permeabilization has greatly facilitated our understanding of nuclear transport. For this purpose, digitonin treatment is the preferred approach.[31] When supplied with proper transport substrates, the treatment maintains the nucleus competent for signal-dependent translocation through NPCs. Since 1998 my laboratory has been exploring a new preparation of naturally occurring cell-free systems. These provide isolated nuclei forming a syncytium. That is, the nuclei float freely in their natural extranuclear fluid. Among these systems we encounter insect embryos and throphoblasts. The remarkable advantage of these systems is that they require no substrate to express cDNA plasmids. The results are very promising and have revealed new properties, some of which I will describe in the electrophysiology section.

Cell/molecular biology methods encompass a broad range of techniques. During a particular project, one may have to resort to a large variety of techniques. For example, the researchers may have to produce antibodies against a particular NPC protein or transfect cells with a particular cDNA to produce sufficient quantities of a protein which are later to be purified and sequence- and structure-analyzed. The DNAs may have to be amplified with the polymerase chain reaction (PCR) technique or the mRNAs produced from DNA templates by reverse transcriptase PCR (RT-PCR). Alternatively, mRNAs and/or proteins may be produced with in vitro transcription and/or transcription-translation systems. The mRNA templates may be inserted into naturally non-expressing systems like *Xenopus laevis* oocytes. Other general techniques include high-pressure liquid chromatography (HPLC) for separation with different columns, electrophoresis for immunoblotting, centrifugation for subcellular and particle fractionation (e.g., NPCs), etc. I shall not describe any of these due to space limitations and their ready availability in the literature (e.g., refs. 32, 33) as well as elsewhere in this book.

Mass Spectrometry

Mass spectrometry has been successfully used in the nuclear pore field. Mass spectrometers are named according to their type of ionization source and mass analyzer.[34] The sample must be charged/ionized and dried. For peptides and proteins, this is usually done with matrix-assisted laser desorption ionization (MALDI) and electrospray ionization (ESI). In MALDI the sample is embedded in a matrix, then dried and then volatilized under UV in a vacuum chamber. With each laser shot, only a small portion of the sample is ablated. The analyzer commonly used with MALDI is the time-of-flight (TOF). Based on simple electrodynamics principles, the TOF analyzer measures the elapsed time from acceleration of the charged (ionized) molecules through a field-free drift region. When ESI is used for ionization, the sample is sprayed under vacuum and voltage, using a principle similar to that used for the preparation of SEM samples. Usually, the spray is taken from a reversed-phase HPLC (RP-HPLC) column or a nanospray device similar to a microinjection needle. During this process, the droplets containing the sample material are dried and ionized. The ionized material is directed into the mass analyzer consisting of one of the following: a triple-quadrupole, an ion trap, a Fourier-transform ion cyclotron resonance (FT-ICR), or a hybrid quadrupole TOF (Qq-TOF).[34]

Mass spectrometry using MALDI and TOF has been used to define the spatial organization of multi-protein complexes such as the NPC-associated proteins Nup50[32] and Nup85p.[35] For this approach, the components are purified via molecular interactions using an affinity tagged member and the purified complex is then partially cross-linked. The products are then separated by gel electrophoresis and their constituent components identified by mass spectrometry.[35] Mass spectrometry may identify protein constituents even at levels of a few hundred femtomoles.[35] Based on these results, a model of the spatial organization of Nup85pp was obtained.[35] The model was supported by biological experiments.[35] Mass spectrometry was also used, in combination with HPLC, to identify the molecular architecture of the NPC.[36] The mammoth task was assisted with a well planned flow-chart diagram.[36] The result was a map of many NPC components and a model of macromolecular transport that suggests Brownian motion along the NPC channel.[36]

Pharmacology

Despite reports on the participation of NE and NPC anomalies in several diseases (e.g., refs. 37-45), both structures have received less attention as pharmacological targets. This is probably due to the limited information on the components of these structures. However, the few papers that have appeared are encouraging. Let us take the example of the work on the vasoconstrictor angiotensin. The hormone angiotensin II (AII) is produced by the action of ACE (angiotensin converting enzyme) on angiotensin I. As a result, either the AII receptors or ACE are pharmacological targets in the clinical treatment of hypertension. In neurons, AII interaction with its type I receptor (AT_1) results in the chronic stimulation of neuromodulation that involves the expression of norepinephrine transporter (NET) and tyrosine hydroxylase (TH). Recent investigations[46] on this G protein-coupled receptor have shown that AII induces nuclear sequestration of the NLS-containing AT_1 and that the sequestration may have implications on AII-induced expression of NET and TH genes. AII caused a time- and dose-dependent stimulation of P62 phosphorylation (a glycoprotein of the NPC).[46] The 6-fold stimulation caused by 15 min of 100 nM AII was completely blocked by the AT_1 clinical blocker losartan but not by PD123,319, a blocker of the type 2 receptor (AT_2).[46] These observations have been confirmed[47] and seem to explain my own observations of AII upregulation of NPC channel gating (Bustamante, unpublished). We have briefly reviewed the short list of plasmalemma-like receptors.[48,49]

Ever since the proposal by Hodgkin and Huxley[50] of the existence of ion channels as independent functional structures, they have been named according to their ion selectivity. As a rule, ion channels are named according to the ions they conduct with most ease (e.g., Na^+-channel is one that has a distinct preference for the conduction of Na^+ over any other ion species). Two advances have greatly facilitated the study of ion channels in all cell types: patch-clamp and reliable single cell preparations. However, before the availability of both, drugs were used in the identification of ion channels participating in a particular mechanism. With the availability of viable single cells and patch-clamp, it is now simple to identify channel selectivity, and thus the ionic nature of the channel, by a procedure called ion substitution or bi-ionic potentials (e.g., 51). In this procedure, one ion species is substituted by another and the permeability ratios are calculated according to a simple Nernst electrochemical equation. When ion selectivity is not shown by the channel, the ion channel is said to be non-selective whereas when only cations pass through, one says that the channel is cation-selective, and so on.

The importance of the concept of ion channel selectivity becomes apparent when studying a very important class of channels: the IP_3-receptor Ca^{2+}-channels. A few patch-clamp publications have demonstrated NE ion channel activity that is activated by IP_3 and blocked by heparin, from what was concluded that these were IP_3-receptor channels (e.g., refs. 52-54). These studies are major contributions to the field. However, there are a few points worth mentioning. First, IP_3-receptors have been only observed with biochemical methods in the inner, rather than the outer nuclear membrane of the NE.[55] The same observation has been made for the cyclic ADP-ribose dependent Ca^{2+} signaling pathway.[56] This contrasts with the fact that these patch-clamp recordings were claimed to correspond to IP_3-receptors on the cytosolic side of the NE because the patch-clamp pipette was attached to the cytosolic side of the NE.[52-54] Second, heparin is a wide spectrum agent that stabilizes the NPC plug (e.g., ref. 57) and modifies the behavior of many ion channels (e.g., ref. 58). Third, as I discuss in the section on electrophysiology, one must make sure that the NPCs do not contribute to patch-clamp recordings and, therefore, that the observed effect is not a result of NPC modulation by the Ca^{2+}-mobilizing agent such as IP_3.

Chloride channels were also reported in patch-clamp studies of isolated nuclei (e.g., refs. 59, 60). These channels were identified by replacement of Cl^- with glutamate, aspartate or gluconate[59,60] and by the blockade of the recorded current with the Cl^- channel blocker DIDS.[59] In contrast, cationic channels did not respond to known blockers (e.g., TTX) of cationic channels.[51,61]

Electrophysiology

I have left the electrophysiological methods for last for two reasons. First, it is the only field that so far promises to measure new features of NPC channel opening in terms of probabilities and other quantities familiar to the ion channel field. Second, it is the area in which I have carried out my research since 1972 when I finished my university studies in physics.

The 1960s saw the birth of the application of electrophysiological techniques to the study of nuclear transport (reviewed in ref. 62). During that decade, Loewenstein and collaborators measured the NE resistance, R_{NE}, with two microelectrodes inserted at both sides of the living NE.[62] By applying a voltage across the NE, V_{NE}, and measuring the resulting current, I_{NE}, it was possible to obtain NE from Ohm's law:

$$R_{NE} = V_{NE} / I_{NE} \tag{4}$$

The principle of operation of the microelectrode measuring system is similar to that of any current-recording instrument (including patch-clamp). One applies a voltage, V_{NE}, across the resistive element, R_{NE}, of unknown value and measures the current, I_{NE}, by using a resistor of known value (R_{known}). The measured current is calculated according to the following formula:

$$I_{NE} = V_{NE} / R_{known} \tag{5}$$

The value of the current is then used to determine the resistance of the resistive element according to (4).

Due to insurmountable technical questions, these measurements were not taken seriously and the electrophysiological approach was soon abandoned. It had to pass three decades in order for electrophysiology to reappear as a suitable methodology (reviewed in ref. 9). The decisive factor was the wide acceptance of patch-clamp and the availability of better methods for the study of nuclear transport. Patch-clamp has given us the first glimpses of NE ion channel activity (see Fig. 3). Various types of activities have been recorded and attributed to NPCs as well as to structures other than NPCs. The principle of operation of patch-clamp is similar to that described in equations (4) and (5). As Figure 4 shows, two electrodes are used to impose/clamp the voltage of the NE and the same are used to record the current passing across the NE. Thus, for example, if we applied a voltage (V) through a known resistor (R) of say 100 GΩ (10^2 x 10^9 Ω) and if this voltage causes a current (I) of 20 pA (20 x 10^{-12} A) to flow through the NE patch, the output of the voltage output of the apparatus would be, by Ohm's law, $V = I$ x $R = 2$ V. That is to say, we can always measure what the current flow is. However, to calculate the actual voltage applied to the patch, one has to know the voltage at the other side of the patch (i.e., the nucleus). Measuring the intranuclear potential may be very difficult due to its Donnan potential nature.[63] This was a significant problem for microelectrode studies because they depended heavily on this measurement. Fortunately, the very high permeability of NPCs to monoatomic ions seems to short-circuit the nuclear interior with the bath electrode used as reference (i.e., voltage set to zero).[51,61] Therefore, the applied voltage is roughly the same as the negative of the voltage across the NE patch. [51,61] Consequently, if in our example, we had applied a pipette voltage of -20 mV (i.e., NE voltage of +20 mV) and if the current was +20 pA, then the resistance is 1 GΩ (20 mV/20 pA). That is, the cannel conductance, γ, is 1 nS.

Another point I should like to discuss is the nomenclature inward and outward-going ion currents. If the voltage inside the pipette is positive, then positive ions move out from the pipette and negative ions get into the pipette. Since by definition, electrical current is flow of positive charges, it is said that these positive and negative ion movements are inward currents because both result in a current that goes into the nucleus. If the voltage inside of the pipette is made negative, then the opposite happens. Outward currents are those that go away from the nucleus and into the pipette. Thus, if the pipette electrode is made negative, positive charges go into the pipette and negative charges go out of it. By convention, inward currents are negative

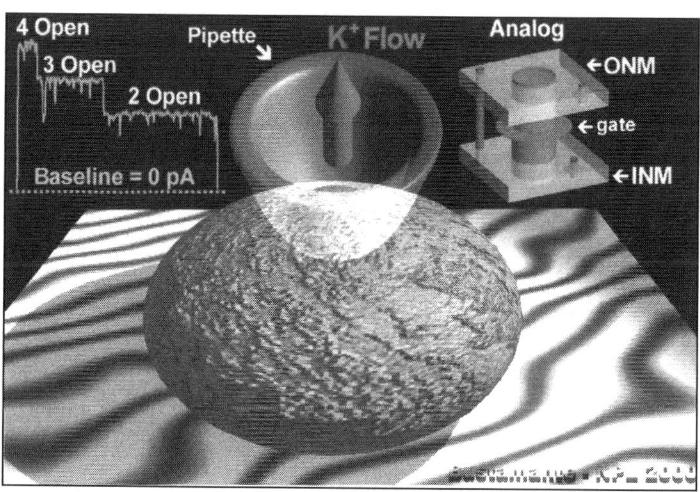

Figure 3. Patch-clamp recording. A nucleus is shown with a patch-clamp pipette attached to it. The arrow depicts a K^+ current going outwardly as a result of a negative voltage applied to the pipette interior. On the top-left a current record is shown with 4, 3 and 2 current levels taken to represent a population of 4 ion-conducting NPCs. On the top-right, a model of ion conduction is shown. Here, the NPC is represented as the cylinder with largest diameter with an on-off switching mechanism in the middle. A parallel conduit, similar to the putative NPC peripheral channels,[10,13,68] is shown without the switching mechanism. Also shown are switching ion channels at the inner and outer nuclear membranes, INM and ONM, respectively.

(e.g., Na^+-channel current[64]). This is done to remind the reader that the current is flowing from the pipette and into the cell. For this reason, it seems erroneous to say that inward movement of negative ions corresponds to inward current (e.g., Fig. 1 in ref. 60.). Perhaps a universal convention, like that accepted for Na^+-channels,[64] is missing in this field.

Several confounding problems do exist. Some of these I previously discussed.[65] The double-membrane nature of the NE poses a potential problem because most patch-clamp recordings are made at the ONM. The pipette is placed on the ONM to avoid destroying the natural NPC environment. It is hard to imagine NPCs working normally in patches excised from their natural environment. The fact that the ONM is continuous with the endoplasmic reticulum, ER also poses the potential problem of contamination (i.e., does the recording corresponds to the NE or ER?). Finally, one more confusing situation is the fact that patch-clamp requires a gigaohm seal (i.e., the gigaseal) between the pipette tip and the membrane whereas it appears that having several NPC channels may prevent a gigaseal from forming.

Let us take the gigaseal issue first. We note that the gigaseal resistance is in parallel with the NPCs resistance. One of the Kirchoff laws for electrical circuits tells us that, for a circuit of parallel resistors, the total resistance can never be greater than the minimal resistance. Thus, if we had 10 NPCs of 1 nano-Siemens (1 nS, which is the inverse of a resistance of gigaohm or $G\Omega$) simultaneously open, we would have a total NPC resistance of 0.1 $G\Omega$ (100 $M\Omega$). In this case, even if a gigaseal of 10 $G\Omega$ were formed, we would never be able to see it while all or any of the channels were open. How then can we say we have a gigaseal? Simply enough... We just wait that all NPCs close. In my experience, although the wait may be long (1 in 10^4 events), this is sufficient to observe the value of the gigaseal (see, for example, refs. 61-67).

The fact that the ONM is continuous with the ER may be used to argue that the ion channel recording may derive from the ER and not the NE. To rule out ER contamination, one simply uses nuclear-targetted molecules such as B-phycoerythrin conjugated to NLS or

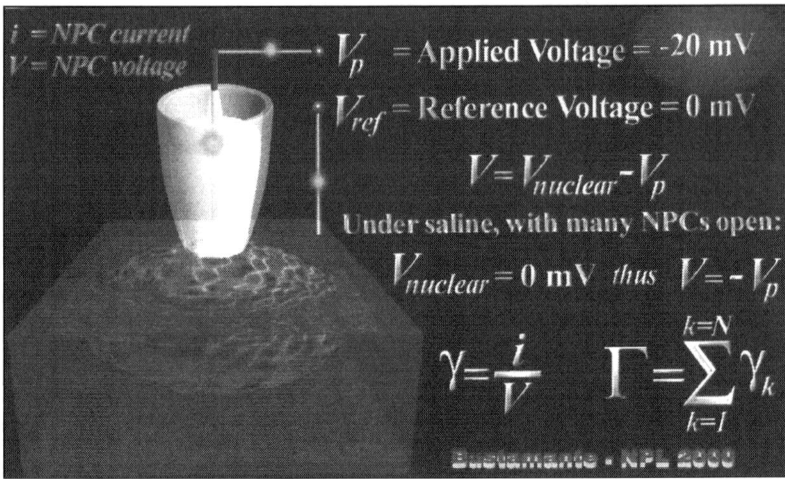

Figure 4. Patch-clamp setup. Schematics of nucleus-attached recording mode. The diffuse circles depict positive electrical charges going into the negative electrode. Under saline conditions, it is known that the NE is quite permeable to small monoatomic ions (e.g., K[+]).[51,61] Consequently, the voltage at the bath electrode goes directly into the nucleoplasmic side of the NE. Thus, the voltage difference across the NE, and thus the NPCs, is the negative of the voltage applied to the pipette electrode. The patch conductance, Γ, equals the sum of each and every single NPC conductance, γ. To measure this conductance, a potential must be applied to the pipette interior (e.g., -20 mV).

even nuclear molecules such as transcription factors (e.g., refs. 18, 66). The addition of these molecules should have no effect on ER channels. Instead, one sees that they do indeed depress the ion channel current.[18,66] We have suggested that this depression is caused by displacement of the electrolyte inside the NPC channel (e.g., refs. 18, 66). That the ion channel activity derives from NPC channel opening and closing is further supported by the ion channel current blockage caused by the addition of the NPC monoclonal antibody mAb414.[18] This antibody, and no other (out of a panel of dozen antibodies against NPC proteins as well as control antibodies), caused such blockade in my experiments.[18]

It may be argued that because the pipette tip is attached to the ONM, the recorded ion channel activity derives from ion channels at the ONM. This would be true only if all NPCs were closed all the time. Instead, all passive transport assays with fluorescence microscopy show that fluorochromes (e.g., Fluo-3, see ref. 24) readily diffuse into the nucleus. It is for this reason that it is hard to think of the NPC as a barrier to ion flow. Therefore, only when all NPCs are plugged by translocating macromolecules one could say that the recorded activity in nucleus-attached patch-clamp corresponds to ion channels other than NPCs. In over a decade o patch-clamping the cell nucleus, I have never observed this phenomenon. Instead, the recorded ion current remains at a zero value as if the electrical circuit was permanently open due to NPC channel plugging. That is, no ion channel opening is seen. The concept behind the ion channel model of the NPC is that of the molecular Coulter counter principle (see, e.g., refs. 18, 67). Figure 5 illustrates this macromolecule conducting ion channel paradigm.

I believe that ion channels must exist in both the ONM and the inner nuclear membranes, INM. Indeed, the experience of several laboratories, including my own, is that patches excised from the NE display channel activity. Figure 6A shows a hydrodynamic analog for all possible channels. Since NPC transport properties depend of the NE cisterna, it is likely that the NPCs in these excised patches are functionally impaired. But impaired does not mean that

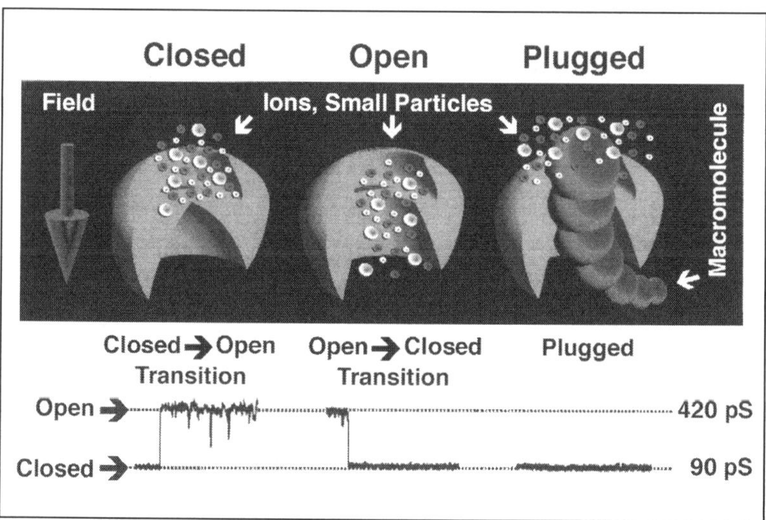

Figure 5. Macromolecule conducting ion channel paradigm for NPCs. When macromolecules are not traversing the inner channel of the NPC, the free space is filled with electrolyte and small particles. Once macromolecules traverse the NPC length, they exclude the electrolyte and the conductance is very small. Note that the value of the conductance is not necessarily zero. This is due in part to unavoidable experimental errors and to imperfect gigaseal formation. In the illustration, when all channels are closed, the NE patch has a conductance of 90 pS ($9 \times 10^{-11} \ \Omega^{-1}$). This value corresponds to a gigaseal of 11 GΩ.

all NPCs are closed because, as Figure 6B illustrates, there is the possibility that some NPCs were transporting macromolecules while others not. One can close all NPCs by adding nuclear-targetted macromolecules (such as large transcription factors) under conditions that support macromolecular transport and then stopping the transport, say with switching to a simple saline solution, so as to catch macromolecules inside the NPC channel (thus plugging it). To demonstrate that the NPCs are closed, it is sufficient to show that small molecules (e.g., fluorescently-tagged dextran) do not get into the nucleus.

Recently, a macroscopic approach has been applied to studying NPC ion permeability.[10,68] The technique deserves discussion because it has reached conclusions not supported by patch-clamp. Like several techniques that were developed simultaneously to patch-clamp (e.g., refs. 69, 64), the technique uses two electrodes to apply a known current and two electrodes to record the voltage drop produced by the presence of the isolated nucleus inside a capillary. The resistance of the NPC population is inferred from a model that makes several assumptions.

First, the nucleus resistance, R_{Nuc}, is made up of three components: the electrical resistances of the upper and lower NE surfaces, R_{NE1} and R_{NE2}, plus the electrical resistance of the nucleoplasm, $R_{Chromatin}$. It is then assumed that the experimenter can make both surfaces of equal area and, therefore, that $R_{NE1} = R_{NE2}$. From this, the electrical resistance of one single NE surface is given as:

$$R_{NE1} = R_{NE2} = (\, (R_{Nuc} - R_{Chromatin}) \, / \, 2 \,) \tag{6}$$

Second, as the authors pointed out (and as techniques contemporaneous to patch-clamp proved—e.g., ref. 64), eq. (6) does not consider the fact that there is a shunt current that bypass the nucleus and that must be considered.[68] They then estimated the shunt effects by measuring the shunt resistance, R_{Shunt}, in the oocyte rather than in its nucleus.[68] While their rationale of using a different preparation to estimate R_{Shunt} may seem justifiable from a

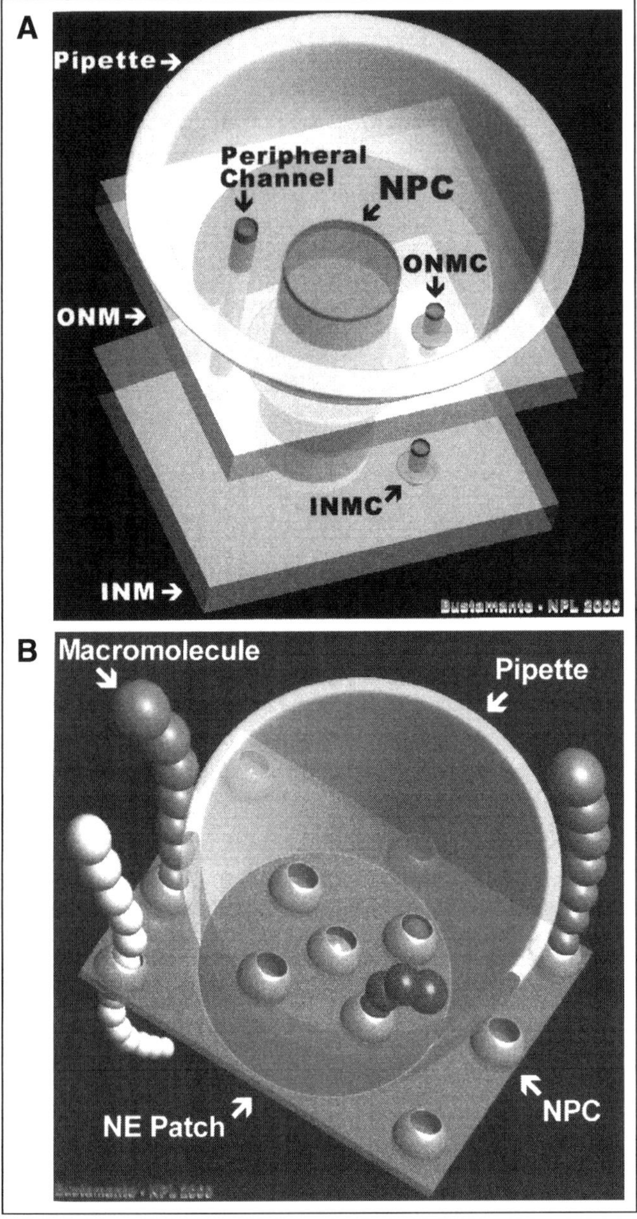

Figure 6. Channels and conduits of the NE. A) Hydrodynamic model of a nucleus-attached patch. The recording pipette isolates a patch of the NE from the rest of the NE. The only physical communication among NPCs is through the NE cisterna. On-off switching ion channels are shown with a gate/valve in the middle: INMC and ONMC are ion channels of the INM and ONM, respectively. The peripheral channel is shown as a non-switching conduit as was originally proposed.[13] B) Macromolecule transporting NPCs are transparent to patch-clamp because they have no electrical current associated with them: they are plugged. In contrast, unplugged NPCs readily transport monoatomic ions such as K^+, which carry the current detected with patch-clamp.

mathematical point of view, it suffers from one major physical problem: not all surfaces stick with the same force to glass. I think that a better justification for disregarding R_{Shunt} is to show that small fluorescent probes do not pass between the NE and the capillary wall.[17] One additional point should be made. Since the macroscopic currents produced by ion flow along NPCs do not have the distinguishing features of classical ion channel currents (e.g., voltage-dependent Na^+-channel[64]), it does not follow that any passive ion flow can be ascribed to ion channel activity. The issue of leak current has been a major consideration ever since the introduction of voltage-clamp (the predecessor of patch-clamp) by Nobel laureates Hodgkin and Huxley.[50] Indeed, patch-clamp gained acceptance over existing techniques thanks to the possibility of forming a tigh seal (i.e., gigaseal) between the pipette tip and the membrane. This very reason resulted in the Nobel Prize award to the patch-clamp creators: Neher and Sakmann.[70]

The same investigators used their macroscopic technique to conclude that the recorded macroscopic currents correspond (rather than to shunt currents) to ion flow along putative channels peripheral to the known NPC channel. These putative ion conduits (8 in general, corresponding to the general 8-fold geometry of the NPC) were introduced through mathematical image reconstruction of cryo-EM data[13] and have found no support with EM[71] and fluorescence microscopy.[26] As introduced, these channels are not ion channels but mere conduits.[13] In over a decade, I have never recorded with patch-clamp any signal that may be ascribed to these putative conduits (e.g., refs. 14, 17, 61-67). If these NPC peripheral conduits existed, then (since they are in parallel to the known NPC central channel) they should be seen as a constant leak current when all channels are closed. Instead, for over a decade all I have measured is gigaseals when all channels close. Moreover, if they were ion channels that open and close (rather than permanently open conduits), then one would see in all the patch-clamp studies reported so far (reviewed in ref. 9) the 8-fold statistics that one should expect. However, this 8-fold statistics has never been reported (see, for example, ref. 61). The only explanation would be that these particular conduits are very peculiar in that they would all open and all close simultaneously—a phenomenon never reported so far in the history of switching ion channels.

The authors also claimed errors of 2% for their macroscopic approach[68] and, more recently, 6%.[10] To judge the significance of their estimates, I refer the reader to the discussion I gave earlier about computing experimental errors. Clearly, their estimate of 2-6% needs reconsideration. Finally, these investigators have suggested that the putative peripheral channels are sensitive to Ca^{2+} and ATP.[10] Once again, their observations are explainable by the effects of these substances on the shunt pathway just as these effects can be seen with techniques contemporaries of the patch-clamp (see, for example, ref. 64). Figure 7 shows the peripheral channels paradox. Namely, patch-clamp experiments report neither leak current nor 8-fold statistics of channel openings.

I like to end this section with a comment on the new preparation that I have been working since 1998: the syncytial nuclei (J.O. Bustamante, in preparation). When I patch-clamp nondividing syncytial nuclei, I observe no ion current for as long as the nuclei are in their natural environment. My observation is in agreement with the abundant reports indicating that macromolecular traffic may be quite heavy in certain phases of the cell cycle. In contrast, when I replaced the natural liquid with physiological saline containing no macromolecular substrates, I observed the development of ion channel activity. As time progresses, the number of ion conducting channels increases. This additional information supports the contention that ion channel activity is possible only in NPCs that are not transporting macromolecules. My experiments, therefore, also confirm the preceding discussion on gigaseals, ER contamination and peripheral channels and offer the framework within which the applicability of patch-clamp data can be evaluated.

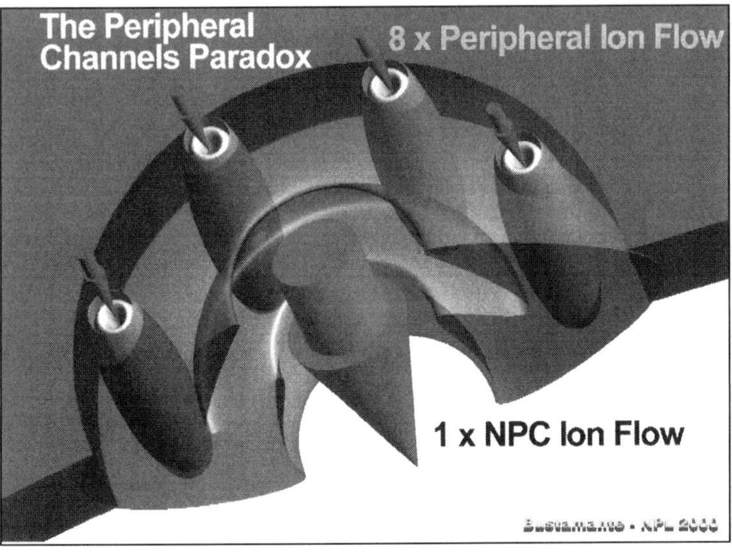

Figure 7. The NPC peripheral channels paradox. A top view showing the channels peripheral to the known NPC channel. The arrows show ion currents through these channels. The concept of peripheral channels was introduced in a cryo-EM image reconstruction paper.[13] As originally conceived,[13] they are non-gated fluid conduits instead of on-off gated ion channels. Therefore, they should contribute a constant current (i.e., one that does not switch on and off). Patch-clamp experiments do not record such gated current or the expected 8-fold statistics for channel openings. One can not interpret the imperfect gigaseal as corresponding to the putative peripheral channels because the range of gigaseal values is very wide under the same conditions. That only a single conduction channel exists within the NPC is supported by independent fluorescence and EM experiments.[26,71]

Other Methods

There are other important methods that are less frequently used for one or more reasons. Some of the methods require instruments that are only accessible to few researchers due to their high cost (e.g., synchrotron). Others methods have been made possible thanks to the recent affordability of the instruments (as was the case for computers). No person in this new millenium can escape hearing about genomics[72] and proteomics[73] for they will invade many corners of our lives. These new fields have begun to contribute to the field of nuclear transport.[74] Thus, functional proteomics (which can report altered protein posttranslational modification and expression) was used to identify 25 cellular targets of the MKK/ERK signaling cascade, some of which suggested novel roles in nuclear transport.[75] Another important technique related to structural determination includes x-ray crystallography assisted with synchrotron radiation (see, for example, ref. 76). The technique yields true atomic resolution at the expense of the sometimes difficult process of producing a crystal for the investigated structure.

Conclusions

Since the 1990s, our knowledge of nuclear pore structure and function has exploded.[30,74] This has been the result of the commercial availability of advanced instruments like patch-clamp setups, compact laser-scanning confocal microscopes, CCDs of high sensitivity, AFMs, genome and proteome microarray analyzers, etc. As our learning has progressed, the veil on the variables and parameters that determine nuclear transport has begun to fall. Our very

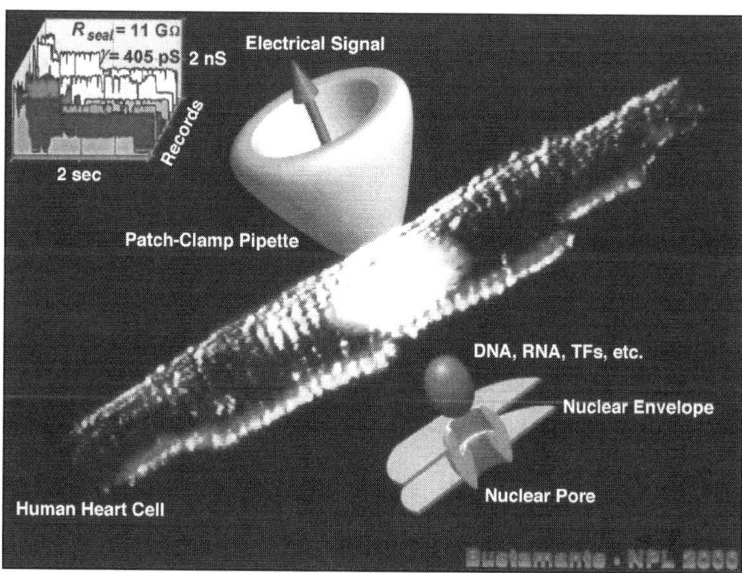

Figure 8. A futuristic view of some integrated methodologies. At the center, an isolated, intact human cardiac myocyte microphotograph is shown with a superimposed isolated nucleus expressing the enhanced green fluorescence protein (e.g., refs. 17, 64, 67). On the upper left corner is shown a set of electrical signals recorded from a nucleus-attached patch. On the lower right corner is shown a simplified diagram of the NPC with a gate. DNAs, RNAs, transcription factors (TFs) and other important macromolecules use the NPC to go into and out of the nucleus. Patch-clamp detects the movement of these molecules because they interrupt the flow of electrical charge carriers. These carriers consist mainly of monoatomic ions such as K^+ and Na^+.[51,61]

knowledge of the nuclear pore workings may be helpful in the development of techniques that will be applied back to genomics.[77,78] We stand witness of the diversity of mechanisms regulating the translocations of particles along the nuclear pores—some complex, others not. This diversity, in turn, enhances our awareness of the mechanisms controlling gene activity and expression. These mechanisms play vital roles in the development and maintenance of various diseases (e.g., refs. 40-45). Therefore, one should not be surprised to find in the future microarray diagnostic chips with markers for such mechanisms. We are now beginning to see the trees within the forest. I anticipate that our increased knowledge of nuclear transport mechanisms will result in a better understanding of the significance of genomics and proteomics, for a gene without expression is a null structure. I also expect that our contributions to the nuclear electrophysiology field will provide the scientific basis for the empirical use of electricity in the cloning of living beings.[79] In Figure 8 I give a futuristic view of the integration of several methodologies for cell diagnostic and therapy.

Acknowledgements

The author is currently a Senior Research Fellow of the National Council for Research Support (CNPq), Brazil. This work was made possible thanks to the Millenium Science Initiative of the Brazilian Ministry of Science and Technology (MCT) through the National Council for Research Support (CNPq): Millenium Institute of Nanosciences.

References

1. Stauber RH. Methods and assays to investigate nuclear export. Curr Top Microbiol Immunol 2001; 259:119-128.
2. PubMed: A service from the National Center for Biotechnology Information (NCBI), National Library of Medicine (NLM). National Institutes of Health (NIH), US government. NCBI Homepage
3. Fahrenkrog B, Stoffler D, Aebi U. Nuclear pore complex architecture and functional dynamics. Curr Top Microbiol Immunol 2001; 259:95-117.
4. Stoffler D, Fahrenkrog B, Aebi U. The nuclear pore complex: from molecular architecture to functional dynamics. Curr Opin Cell Biol 1999; 11(3):391-401.
5. Allen TD, Cronshaw JM, Bagley S et al. The nuclear pore complex: mediator of translocation between nucleus and cytoplasm. J Cell Sci 2000; 113(Pt 10):1651-1659.
6. Kiseleva E, Goldberg MW, Cronshaw J et al. The nuclear pore complex: structure, function, and dynamics. Crit Ver Eukaryot Gene Expr 2000; 10(1):101-112.
7. Hitachi's line of electron microscopes. Hitachi's EM Line
8. Fruhwirth R, Regler M, Bock RK et al, eds. Data Analysis Techniques for High-Energy Physics. 2nd ed. Cambridge: Cambridge University Press, 2000:408.
9. Mazzanti M, Bustamante JO, Oberleithner H. Electrical dimension of the nuclear envelope. Physiol Ver 2001; 81(1):1-19.
10. Shahin V, Danker T, Enss K et al. Evidence for Ca^{2+}- and ATP-sensitive peripheral channels in nuclear pore complexes. FASEB J 2001(11); 15:1895-1901.
11. Wang H, Clapham DE. Conformational changes of the in situ nuclear pore complex. Biophys J 1999; 77(1):241-247.
12. Stoffler D, Goldie KN, Feja B et al. Calcium-mediated structural changes of native nuclear pore complexes monitored by time-lapse atomic force microscopy. J Mol Biol 1999; 287(4):741-752.
13. Hinshaw JE, Carragher BO, Milligan RA. Architecture and design of the nuclear pore complex. Cell 1992; 69(7):1133-1141.
14. Bustamante JO, Liepins A, Prendergast RA et al. Patch clamp and atomic force microscopy demonstrate TATA-binding protein (TBP) interactions with the nuclear pore complex. J Membr Biol 1995; 146:263-272.
15. Monnè L. Permeability of the nuclear membrane to vital stains. Proc Soc Exptl Biol Med 1935; 32:1197-1199.
16. Molecular Probes' line of fluorescent probes. Molecular Probes Homepage
17. Bustamante JO, Michelette ER, Geibel JP et al. Dendrimer-assisted patch-clamp sizing of nuclear pores. Pflügers Arch 2000; 439(6):829-837.
18. Bustamante JO, Hanover JA, Liepins A. The ion channel behavior of the nuclear pore complex. J Membr Biol 1995; 146:239-251.
19. Clontech's lines of products related to GFP. Clontech's Homepage
20. Ayoob JC, Sanger JM, Sanger JW. Visualization of the expression of green fluorescent protein (GFP)-linked proteins. Methods Mol Biol 2000; 137:153-157.
21. Tomura A, Goto K, Morinaga H et al. The subnuclear three-dimensional image analysis of androgen receptor fused to green fluorescence protein. J Biol Chem 2001; 276:28395-28401.
22. Terskikh A, Fradkov A, Ermakova G et al. "Fluorescent timer": Protein that changes color with time. Science 2000; 290(5496):1585-1588.
23. Stage-Zimmermann T, Schmidt U, Silver PA. Factors affecting nuclear export of the 60S ribosomal subunit in vivo. Mol Biol Cell 2000; 11:3777-3789.
24. Perez-Terzic C, Stehno-Bittel L, Clapham DE. Nucleoplasmic and cytoplasmic differences in the fluorescence properties of the calcium indicator Fluo-3. Cell Calcium 1997; 21:275-282.
25. Badminton MN, Kendall JM, Rembold CM et al. Current evidence suggests independent regulation of nuclear calcium. Cell Calcium 1998; 23:79-86.
26. Keminer O, Peters R. Permeability of single nuclear pores. Biophys J 1999; 77:217-228.
27. Keminer O, Siebrasse JP, Zerf K et al. Optical recording of signal-mediated protein transport through single nuclear pore complexes. Proc Natl Acad Sci USA 1999; 96:11842-11847.

28. Kues T, Peters R, Kubitscheck U. Visualization and tracking of single protein molecules in the cell nucleus. Biophys J 2001; 80:2954-2967.

29. Radtke T, Schmalz D, Coutavas E et al. Kinetics of protein import into isolated *Xenopus* oocyte nuclei. Proc Natl Acad Sci USA 2001; 98:2407-2412.

30. Cullen BR. Journey to the center of the cell. Cell 2001; 105:697-700.

31. Liu J, Xiao N, DeFranco DB. Use of digitonin-permeabilized cells in studies of steroid receptor subnuclear trafficking. Methods 1999; 19:403-409.

32. Guan T, Kehlenbach RH, Schirmer EC et al. Nup50, a nucleoplasmically oriented nucleoporin with a role in nuclear protein export. Mol Cell Biol 2000; 20:5619-5630.

33. Petersen JM, Her LS, Dahlberg JE. Multiple vesiculoviral matrix proteins inhibit both nuclear export and import. Proc Natl Acad Sci USA 2001; 98:8590-8595.

34. Patterson SD. Mass spectrometry and proteomics. Physiol Genomics 2000; 2(2):59-65.

35. Rappsilber J, Siniossoglou S, Hurt EC et al. A generic strategy to analyze the spatial organization of multi-protein complexes by cross-linking and mass spectrometry. Anal Chem 2000; 72:267-275.

36. Rout MP, Aitchison JD, Suprapto A et al. The yeast nuclear pore complex: Composition, architecture, and transport mechanism. J Cell Biol 2000; 148(4):635-651.

37. Funakoshi M, Tsuchiya Y, Arahata K. Emerin and cardiomyopathy in Emery-Dreifuss muscular dystrophy. Neuromuscul Disord 1999; 9(2):108-114.

38. Haraguchi T, Koujin T, Hayakawa T et al. Live fluorescence imaging reveals early recruitment of emerin, LBR, RanBP2, and Nup153 to reforming functional nuclear envelopes. J Cell Sci 2000; 113 (Pt 5):779-794.

39. Ahuja HG, Popplewell L, Tcheurekdjian L et al. NUP98 gene rearrangements and the clonal evolution of chronic myelogenous leukemia. Genes Chromosomes Cancer 2001; 30(4):410-415.

40. Burke B, Mounkes LC, Stewart CL. The nuclear envelope in muscular dystrophy and cardiovascular diseases. Traffic 2001; 2(10):675-683.

41. Hegele RA, Yuen J, Cao H. Single-nucleotide polymorphisms of the nuclear lamina proteome. J Hum Genet 2001; 46:351-354.

42. Invernizzi P, Podda M, Battezzati PM et al. Autoantibodies against nuclear pore complexes are associated with more active and severe liver disease in primary biliary cirrhosis. J Hepatol 2001; 34(3):366-372.

43. Jaju RJ, Fidler C, Haas OA et al. A novel gene, NSD1, is fused to NUP98 in the t(5;11)(q35;p15.5) in de novo childhood acute myeloid leukemia. Blood 2001; 98(4):1264-1267.

44. Nesher G, Margalit R, Ashkenazi YJ. Anti-nuclear envelope antibodies: Clinical associations. Semin Arthritis Rheum 2001; 30(5):313-320.

45. Perez-Terzic C, Gacy AM, Bortolon R et al. Directed inhibition of nuclear import in cellular hypertrophy. J Biol Chem 2001; 276(23):20566-20571.

46. Lu D, Yang H, Shaw G, Raizada MK. Angiotensin II-induced nuclear targeting of the angiotensin type 1 (AT1) receptor in brain neurons. Endocrinology 1998; 139(1):365-375.

47. Bhattacharya M, Peri KG, Almazan G et al. Nuclear localization of prostaglandin E2 receptors. Proc Natl Acad Sci USA 1998; 95(26):15792-157927.

48. Bustamante JO. Nuclear electrophysiology. J Membr Biol 1994; 138(2):105-112.

49. Bustamante JO, Liepins A, Hanover JA. Nuclear pore complex ion channels. Mol Membr Biol 1994; 11(3):141-150.

50. The Nobel Prize in Physiology or Medicine 1963: Nobel Lectures of Hodgkin and Huxley. Nobel Prize for 1963

51. Bustamante JO. Restricted ion flow at the nuclear envelope of cardiac myocytes. Biophys J 1993; 64(6):1735-1749.

52. Stehno-Bittel L, Luckhoff A, Clapham DE. Calcium release from the nucleus by InsP$_3$ receptor channels. Neuron 1995; 14(1):163-167.

53. Mak DO, Foskett JK. Single-channel kinetics, inactivation, and spatial distribution of inositol trisphosphate IP$_3$ receptors in Xenopus oocyte nucleus. J Gen Physiol 1997; 109(5):571-587.

54. Mak DO, McBride S, Foskett JK. ATP regulation of type 1 inositol 1,4,5-trisphosphate receptor channel gating by allosteric tuning of Ca^{2+} activation. J Biol Chem 1999; 274(32):22231-22237.

55. Humbert JP, Matter N, Artault JC et al. Inositol 1,4,5-trisphosphate receptor is located to the inner nuclear membrane vindicating regulation of nuclear calcium signaling by inositol 1,4,5-trisphosphate. Discrete distribution of inositol phosphate receptors to inner and outer nuclear membranes. J Biol Chem 1996; 271(1):478-485.

56. Khoo KM, Han MK, Park JB et al. Localization of the cyclic ADP-ribose-dependent calcium signaling pathway in hepatocyte nucleus. J Biol Chem 2000; 275(32):24807-24817.

57. Strambio-de-Castillia C, Blobel G, Rout MP. Isolation and characterization of nuclear envelopes from the yeast Saccharomyces. J Cell Biol 1995; 131(1):19-31.

58. Lacinova L, Cleemann L, Morad M. Ca^{2+} channel modulating effects of heparin in mammalian cardiac myocytes. J Physiol 1993; 465:181-201.

59. Tabares L, Mazzanti M, Clapham DE. Chloride channels in the nuclear membrane. J Membr Biol 1991; 123(1):49-54.

60. Franco-Obregon A, Wang HW, Clapham DE. Distinct ion channel classes are expressed on the outer nuclear envelope of T- and B-lymphocyte cell lines. Biophys J 2000; 79(1):202-214.

61. Bustamante JO. Nuclear ion channels in cardiac myocytes. Pflügers Arch 1992; 421(5):473-485.

62. Loewenstein WR, Kanno Y, Ito S. Permeability of nuclear membranes. Ann NY Acad Sci 1966; 137(2):708-716.

63. Overbeek JThG. The Donnan equilibrium. Progr Biophys Biophys Chem 1956; 6:57-84.

64. Bustamante JO, McDonald TF. Sodium currents in segments of human heart cells. Science 1983; 220(4594):320-321.

65. Bustamante JO, Varanda WA. Patch-clamp detection of macromolecular translocation along nuclear pores. Braz J Med Biol Res 1998; 31(3):333-354.

66. Bustamante JO, Oberleithner H, Hanover JA et al. Patch clamp detection of transcription factor translocation along the nuclear pore complex channel. J Membr Biol 1995; 146(3):253-261.

67. Bustamante JO, Michelette ER, Geibel JP et al. Calcium, ATP and nuclear pore channel gating. Pflügers Arch 2000; 439(4):433-444.

68. Danker T, Schillers H, Storck J et al. Nuclear hourglass technique: An approach that detects electrically open nuclear pores in Xenopus laevis oocyte. Proc Natl Acad Sci USA 1999; 96(23):13530-13535.

69. Kostyuk PG. Intracellular perfusion. Annu Rev Neurosci 1982; 5:107-120.

70. The Nobel Prize in Physiology or Medicine 1991: Nobel Lectures of Neher and Sakmann. Nobel Prize for 1991

71. Feldherr CM, Akin D. The location of the transport gate in the nuclear pore complex. J Cell Sci 1997; 110(Pt 24):3065-3070.

72. Celera's animated primer on genomics. Celera's Educational Link

73. Allen NP, Huang L, Burlingame A et al. Proteomic analysis of nucleoporin interacting proteins. J Biol Chem 2001; 276(31):29268-29274.

74. Adam SA. The nuclear pore complex. Genome Biol 2001; 2:REVIEWS0007.

75. Lewis TS, Hunt JB, Aveline LD et al. Identification of novel MAP kinase pathway signaling targets by functional proteomics and mass spectrometry. Mol Cell 2000; 6:1343-1354.

76. Bayliss R, Kent HM, Corbett AH et al. Crystallization and initial X-ray diffraction characterization of complexes of FxFG nucleoporin repeats with nuclear transport factors. J Struct Biol 2000; 131(3):240-247.

77. Kasianowicz JJ, Brandin E, Branton D et al. Characterization of individual polynucleotide molecules using a membrane channel. Proc Natl Acad Sci USA 1996; 93(24):13770-13773.

78. Deamer DW, Akeson M. Nanopores and nucleic acids: prospects for ultrarapid sequencing. Trends Biotechnol 2000; 18(4):147-151.

79. Wilmut I, Schnieke AE, McWhir J et al. Viable offspring derived from fetal and adult mammalian cells. Nature 1997; 385(6619):810-813.

Index

A

A1 protein 55, 119, 124, 149, 190
Adenovirus 13, 177, 189, 197, 198
Agrobacterium 64, 66, 70, 73, 77, 83-94, 103, 189, 196, 199
A-kinase anchoring protein (AKAP) 42
Alkylating agent N-ethylmaleimide (NEM) 8
AML transcription factors 39, 166
Arabidopsis 17, 30, 63, 64, 66-69, 71-73, 75, 89-91, 93, 103, 105-111
ARM repeats 149-151
At-IMP 66-68, 70, 74, 103
AtKAP 66, 67, 69, 70, 84, 90-93
Atomic force microscopy (AFM) 3, 4, 12, 132, 206, 209-211, 220

B

Barr body 131
Basic leucine-zipper (bZIP) 71, 72, 91, 105, 108-112, 144
Basic type NLS 51-54, 177, 182, 183
Bipartite basic amino acid 40, 51, 52, 63, 64, 66, 67, 71, 100, 110, 138, 148, 178
BL1 73, 75
Blue/UV-A receptor 105 *see also* Phototropin
BR1 73-75

C

C. elegans 30, 31, 149, 150, 166
CAK2M 84, 90, 91, 93
Calreticulin 123-126
cAMP-dependent protein kinase (PKA) 38, 42, 43, 53, 54, 102, 156, 177, 210
CBF transcription factors 39
Cell-free extract 31
Cell nucleus 64, 87-94, 137, 201, 216
cGMP 108
Chromatin 1, 2, 8-11, 17, 31, 35, 45, 84, 118, 131, 132
CMV 191, 196

Common plant regulatory factor (CPRF) 72, 73, 109

Constitutive photomorphogenesis (COP1) 71, 72, 105, 110-112
Constitutive transport element (CTE) 165, 180
CRM1 13, 53-55, 66, 75, 102, 121, 125, 127-130, 139, 162-164, 179-181
Cryo-electron microscopy 5
Cryogenic EM 208, 209, 219, 220
Cryptochrome (CRY) 44, 72, 109, 110, 111
  Blue light 44, 72, 73, 109-111
  CRY1 109-111
  CRY2 44, 109-111
Cyclophilin 84, 89, 90, 93

D

DEAD-box 131, 166, 167
Dexamethasone 43, 122
DNA-binding domain (DBD) 36, 38-40, 42, 43, 45, 53, 125, 190, 191
DNA binding protein replication protein A (RPA) 149
Drosophila 8, 10, 30, 31, 42, 64, 70, 89, 109, 132, 149, 151, 153, 165, 168, 169

E

EBNA-LP *see* Nuclear antigen leader protein
Egg extract 4, 7-13, 17, 74
Electron microscopy (EM) 2-7, 9-16, 18, 62, 87, 132, 179, 182, 206-209, 219, 220
Electrophysiology 133, 209, 212-214, 221
Electrospray ionization (ESI) 212
Epstein-Barr virus (EBV) 41, 147
ER 28-32, 38, 39, 215, 216, 219
Exon-exon junction (EEJ) 131, 164, 166-169
Exon-exon junction complex (EJC) 131, 164, 166-169
Exportin 1 66, 75, 121, 123, 128, 162-164
  see also CRM1, XPO1 and Xpo1p

F

FG repeats 14-17, 30, 119, 125, 128, 131, 138, 139, 151, 163-167
FT-ICR *see* Ion cyclotron resonance

G

GAL4 36, 39, 144, 153-155
GBF2 73, 109-111
G-box 70-72, 109-111, 126, 144
Geminivirus 73
Gene delivery 187, 188
Genetic transformation 83-85, 90, 91
G-protein 108, 213
GTPase 2, 12, 51, 64, 65, 67, 68, 93, 100, 102, 103, 112, 125, 127, 138, 156, 161-163, 165, 166, 170
GTPase activating protein (GAP) 2, 12, 51, 68, 69, 138, 156
GTPase Ran 2, 12, 64, 67, 68, 93, 100, 102, 103, 125, 127, 138, 156, 161, 162, 166
GTP-binding protein 28, 127, 162
GTPγS *see* Nonhydrolyzable GTP

H

HEAT repeat 15, 52, 124, 125, 150, 151
HeLa 67, 69, 70, 89, 121, 122, 151, 177, 188, 193
Hemagglutinin(A) 32
Herpes simplex virus (HSV) 182, 189, 193, 197, 198
Herpes simplex virus thymidylate kinase 193
Heterogeneous nuclear ribonucleoprotein (hnRNP) 38, 39, 52, 53, 55, 56, 66, 119, 124, 130, 131, 140-143, 145, 148, 150, 151, 161, 162, 164-166, 168
High irradiance response (HIR) 72, 105, 106
HIV Rev 75, 119, 121, 162
Homeodomain proteins 40
Human androgen receptor 36
Human glucocorticoid receptor (hGR) 36-38
Human immunodeficiency virus (HIV-1) 52, 54, 119, 140, 143, 147, 150, 163, 176, 177, 178, 179, 180, 181, 182

I

IκBα 52, 56, 154
Importin 2, 11, 12, 15-17, 28, 50-53, 55, 56, 62, 64-70, 73,-77, 100-104, 112, 121, 123-125, 127-132, 137-143, 145-156, 161-164, 179, 190, 192, 195, 197, 199, 200
Importin α (IMPα) 2, 11, 12, 17, 52, 62, 64-67, 69, 73-76, 101, 103, 112, 123, 124, 132, 138-142, 147-155
    see also Karyopherin α
Importin β (IMPβ) 2, 11, 12, 15, 16, 50-53, 55, 56, 62, 64-69, 74, 77, 101, 103, 123-125, 127, 128, 130, 138-143, 145-155, 197 *see also* Karyopherin β
Importin β binding domain (IBB domain) 52, 64, 101, 148, 150, 151
Inner nuclear membrane (INM) 1, 28, 30, 181, 182, 206, 215, 216, 218
Integral membrane protein 28-30 *see also* POM
Integrase (IN) 140, 143, 147, 176-179, 198, 199
Ion channel 87, 206, 207, 213-220
Ion cyclotron resonance (FT-ICR) 212

K

Kap 15-17, 51, 91, 123, 139, 143-146, 150, 153-155, 163
    Kap60p 15, 16, 139, 143
Karyopherin α 2, 90-93, 177-179, 182
    see also Importin α
Karyopherin β 2, 50, 90, 93, 140, 145, 178-180 *see also* Importin β
Kinase adapter 36, 42
KNS sequence 56
Kruppel-associated box (KRAB) 37, 41

L

Laser desorption ionization (MALDI) 108, 212
Leptomycin B (LMB) 53, 54, 75, 121, 123, 163, 180, 181
Light microscopy 210
Light signaling 71, 72, 100, 104, 105, 107, 108, 110, 111
Low fluence response (LFR) 72, 105, 106
Lysine/arginine 51

M

M9 sequence 55, 56, 145
M9 signal 162
MALDI *see* Laser desorption ionization
Mass spectrometry 29, 212
Matα2 NLS 64
Matrix (MA) 1, 35, 36, 37, 38, 39, 40, 41,
 42, 43, 44, 45, 46, 62, 140, 176-180,
 198, 199, 201, 212
Matrix attachment regions (MARs) 35, 36,
 38, 39, 44, 45
Mazon-Pfizer monkey virus (MPMV) 180
Messenger ribonucleoprotein particle (MRNP)
 93, 131, 164-168, 170
Method 3, 7, 14, 62, 68, 70, 118-120, 200,
 201, 206-209, 212-214, 220
Mitogen-activated protein kinase (MAPK) 54,
 146, 156, 177
Monopartite 63, 66, 100
MRNA export 2, 13, 53, 56, 130, 131,
 161-168, 170, 201

N

NEM *see* Alkylating agent N-ethylmaleimide
Non-basic type NLS 52, 53
Nonhydrolyzable GTP (GTPγS) 8, 9
Nonsense-mediated mRNA decay (NMD)
 131, 168, 169
NTF2-related export protein 1 (NXT1) 103,
 128, 165
Nuclear antigen leader protein (EBNA-LP)
 41, 42
Nuclear envelope (NE) 1-4, 6-13, 17, 28-31,
 35, 50, 54, 61-63, 66-68, 70, 75, 76,
 100, 112, 118, 119, 131, 133, 135, 137,
 161, 162, 175, 176, 179, 187-189, 197,
 201, 206, 207, 209-211, 213-219
Nuclear export factor (NXF) 165, 179, 180
Nuclear export signal (NES) 28, 30, 50, 51,
 53-56, 66, 69, 74, 75, 101-104, 110,
 119, 121-123, 125, 162-164, 175,
 177-181
Nuclear import and export 50, 55, 56, 91,
 100, 102, 103, 118, 119, 125, 162, 175,
 176, 182
Nuclear lamina 17, 28, 35, 40
Nuclear localization signal (NLS) 2, 12, 13,
 16, 17, 28, 30, 36-43, 50-56, 61-68,
 70-75, 77, 88-90, 93, 100, 101, 104,
 105, 108-110, 126, 138, 139, 144,
 147-156, 162, 175, 177-179, 182, 183,
 187, 190-192, 194-198, 199, 201, 210,
 213, 215
 NLS C 63
 NLS M 63
Nuclear matrix 35-46, 62, 180, 201
Nuclear matrix targeting signal (NMTS)
 36-43, 45
Nuclear pore complex (NPC) 1-9, 11-15, 17,
 28-32, 35, 50, 51, 56, 61-64, 66, 68, 70,
 74, 76, 77, 87, 100-103, 112, 118-121,
 124, 127, 128, 131-133, 137-139, 150,
 151, 155, 156, 161-163, 165-168, 175,
 178-181, 189, 190, 195-199, 201,
 206-221
Nuclear protein import 65, 66, 69, 71, 76,
 137-139, 150, 161
Nuclear targeting sequence 153, 191, 193
Nuclear transport factor 2 (NTF2) 52, 56, 64,
 66, 103, 127, 128, 131, 137-139, 148,
 156, 165, 166
Nucleocapsid (NC) 178, 179, 182, 198
Nucleocytoplasmic transport 2, 8, 13, 28,
 118, 161, 162, 165, 206
Nucleoplasmic ring 5-7, 12
Nucleoporin (Nup) 11-17, 28-31, 50, 56, 62,
 63, 100, 102, 104, 119, 120, 128, 131,
 132, 138, 139, 151, 156, 163-167,
 178-181, 212

O

Oligonucleotide 12, 196, 199-201
Oocyte 3, 6, 12, 14, 61, 68, 89, 119, 151,
 166, 180, 199, 212, 217
Outer nuclear membrane (ONM) 28, 29, 32,
 182, 206, 213, 215, 216, 218

P

Papillomavirus E2 protein 42
Papovavirus 176, 197
Parathyroid hormone-related protein (PTHrP)
    144, 150, 151, 153, 156
Patch-clamp 206, 211, 213-221
Photomorphogenesis 71, 72, 104, 105, 107,
    110, 111
Phototropin 105, 109
Phytochrome (Phy) 44, 72, 105-112
    PhyA 105, 106
    PhyE 105
    Red light 44, 72, 105-107, 109-111
Piggy-back 56, 67, 91, 93, 168, 190
PKA *see* cAMP-dependent protein kinase
Plants 1, 6, 8, 12, 13, 17, 28, 31, 35, 43, 44,
    61-64, 66-77, 83-92, 94, 100, 103-111,
    199, 201
Plasmid nuclear import 189-192, 194, 196,
    197
POM 11, 17, 28-30, 32 *see also* Integral
    membrane protein
PPIase 89, 90
Preintegration complex (PIC) 176-179, 190,
    198, 199

R

Ran 2, 10-12, 17, 28, 51-56, 62, 64-69, 74,
    75, 93, 100-103, 112, 124-132, 137-139,
    145, 146, 148, 150-152, 155, 156,
    161-167, 170, 181, 195, 199, 201
RanGAP 2, 12, 17, 65, 68, 69, 101-103, 112,
    127, 128, 138, 156, 162
Ran-binding protein (RanBP) 12, 17, 51, 68,
    69, 75, 101-103, 112, 127, 128, 138,
    145, 146, 150, 151, 156, 162
    RanBP1 51, 68, 69, 75, 101-103, 112,
        127, 128, 138, 156, 162
Ran-GDP 64-68, 103, 127, 138, 139, 162,
    163
Ran-GTP 2, 11, 51, 53, 55, 64-67, 74,
    101-103, 124, 125, 127, 128, 130, 131,
    138, 139, 148, 151, 155, 161-163, 165,
    167, 170, 181
Ran GTPase *see* GTPase Ran

Ran GTPase activating protein 1 (RanGAP1)
    65, 68, 69, 102, 103, 138, 156
Ran-interacting factor NTF2 64
Red/far-red 105, 106, 109
Red/far-red reversible 105 *see also* Low
    fluence response
Retroviridae 175, 176
Reverse transcriptase (RT) 176, 178, 179,
    183, 212
Reverse transcription complex (RTC) 176,
    178, 179, 198
Reverse-transcribed viral genome (rtDNA)
    187
Rev-response element (RRE) 163, 179-181
Ribonucleoprotein complexes 39, 118, 180
    see also snRNP
Ribosome 1, 28, 118, 128, 129, 161, 168,
    169, 180
RNA helicase (RH) 42, 131, 166
RNA-protein complex (RNP) 38, 39, 52-56,
    66, 100, 119, 123, 124, 130, 131,
    140-143, 145, 147, 148, 150, 151, 161,
    162, 164-168, 170, 175, 180, 181
Rough endoplasmic reticulum see ER

S

Saccharomyces cerevisiae 28, 29, 50, 54, 103,
    139, 161, 165-168
Scanning electron microscopy (SEM) 3, 4, 7,
    9-12, 207, 208, 212
Signal transducer and activator of transcription
    (STAT) 52, 53, 142, 147, 148, 153, 154
    STAT1 52, 53, 142, 148, 153
Simian virus 40 (SV40) 43, 51, 52, 63, 64,
    67, 100, 138, 148, 156, 162, 176, 177,
    187, 189-193, 195-198, 201
Small nuclear RNA (snRNA) 12, 54, 128,
    130, 132, 163, 195, 201
snRNP 54, 66, 100, 130, 143, 147, 148, 151,
    166
Squash leaf curl virus (SqLCV) 73, 74
SRP1 66, 139, 140, 142, 143
Srp1 90, 91
SUMO-1 128
SV40 large tumor antigen 43, 51, 52, 67, 138,
    162, 177 *see also* T-ag

T

T-ag 138, 140-143, 148, 151, 153, 156
TAP 131, 132, 164-169, 180
TATA box-binding protein (TBP) 84, 90, 91, 93, 147
T-cell protein tyrosine phosphatase (TCPTP) 143, 150, 151, 156
T-complex 83-89, 91-94
Three-dimensional structure 3, 7, 87
Three-dimensional structure 53, 193
Time-of-flight (TOF) 108, 212
Ti plasmid 84-86, 199
Tomographic reconstruction 5, 7, 14
Transcription-coupled repeat (TCR) 91
Transferred DNA (T-DNA) 64, 67, 83-92, 94, 196, 199
Transmission electron microscopy (TEM) 3, 4, 6, 14, 87, 207, 208
tRNA 12, 63, 66, 100, 118, 128-130, 132, 163, 201
T-strand 84-89, 90-93

U

U small nuclear RNA (U snRNA) 54, 163, 195, 201
UV-B 72, 105
UV-B receptor 105

V

Very low fluence response (VLFR) 72, 105, 106
VIP1 67, 84, 91-93
Viral nuclear import 175, 176, 197
Viral protein R (Vpr) 176-179, 182, 198, 199
Viral RNP (vRNP) 180, 181
VirD2 64, 66, 69, 70, 84-86, 88-94, 103, 199
VirE1 87, 92
VirE2 64, 67, 68, 70, 84-89, 91-94, 199
VirE3 84, 92, 93

W

WD-40 repeat 71, 110
Wheat germ agglutinin (WGA) 9, 13, 30, 62, 70, 88, 103, 123, 126, 189, 190, 195, 197, 200, 201
WPP domain 68

X

Xenopus 3-14, 17, 31, 61, 64, 68, 70, 74, 89, 119, 120, 146, 151, 166, 180, 187, 195, 199, 212
XPO1 101-103
Xpo1p 53, 121, 124, 129, 130

Z

Zinc finger transcription factors (Zn finger) 12, 37, 38, 40
see also Basic leucine-zipper